Techniques in PROTEIN MODIFICATION

Roger L. Lundblad

CRC

CRC Press
Boca Raton Ann Arbor London Tokyo

Library of Congress Cataloging-in-Publication Data

Lundblad, Roger L.
Techniques in protein modification / author, Roger L. Lundblad.
p. cm.
Includes bibliographical references and index.
ISBN 0-8493-2606-0
1. Proteins—Chemical modification. I. Title.
QP551.L883 1994
574.19'245—dc20
DNLM/DLC
for Library of Congress 94-11646
CIP

Direct all inquiries to CRC Press, Inc., 2000 Corporate Blvd., N.W., Boca Raton, Florida 33431.

International Standard Book Number 0-8493-2606-0
Library of Congress Card Number 94-11646
Printed in the United States of America 2 3 4 5 6 7 8 9 0
Printed on acid-free paper

PREFACE

This book is an extension of the previous editions of *Chemical Reagents for Protein Modification* (CRC Press, Boca Raton, FL). Since it has been the author's experience that the previous editions were extensively used in the laboratory, this current version has been designed for the laboratory worker. Thus, a majority of the illustrations from the previous volumes have been deleted, as the author did not consider them essential to the laboratory research phase. Material regarding sources of materials and information have been included.

While a consideration of major biochemical journals would suggest that the use of chemical modification to study the relationship between structure and function in proteins is no longer an active area, the reader is reminded that the chemical characterization of protein is still a major concern for biopharmaceutical companies, and there is an active area in the development of site-specifically modified proteins as therapeutics.

A major portion of this book has been written in various airports and at various altitudes. The support of numerous colleagues is greatly appreciated. In particular, the author is indebted to Professor Charles Craik of the University of California at San Francisco and Professor Bryce Plapp of the University of Iowa, Iowa City for many useful discussions regarding protein engineering and functional group reactivity. The author is also indebted to Professor Ralph Bradshaw of the University of California at Irvine for many entertaining hours of discussions concerning both protein chemistry and the theoretical aspects of Duke-Carolina basketball. Finally, the fifth floor of Flexner Hall at Rockefeller University, New York continues to be a dominant force.

THE AUTHOR

Roger L. Lundblad, Ph.D., is Director of Science and Technology Development in the Hyland Division of Baxter Biotech Group of Baxter Healthcare in Duarte, California. He is also Adjunct Professor of Pathology in the School of Medicine and Adjunct Professor of Periodontics in the School of Dentistry at the University of North Carolina at Chapel Hill.

Dr. Lundblad received his undergraduate education at Pacific Lutheran University in Tacoma, Washington and his Ph.D. degree in biochemistry at the University of Washington in 1965. Before joining Baxter Biotech in 1988, Dr. Lundblad was a full-time faculty member of the University of North Carolina at Chapel Hill. Prior to the University of North Carolina, Dr. Lundblad was a Research Associate in the Moore-Stein group at the Rockefeller University in New York City. Dr. Lundblad has received awards from Pacific Lutheran University, the University of North Carolina, and Baxter Biotech in recognition of his research accomplishments.

Dr. Lundblad's research interests are in the manufacture and characterization of protein biotherapeutics, the role of proteases in biological regulation and wound healing, and the solution chemistry of proteins.

TABLE OF CONTENTS

Chapter 1

SITE-SPECIFIC CHEMICAL MODIFICATION OF PROTEINS

This book is intended to describe laboratory techniques in the site-specific chemical modification of proteins. While such techniques are not used in current protein research as extensively as 20 years ago, such technology is still useful in the characterization of proteins.[1–20] In addition, site-specific chemical modification has been used to produce biotherapeutics.[21,22] The current presentation is not intended to be encyclopedic, and the reader is referred to other more comprehensive works.[23–27] In addition, there are several volumes of *Methods in Enzymology*[28–30] which are extremely useful. The reader is also referred to Appendix I for information concerning the availability of various reagents, addresses of suppliers, and a list of current journals active in publishing the results of site-specific chemical modification studies.

The specific purpose of this chapter is to briefly introduce the concept of site-specific chemical modification including methods of characterizing the product of chemical modification reactions. *Site-specific chemical modification is strictly defined as a process which yields a stoichiometrically altered protein with the quantitative covalent derivatization of a single, unique amino acid residue without either modification of any other amino acid residue or conformational change.* In fact, this objective is rarely obtained because there are several major problems which confound this goal. First, few reagents are specific for the modification of a single functional group (Table 1). Second, even with a reagent which appears to be functional group specific, modification of only one of several residues within a functional group class is unlikely except where the specific residue is uniquely reactive. An example of this is provided by the reactivity of histidine residues at the active sites of enzymes.[31–45] There are also examples where only a small fraction of a given amino acid residue will react. An example of this is provided by the modification of lysyl residues by pyridoxal-5′-phosphate.[46–50] Finally, it is unlikely that the site-specific modification of a protein can be achieved without any conformational change.[51–57]

Establishing the stoichiometry of modification is a relatively straightforward process. First, the molar quantity of modified residue is established by analysis. This could be spectrophotometric as, for example, with the trinitrophenylation of primary amino groups, the nitration of tyrosine with tetranitromethane, or the alkylation of tryptophan with 2-hydroxy-5-nitrobenzyl bromide or by amino acid analysis to determine either the loss of a residue as, for example, in photooxidation of histidine and the oxidation of the indole ring of tryptophan with *N*-bromosuccinimide or the appearance of a modified residue such as with *S*-carboxymethylcysteine or N^1- or N^3-carboxymethylhistidine. In the situation where spectral change or radiolabel incorporation is used to

Table 1
DISSOCIATION CONSTANTS FOR NUCLEOPHILES IN PROTEINS

Potential nucleophile	pKa
Carboxyl	4.6
Imidazole	7.0
Sulfhydryl	7.0
α-Amino	7.8
Phenolic hydroxyl	9.6
ε-Amino	10.2

establish stoichiometry, analysis must be performed to determine that there is not a reaction with another amino acid. For example, the extent of oxidation of tryptophan by *N*-bromosuccinimide can be determined spectrophotometrically, but amino acid analysis is *required* to determine if modification has also occurred with another amino acid such as histidine.

In case of the site-specific chemical modification of a protein, it must be established that the modification of one residue mole per mole of protein (or functional subunit) has occurred without modification of another amino acid (e.g., modification has only occurred with lysine and not with tyrosine). The reaction pattern of a given reagent with free amino acids or amino acid derivatives does not necessarily provide the basis for reaction with such amino acid residues in protein. Furthermore, the reaction pattern of a given reagent with one protein cannot necessarily be extrapolated to all proteins. The results of a chemical modification can be markedly affected by reaction conditions (e.g., pH, temperature, solvent and/or buffer used, degree of illumination, etc.). Establishment of stoichiometry does not necessarily mean that this modification has occurred at a unique residue (unique in terms of position in the linear peptide chain — not necessarily unique with respect to reactivity). It is, of course, useful if there is a change in biological activity (catalysis, substrate binding, ion binding, etc.) which occurs concomitant with the chemical modification. Ideally, one would like to establish a direct relationship (i.e., 0.5 mol/mol of protein with 50% activity modification; 1.0 mol/mol of protein with 100% activity modification). More frequently, one might have the situation where there are several moles of a given amino acid modified.[58] In some of these situations it is possible to fractionate the protein into uniquely modified species. The separation of carboxy-methyl-HIS^{12}-pancreatic ribonuclease from carboxy-methyl-HIS^{119}-pancreatic ribonuclease is a classic example of this type of a situation.[58] More recently, it has been possible to separate various derivatives of lysozyme obtained from the modification of carboxyl groups.[59] Frequently, however, while there is good evidence that multiple modified species are obtained as a result of the reaction, it is not possible to separate apparently uniquely modified species. In the reaction of tetranitromethane with thrombin,[60] apparent stoichiometry of inactivation was obtained with equivalent modification of two separate tyrosine residues (Tyr 71 and Tyr 85 in the B chain) and it was not possible to separate these derivatives.

Indirect support of a site-specific modification can be obtained from a consideration of the functional consequences. In a situation where there are clearly multiple sites of reaction which can be distinguished by analytical techniques, the approach advanced by Ray and Koshland can still be useful.[61] This analysis is based on establishing a relationship between the rate of loss of biological activity and the rate of modification of a single residue. With current technology, it would probably be faster and easier to systematically examine the effect of alanine scanning.[62,63]

The statistical approach advanced by Tsou[64–66] is based on establishing a relationship between the number of residues modified and the change in biological activity. This approach is quite valuable when it is difficult to accurately determine the rate of functional group modification as for example with *N*-bromosuccinimide or 2-hydroxy-5-nitrobenzyl bromide. In Tsou's least complicated example, the biologically essential groups are all of the same type and both essential and nonessential groups are modified at the same rate. Assuming that the modification of any essential group results in the loss of activity, the fraction of biologically active protein remaining will be equal to the fraction of activity remaining (denoted as a). In the situation where there is a single essential group, the fraction of essential groups remaining after any period of modification (denoted as x_c) will be equal to a. In the situation where the number of essential groups is i (by definition greater than 1) among all functional groups of type X, the fraction of each essential group remaining after a period of modification will be x_c. Only those proteins which have all their essential groups intact will retain full activity. Therefore,

$$a = x_c^i \quad \text{or} \quad a^{1/i} = x$$

When all groups of type X react at the same rate, then x_c will be equal to the fraction of the overall fraction of unmodified type X groups and

$$a^{1/i} = x$$

The use of this approach requires the plotting of log a vs. log x; the slope of the resulting line yields i. A number of investigators plot a (activity) vs. m (residues modified).

Horiike and McCormick[67] have explored the approach of relating changes in activity to extent of chemical modification. These investigators state that the original concepts which form the basis of this approach are sound, but that extrapolation from a plot of activity remaining vs. residues modified is not

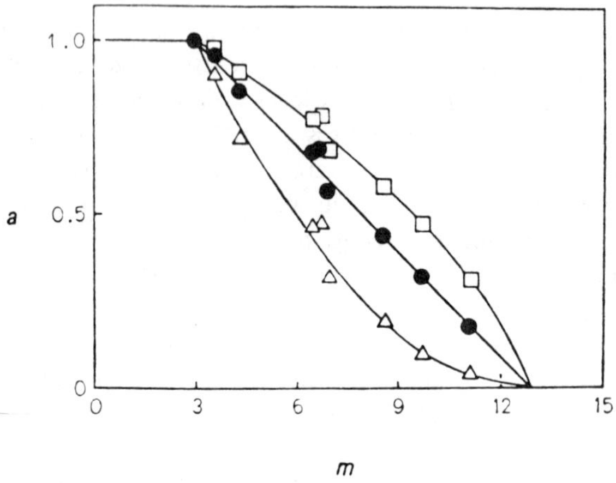

FIGURE 1. Tsou plot for the modification of pepsin with trimethyloxonium fluoroborate. Shown is a plot of (a) vs. (m) where a is the remaining catalytic activity and m is the number of methyl groups incorporated from reaction with trimethyloxonium fluoroborate. The line for i = 2 (●) is the least-squares straight line for these points. The lines for i = 1 (△) and i = 3 (□) are the theoretical curves based upon the i = 2 line. (From Paterson, A. K. and Knowles, J. R., *Eur. J. Biochem.*, 31, 510, 1972. With permission.)

necessarily sound. Such extrapolation is only valid if the "nonessential" residues react much slower (rate at least 100 times slower). Given a situation where all residues within a given group are equally reactive toward the reagent in question, the number of essential residues obtained from such a plot is correct only when the total number of residues is equal to the number of essential residues which is, in turn, equal to 1.0. However, it is important to emphasize that this approach is useful when there is a difference in the rate of reaction of an *essential* residue or residues and all other residues in that class as is the example in the modification of histidyl residues with diethyl pyrocarbonate in lactate dehydrogenase,[68,69] and pyridoxamine-5′-phosphate oxidase.[70] Some examples of the application of Tsou plots to specific chemical modification are presented in Figures 1 to 3.

A major advantage in relating changes in "activity" to a specific chemical modification is being able to demonstrate that the reversal of modification (see Figures 4 and 5) is directly associated with the reversal of the change(s) in biological activity. Demonstrating that the "effects" of a specific chemical modification are reversible lends support *against* the argument that such "effects" are a result of irreversible and "nonspecific" conformational change.

It is useful to consider some factors which influence the reactivity of nucleophilic centers in proteins. From a consideration of the three-dimensional structure of proteins the majority of polar amino acids (i.e., Lys, Arg, Gly, Asp) are located on the exterior surface of the molecule, while the majority of the hydrophobic (nonpolar) residues are located in the interior of these molecules.

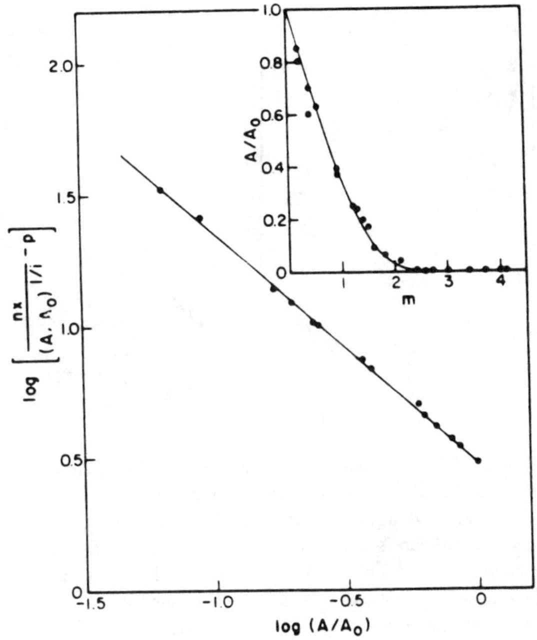

FIGURE 2. Tsou plot for the modification of pyridoxamine phosphate oxidase by diethylpyrocarbonate. The data were plotted using the following equation:

$$\log\left[\frac{nx}{A/A_o^{1/i}} - p\right]$$

$$= \log(n - p) + \left(\frac{\alpha - 1}{i}\right)\log(A/A_o)$$

where A/A_o is the fraction of enzyme molecules retaining full activity, n is the number of modifiable residues of type X consisting of p residues, of which i are essential, react with the reagent at a pseudo-first-order rate constant k_1 and $n - p$ residues which are not essential reacting at a pseudo-first-order rate constant $k_2(-\alpha k_1)$, and x is the number of residues remaining after reaction with reagent. The data are plotted assuming that in the above equation $n = 4$ and $p = i = 1$. The inset describes the relationship between the number of histidyl residues modified per mole of enzyme (m) and A/A_o. (From Horiike, K., Tsuge, H., and McCormick, D. B., *J. Biol. Chem.*, 254, 6638, 1979. With permission.)

Thus, a gradient of polarity (dielectric constant) will exist going from the surface of the protein into the interior. Such a gradient could also be considered to exist in "pocket-like" indentations on the protein surface. For example, the area immediately adjacent to the active site and substrate site S_1 in thrombin is definitely hydrophobic with respect to the surrounding environment and the

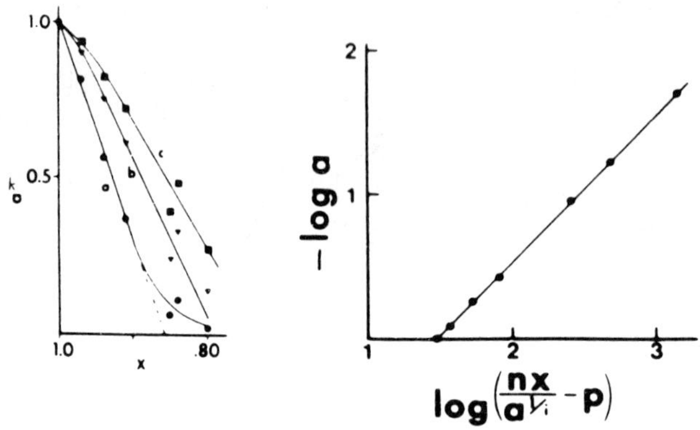

FIGURE 3. The left figure is a Tsou plot for the modification of arginine residues in transketolase by phenylglyoxal. The abscissa is a ratio of unmodified arginine residues to the total number of arginine residues. The ordinate is where a is the fraction of activity remaining and i is a small integer. A linear segment is generated in the case where p residues including i essential residues react at a rate k and n – p residues react with rate αk. The following equation adapted from Tsou:

$$nx = pa^{1/i} + (n - p)\,a^{\alpha/i}$$

reduces to

$$a^{1/i} = \frac{nx - (n - p)}{p}$$

when $\alpha \ll 1$. This gives a straight line with the x intercept equal to n – p/n. This is represented by the extrapolation of the linear portion of the curve to the x axis: a, i = 1, b, i = 2, c, i = 3. The best fit is provided by i – 1 and in this case p = 4 to 5 residues/active site. The right figure describes the determination of α (the constant relating the reaction rates of rapidly and slowly reacting residues). The following equation

$$nx = pa^{1/i} + (n - p)\,a^{\alpha/i}$$

adapted from Tsou can be rearranged as

$$\log\,(nx/a^{1/i} - p) = \log\,(n - p) + [(\alpha - 1)/i]\log a,$$

where i and p are determined as described above. α is determined from the slope of the resulting line. In this situation, α = 0.023 implying that the rapidly reacting residues have a rate constant approximately 40-fold greater than the slowly reacting residues. (From Kremer, A. B., Egan, R. M., and Sable, H. Z., *J. Biol. Chem.*, 255, 2405, 1980. With permission.)

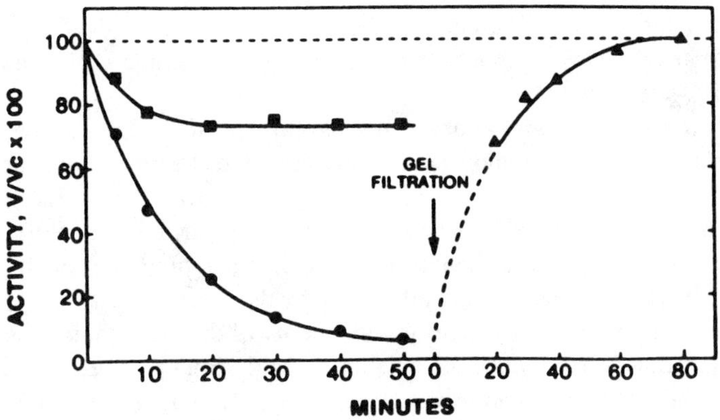

FIGURE 4. The reversible modification of pyridoxamine-5′-phosphate oxidase by 2,3-butanedione. The enzyme (2.1 μ*M*) was incubated with 10 m*M* 2,3-butanedione in the presence of 5 μ*M* flavin mononucleotide (FMN) in either 50 m*M* potassium borate, pH 8.0 (●) or 50 m*M* potassium phosphate, pH 8.0 (■). The reaction mixture in borate was passed over a G-25 Sephadex column and assayed for enzyme activity (▲). The arrow indicates the time at which the reaction mixture in borate was applied to the gel filtration column. (From Choi, J.-D. and McCormick, D. B., *Biochemistry,* 20, 5722, 1981. With permission.)

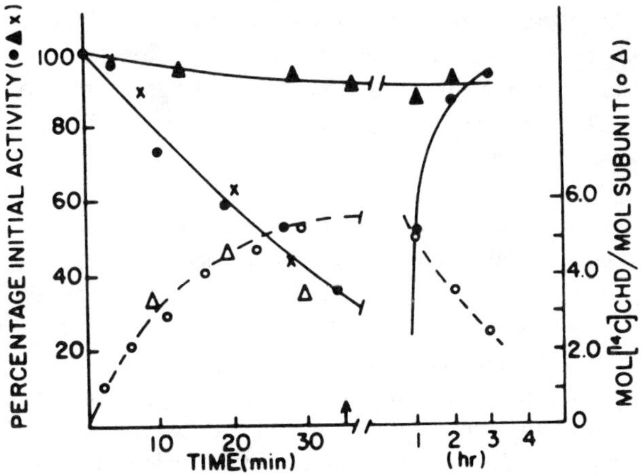

FIGURE 5. The reversible modification of ADP-glucose synthetase by 1,2-cyclohexanedione (CHD). The enzyme was incubated with 10 m*M* [^{14}C] CHD in the presence (●, ○) or absence (x, △) of 50 m*M* sodium tetraborate. The control (▲) was incubated in the absence of CHD. Portions were removed at the indicated times for the determination of incorporated radioactivity (open symbols) or fructose diphosphate-stimulated enzyme activity (closed symbols). The arrow indicated the time of addition of 0.2 *M* hydroxylamine. (From Carlson, C. A. and Preiss, J., *Biochemistry,* 21, 1929, 1982. With permission.)

aqueous solution. This is best demonstrated by the increase in the fluorescence of *N*-α-dansyl-L-arginine-*N*-(3-ethyl-1,5-pentanediyl) amide upon binding to the active-site region.[71]

It should follow from the above discussion that the surface of a globular protein is definitely not homogeneous with respect to electrical charge or, more critically for our consideration, with respect to dielectric constant. As a result of this lack of homogeneity, a variety of surface polarities will surround the various functional groups. The physical and chemical properties of any given functional group will be strongly influenced by the nature (e.g., polarity) of the local microenvironment. Changes in the polarity of the microenvironment can have a profound effect on the dissociation of acids. For example, consider the effect of the addition of an organic solvent, ethyl alcohol, on the pKa of acetic acid. In 100% H_2O, acetic acid has a pKa of 4.70. The addition of 80% ethyl alcohol results in an increase of the pKa to 6.9. In 100% ethyl alcohol the pKa of acetic acid is 10.3. This is particularly important in considering the reactivity of nucleophilic groups such as amino groups, cysteine, carboxyl groups, and the phenolic hydroxyl group. In the case of the primary amines present in protein, these functional groups are not reactive except in the free base form. In other words, the proton present at neutral pH must be removed from the ε-amino group of lysine before this functional group can function as an effective nucleophile. A listing of the "average" pKa values for the various functional groups present in protein is also given in Table 1.

Other factors which can influence the pKa of a functional group in a protein include hydrogen binding with an adjacent functional group, the direct electrostatic effect of the presence of a charged group in the immediate vicinity of a potential nucleophile, and direct steric effects on the availability of a given functional group.

There is another consideration which can in a sense be considered either a cause or consequence of microenvironmental polarity. This has to do with the functional groups/environment immediately around the nucleophilic species in question. These are the "factors" that can cause a "selective" increase (or decrease) in reagent concentration in the vicinity of a potentially reactive species. The most clearly understood example of this is the process of affinity labeling. Another situation can be related to the differences in polarity of the microenvironment around a nucleophilic center. There is also the consideration that a charged reagent can be either attracted to or repelled from the vicinity of a nucleophilic center. This is easily demonstrated by the differences in the comparative rates of modification of the active-site cysteinyl residue by chloroacetic acid and chloroacetamide in papain.[71a]

Another major use of chemical modification has been in the determination of the primary structure of proteins. This includes reagents such as cyanogen bromide for the chemical cleavage of specific peptide bonds, citraconic anhydride for the reversible blocking of lysine residues to restrict tryptic cleavage to arginine residues, and the reversible blocking of arginine residues with 1,2-cyclohexanedione to restrict tryptic cleavage to lysine residues.

The use of specific chemical modification to study changes in the environment has been studied over the past 30 years. The study of Kirtley and Koshland[72] provided the basis for the concept of using "reporter" groups to study changes in the microenvironment surrounding a site of modification. This study used 2-bromoacetamido-4-nitrophenol to modify a limited number of sulfhydryl groups in glyceraldehyde-3-phosphate dehydrogenase. The modified protein has a λ_{max} at 390 nm ($\varepsilon = 7100\ M^{-1}\ cm^{-1}$) between pH 7.0 and 7.6. The addition of the coenzyme NAD caused a marked change in the spectral properties (decrease in absorbance at approximately 375 nm and increase in absorbance at approximately 420 nm) of the modified enzyme which is consistent with a change in the microenvironment around the modified residue (increase in polarity of medium which results in increased formation of the nitrophenolate ion). The reaction of 2-hydroxy-5-nitrobenzyl bromide with tryptophanyl residues to yield the 2-hydroxy-5-nitrobenzyl derivative[73] and the reaction of tetranitromethane with tyrosyl residues[74] to form the 3-nitrotyrosyl derivative was extensively used to study microenvironmental changes in the modified proteins. Spin-labeled reagents have also been useful.[75] One early study used spin-labeled derivatives of diisopropylphosphorofluoridate to study the active site environment of trypsin.[76] Subsequent studies used various spin-label derivatives (piperidinyl nitroxide, pyrrolidinyl nitroxide and pyrrolinyl nitroxide substituent groups) of phenylmethylsulfonyl fluoride to compare microenvironments surrounding the active sites in α-chymotrypsin and trypsin.[77,78] These reagents have been more recently used to study the active site of thrombin.[79,80] The preparation of spin-labeled pepsinogen has been reported.[81] This study used a *N*-hydroxy succinimide ester derivative, 3-[[(2,5-dioxo-1-pyrrolidiny)oxyl] carbonyl]-2,5,-dihydro-2,2,5,5-tetramethyl-1H-pyrrolyl-1-oxy, to modify lysyl residues in pepsinogen. Coupling was accomplished at pH 7.0 (0.1 *M* sodium phosphate) for 7 h at 22°C resulting in the derivatization of approximately three amino groups.

REFERENCES

1. **Nagel-Starczynowska, G. and Kaletha, K.,** AMP-Deaminase from human uterine smooth muscle: the effect of DTNB treatment on kinetic and regulatory properties of the enzyme, *Biochim. Biophys. Acta,* 1164, 261, 1993.
2. **Liao, T.-H., Ho, H.-C., and Abe, A.,** Chemical modification of bovine pancreatic deoxyribonuclease with phenylglyoxal — the involvement of Arg-9 and Arg-41 in substrate binding, *Biochim. Biophys. Acta,* 1079, 335, 1993.
3. **Esmann, M., Sar, P. C., Hideg, K., and Marsh, D.,** Maleimide, iodoacetamide, indanedione, and chloromercuric spin label reagents with derivatized nitroxide rings as ESR reporter groups for protein conformation and dynamics, *Anal. Biochem.,* 213, 336, 1993.
4. **Turk, T. and Macek, P.,** The role of lysine, histidine and carboxyl residues in biological activity of equinatoxin II, a pore forming polypeptide from the sea anemone Actinia equina L, *Biochim. Biophys. Acta,* 1119, 5, 1992.

5. **Turk, T., Macek, P., and Gubensek, F.,** The role of tryptophan in structural and functional properties of equinatoxin II, *Biochim. Biophys. Acta,* 1119, 1, 1992.
6. **Nureki, O., Kohno, T., Sakamoto, K., Miyazawa, T., and Yokoyama, S.,** Chemical modification and mutagenesis studies on zinc binding of aminoacyl-tRNA syntheses, *J. Biol. Chem.,* 268, 15368, 1993.
7. **Xia, C., Meyer, D. J., Chen, H., Reinemer, P., Huber, R., and Ketterer, B.,** Chemical modification of GSH transferase P1-1 confirms the presence of Arg-13, Lys-44 and one carboxylate group in the GSH-binding domain of the active site, *Biochem. J.,* 293, 357, 1993.
8. **Clark, S. and Konstantopoulos, N.,** Sulfydryl agents modulate insulin- and epidermal growth factor (EGF)-receptor kinase via reaction with intracellular receptor domains: differential effects on basal versus activated receptors, *Biochem. J.,* 292, 217, 1993.
9. **Hsieh, P.-C., Shenoy, B. C., Haase, F. C., Jentoft, J. E., and Phillips, N. F. B.,** Involvement of tryptophan(s) at the active site of polyphosphate/ATP glucokinase from Mycobacterium tuberculosis, *Biochemistry,* 32, 6243, 1993.
10. **Sellak, H., Franzini, E., Hakim, J., and Pasquier, C.,** Mechanism of lysozyme inactivation and degradation by iron, *Arch. Biochem. Biophys.,* 299, 172, 1992.
11. **Karlstom, A. R., Shames, B. D., and Levine, R. L.,** Reactivity of cysteine residues in the protease from human immunodeficiency virus: identification of a surface-exposed region which affects enzyme function, *Arch. Biochem. Biophys.,* 304, 163, 1993.
12. **Shen, S. and Strobel, H. W.,** Role of lysine and arginine residues of cytochrome P450 in the interaction between cytochrome P4502B1 and NADPH-cytochrome P450 reductase, *Arch. Biochem. Biophys.,* 304, 257, 1993.
13. **Skawinski, W. J., Adebodum, F., Cheng, J. T., Jordan, F., and Mendelsohn, R.,** Labeling of tyrosines in proteins with [15N]tetranitromethane, a new NMR reporter for nitrotyrosines, *Biochim. Biophys. Acta,* 1162, 297, 1993.
14. **Cutruzzola, F., Ascenzi, P., Barra, D., Bolognesi, M., Menegatti, E., Sarti, P., Schnebli, H.-P., Tomova, S., and Amiconi, G.,** Selective oxidation of Met-192 in bovine alpha-chymotrypsin. Effect on catalytic and inhibitor binding properties, *Biochim. Biophys. Acta,* 1161, 201, 1993.
15. **Siksnys, V. and Pleckaityte, M.,** Role of the reactive cysteine residue in restriction endonuclease Cfr9I, *Biochim. Biophys. Acta,* 1160, 199, 1992.
16. **Thompson, S. A. and Fiddes, J. C.,** Chemical characterization of the cysteines of basic fibroblast growth factor, *Ann. N. Y. Acad. Sci.,* 638, 78, 1991.
17. **Reyes, A. M., Bravo, N., Ludwig, H., Iriarte, A., and Slebe, J. C.,** Modification of Cys-128 of pig kidney fructose 1,6-disphosphatase with different thiol reagents: size dependent effect on the substrate and fructose-2,6-bisphosphate interaction, *J. Prot. Chem.,* 2, 159, 1993.
18. **Chang, G.-G., Satterlee, J., and Hsu, R. Y.,** Essential sulfydryl group of malic enzyme from Escherichia coli, *J. Prot. Chem.,* 12, 7, 1993.
19. **Kobayashi, R., Kanatani, A., Yoshimoto, T., and Tsuru, D.,** Chemical modification of neutral protease from Bacillus subtilis var. amylosacchariticus with tetranitromethane: assignment of tyrosyl residues nitrated, *J. Biochem.,* 106, 1110, 1993.
20. **Alon, R., Bayer, E. A., and Wilchek, M.,** Affinity cleavage of cell surface antibodies using the avidin-biotin system, *J. Immunol. Meth.,* 165, 127, 1993.
21. **Smith, R. A. G., Dewdney, J. M., Fears, R., and Poste, G.,** Chemical derivatization of therapeutic proteins, *Trends Biotechnol.,* 11, 297, 1993.
22. **Poznansky, M. J.,** Soluble enzyme-albumin conjugates: new possibilities for enzyme replacement therapy, *Meth. Enzymol.,* 137, 566, 1988.
23. **Lundblad, R. L. and Noyes, C. M.,** *Chemical Reagents for Protein Modification,* CRC Press, Boca Raton, FL, 1984.
24. **Lundblad, R. L.,** *Chemical Reagents for Protein Modification,* 2nd ed., CRC Press, Boca Raton, FL, 1991.

25. **Walker, J. M., Ed.,** *Methods in Molecular Biology, Vol. 1, Proteins,* Humana Press, Clifton, NJ, 1984.
26. **Shively, J. E., Ed.,** *Methods of Protein Microcharacterization. A Practical Handbook,* Humana Press, Clifton, NJ, 1986.
27. **Oxender, D. L. and Fox, C. F., Eds.,** *Protein Engineering,* Alan R. Liss, New York, 1987.
27a. **Fasman, G. D.,** *Practical Handbook of Biochemistry and Molecular Biology,* CRC Press, Boca Raton, FL, 1989.
27b. **Fasman, G. D.,** *Handbook of Biochemistry and Molecular Biology, Vol. 2, Proteins,* 3rd. ed., CRC Press, Boca Raton, FL, 1976.
27c. **Walker, J. M., Ed.,** *Methods in Molecular Biology, Vol. 3, New Protein Techniques,* Humana Press, Clifton, NJ, 1988.
28. **Hirs, C. H. W., Ed.,** *Methods in Enzymology,* Vol. 2, 197
29. **Hirs, C. H. W. and Timasheff, S. N., Eds.,** *Methods in Enzymology,* Vol. 25, 19.
30. **Hirs, C. H. W. and Timasheff, S. N., Eds.,** *Methods in Enzymology,* Vol. 47, 19.
31. **Blanke, S. R. and Hager, L. P.,** Chemical modification of chloroperoxidase with diethylpyrocarbonate. Evidence for the presence of an essential histidine residue, *J. Biol. Chem.,* 265, 12454, 1990.
32. **Dumas, D. P. and Raushel, F. M.,** Chemical and kinetic evidence for an essential histidine in the phosphotriesterase from *Pseudomonas diminuta, J. Biol. Chem.,* 265, 21498, 1990.
33. **Abdulwajid, A. W. and Wu, F. Y.-H.,** Chemical modification of Escherichia coli RNA polymerase by diethyl pyrocarbonate: evidence of histidine requirement for enzyme activity and intrinsic zinc binding, *Biochemistry,* 25, 8167, 1986.
34. **Sams, C. F. and Matthews, K. S.,** Diethyl pyrocarbonate reaction with the lactose repressor protein affects both inducer and DNA binding, *Biochemistry,* 27, 2277, 1988.
35. **Sokol, P. P., Holohan, P. D., and Ross, C. R.,** Arginyl and histidyl groups are essential for organic anion exchange in renal brush-border membrane vesicles, *J. Biol. Chem.,* 263, 7118, 1988.
36. **Gordillo, E., Ayala, A., F.-Lobato, M., Bautista, J., and Machado, A.,** Possible involvement of histidine residues in the loss of enzymatic activity of rat liver malic enzyme during aging, *J. Biol. Chem.,* 263, 8053, 1988.
37. **Gaikwad, S. M., More, M. W., Vartak, H. G., and Deshpande, V. V.,** Evidence for the essential histidine residue at the active site of glucosexylose isomerase from *Streptomyces, Biochem. Biophys. Res. Commun.,* 155, 270, 1988.
38. **Gacheru, S. N., Trackman, P. C., and Kagan, H. M.,** Evidence for a functional role for histidine in lysyl oxidase catalysis, *J. Biol. Chem.,* 263, 16704, 1988.
39. **Klotz, K.-N., Lohse, M. J., and Schwabe, U.,** Chemical modification of A_1 adenosine receptors in rat brain membranes. Evidence for histidine in different domains of the ligand binding site, *J. Biol. Chem.,* 263, 17522, 1988.
40. **Morishima, I., Shiro, Y., Adachi, S., Yano, Y., and Orii, Y.,** Effect of the distal histidine modification (cyanation) of myoglobin on the ligand binding kinetics and the heme environmental structures, *Biochemistry,* 28, 7582, 1989.
41. **Vangrysperre, W., Ampe, C., Kersters-Hilderson, H., and Tempst, P.,** Single active-site histidine in D-xylose isomerase from *Streptomyces violaceoruber.* Identification by chemical derivatization and peptide mapping, *Biochem. J.,* 263, 195, 1989.
42. **Batt, C. A., Jamieson, A. C., and Vandeyar, M. A.,** Identification of essential histidine residues in the active site of *Escherichia coli* xylose (glucose) isomerase, *Proc. Natl. Acad. Sci. U.S.A.,* 87, 618, 1990.
43. **Kuno, S., Fukui, S., and Toraya, T.,** Essential histidine residues in coenzyme B_{12}-dependent diol dehydrase: dye-sensitized photooxidation and ethoxycarbonylation, *Arch. Biochem. Biophys.,* 277, 211, 1990.
44. **Mattsson, P., Pohjalainen, T., and Korpela, T.,** Chemical modification of cyclomaltodextrin glucanotransferase from Bacillus circulans var. alkalophilus, *Biochim. Biophys. Acta,* 1122, 33, 1992.

45. **Nakayama, T., Tanabe, H., Deyashiki, Y., Shinoda, M., Nara, A., and Sawada, H.,** Chemical modification of cysteinyl, lysyl and histidyl residues of mouse liver 17-beta-hydroxysteroid dehydrogenase, *Biochim. Biophys. Acta,* 1120, 144, 1992.
46. **Kawano, Y., Okubo, K., Tokunaga, F., Miyata, T., Iwanaga, S., and Hamasaki, N.,** Localization of the pyridoxal phosphate binding site at the COOH-terminal region of erythrocyte band 3 protein, *J. Biol. Chem.,* 263, 8232, 1988.
47. **Basu, S., Basu, A., and Modak, M. J.,** Pyridoxal 5′-phosphate mediated inactivation of *Escherichia coli* DNA polymerase I: identification of lysine-635 as an essential residue for the processive mode of DNA synthesis, *Biochemistry,* 27, 6710, 1988.
48. **Jakubowski, H. V. and Owen, W. G.,** Macromolecular specificity determinants on thrombin for fibrinogen and thrombomodulin, *J. Biol. Chem.,* 264, 11117, 1989.
49. **Church, F. C., Pratt, C. W., Noyes, C. M., Kalayanamit, T., Sherrill, G. B., Tobin, R. B., and Meade, J. B.,** Structural and functional properties of human α-thrombin, phosphopyridoxylated α-thrombin, and $gamma_T$-thrombin. Identification of lysyl residues in α-thrombin that are critical for heparin and fibrin(ogen) interactions, *J. Biol. Chem.,* 264, 18419, 1989.
50. **Naoi, M., Ichinose, H., Takahashi, T., and Nagatsu, T.,** Sensitive assay for determination of pyridoxal-5-phosphate in enzymes using high-performance liquid chromatography after derivatization with cyanide, *J. Chromatogr.,* 434, 209, 1988.
51. **Okajima, T., Kawata, Y., and Hamaguchi, K.,** Chemical modification of tryptophan residues and stability changes in proteins, *Biochemistry,* 29, 9168, 1990.
52. **Rao, A. G. and Neet, K. E.,** Tryptophan residues of the gamma subunit of 7S nerve growth factor: intrinsic fluorescence, solute quenching, and *N*-bromosuccinimide oxidation, *Biochemistry,* 21, 6843, 1982.
53. **Winkler, M. A., Fried, V. A., Merat, D. L., and Cheung, W. Y.,** Differential reactivities of lysines in calmodulin complexed to phosphatase, *J. Biol. Chem.,* 262, 15466, 1987.
54. **Morrison, S. A.,** Origin of a fluorescence increase accompanying the limited proteolysis of fluorescein-labeled human prothrombin by Factor Xa, *Int. J. Pept. Protein Res.,* 24, 377, 1984.
55. **Davies, K. J. A.,** Protein damage and degradation by oxygen radicals III. Modification of secondary and tertiary structure, *J. Biol. Chem.,* 262, 9908, 1987.
56. **Suckau, D., Mak, M., and Przybylski, M.,** Protein surface topology-probing by selective chemical modification and mass spectrometric peptide mapping, *Proc. Natl. Acad. Sci. U.S.A.,* 89, 5630, 1992.
57. **Gettins, P. G. W., Fan, B., Crews, B. C., Turko, I. V., Olson, S. T., and Streusand, V. J.,** Transmission of conformational change from the heparin binding site to the reactive center of antithrombin, *Biochemistry,* 32, 8385, 1993.

57a. **Brocklehurst, K., Brocklehurst, S. M., Kowlessur, D., O'Driscoll, M., Patel, G., Salih, E., Templeton, W., Thomas, E., Topham, C. M., and Willenbrock, F.,** Supracrystallographic resolution of interactions contributing to enzyme catalysis by use of natural structural variants and reactivity-probe kinetics, *Biochem. J.,* 256, 543, 1988.

57b. **Buechler, J. A., Vedvick, T. A., and Taylor, S. S.,** Differential labeling of the catalytic subunit of cAMP-dependent protein kinase with acetic anhydride: substrate-induced conformational changes, *Biochemistry,* 28, 3018, 1989.

57c. **Mykkanen, H. M. and Wasserman, R. H.,** Reactivity of sulfhydryl groups in the brush-border membranes of chick duodena is increased by 1,25-dihydroxycholecalciferol, *Biochim. Biophys. Acta: Gen. Subjects,* 1033, 282, 1990.

57d. **Landfear, S. M., Evans, D. R., and Lipscomb, W. N.,** Elimination of cooperativity in aspartate transcarbamylase by nitration of a single tyrosine residue, *Proc. Natl. Acad. Sci. U.S.A.,* 75, 2654, 1978.

57e. **Kumar, G. K., Beegen, N., and Wood, H. G.,** Involvement of tryptophans at the catalytic site and subunit-binding domains of transcarboxylase, *Biochemistry,* 27, 5972, 1988.

57f. **Okajima, T., Kawata, Y., and Hamaguchi, K.,** Chemical modification of tryptophan residues and stability changes in proteins, *Biochemistry,* 29, 9168, 1990.

58. **Crestfield, A. M., Stein, W. H., and Moore, S.,** Alkylation and identification of the histidine residues at the active site of ribonuclease, *J. Biol. Chem.,* 238, 2418, 1963.
59. **Yamada, A., Imoto, T., Fujita, K., Okasaki, K., and Motomura, M.,** Selective modification of aspartic acid-101 in lysozyme by carbodiimide reaction, *Biochemistry,* 20, 4836, 1981.
60. **Lundblad, R. L., Noyes, C. M., Featherstone, G. L., Harrison, J. H., and Jenzano, J. W.,** The reaction of bovine alpha-thrombin with tetranitromethane. Characterization of the modified protein, *J. Biol. Chem.,* 263, 3729, 1988.
61. **Ray, W. J., Jr. and Koshland, D. E., Jr.,** A method for characterizing the type and numbers of groups involved in enzyme action, *J. Biol. Chem.,* 236, 1973, 1961.
62. **Nagashima, M., Lundh, E., Leonard, J. C., Morser, J., and Parkinson, J. F.,** Alanine-scanning mutagenesis of the epidermal growth factor-like domains of human thrombomodulin identifies critical residues for its cofactor activity, *J. Biol. Chem.,* 268, 2888, 1993.
63. **Maher, D. W., Strawn, L. M., and Donoghue, D. J.,** Alanine mutagenesis of conserved residues in the platelet-derived growth factor family: identification of residues necessary for dimerization and transformation, *Oncogene,* 8, 533, 1993.
64. **Tsou, C.-L.,** Relation between modification of functional groups of proteins and their biological activity. I. A graphical method for the determination of the number and type of essential groups, *Sci. Sinica,* 11, 1535, 1962.
65. **Tsou, C.-L.,** Kinetics of substrate reaction during irreversible modification of enzyme activity, *Adv. Enzymol.,* 61, 381, 1988.
66. **Zhou, J.-M., Liu, C., and Tsou, C.-L.,** Kinetics of trypsin inhibition by its specific inhibitors, *Biochemistry,* 28, 1070, 1989.
67. **Horiike, K. and McCormick, D. B.,** Correlations between biological activity and the number of functional groups chemically modified, *J. Theoret. Biol.,* 79, 403, 1979.
68. **Holbrook, J. J. and Ingram, V. A.,** Ionic properties of an essential histidine residue in pig heart lactate dehydrogenase, *Biochem. J.,* 131, 729, 1973.
69. **Bloxham, D. P.,** The chemical reactivity of the histidine-195 residue in lactate dehydrogenase thiomethylated at the cysteine-165 residue, *Biochem. J.,* 193, 93, 1981.
70. **Horiike, K., Tsuge, H., and McCormick, D. B.,** Evidence for an essential histidyl residue at the active site of pyridoxamine (pyridoxine)-5′-phosphate oxidase from rabbit liver, *J. Biol. Chem.,* 254, 6638, 1979.
71. **Nesheim, M. E., Prendergast, F. G., and Mann, K. G.,** Interactions of a fluorescent active-site-directed inhibitor of thrombin: dansylarginine *N*-(3-ethyl-1,5-pentanediyl) amide, *Biochemistry,* 18, 996, 1979.

71a. **Chaiken, I. M. and Smith, E. L.,** Reaction of chloroacetamide with the sulfhydryl groups of papain, *J. Biol. Chem.,* 244, 5087, 1969.

72. **Kirtley, M. E. and Koshland, D. E., Jr.,** The introduction of a "reporter" group at the active site of glyceraldehyde-3-phosphate dehydrogenase, *Biochem. Biophys. Res. Commun.,* 23, 810, 1966.
73. **Loudon, G. M. and Koshland, D. E., Jr.,** The chemistry of a reporter group: 2-hydroxy-5-nitrobenzyl bromide, *J. Biol. Chem.,* 245, 2247, 1970.
74. **Riordan, J. R., Sokolovsky, M., and Vallee, B. L.,** Environmentally sensitive tyrosyl residues. Nitration with tetranitromethane, *Biochemistry,* 6, 358, 1967.
75. **Berliner, L. J., Ed.,** *Spin Labeling: Theory and Applications,* Academic Press, New York, 1975.
76. **Berliner, L. J. and Wong, S. S.,** Evidence against two "pH locked" conformations of phosphorylated trypsin, *J. Biol. Chem.,* 248, 1118, 1973.
77. **Berliner, L. J. and Wong, S. S.,** Spin-labeled sulfonyl fluorides as active site probes of protease structure. I. Comparison of the active site environments in α-chymotrypsin and trypsin, *J. Biol. Chem.,* 249, 1668, 1974.
78. **Wong, S. S., Quiggle, K., Triplett, C., and Berliner, L. J.,** Spin-labeled sulfonyl fluorides as active site probes of protease structure. II. Spin label synthese and enzyme inhibition, *J. Biol. Chem.,* 249, 1678, 1974.

79. **Berliner, L. J. and Shen, Y. Y.,** Probing active site structure by spin label (ESR) and fluorescence methods, in *Chemistry and Biology of Thrombin,* Lundblad, R. L., Fenton, J. W., II, and Mann, K. G., Eds., Ann Arbor Science, Ann Arbor, MI, 1977, 197.
80. **Berliner, L. J., Bauer, R. S., Chang, T.-L., Fenton, J. W., II, and Shen, Y. Y. L.,** Active-site topography of human coagulant (α) and noncoagulant (γ) thrombins, *Biochemistry,* 20, 1831, 1981.
81. **Twining, S. S., Sealy, R. C., and Glick, D. M.,** Preparation and activation of spin-labelled pepsinogen, *Biochemistry,* 20, 1267, 1981.

Chapter 2

AMINO ACID ANALYSIS

The critical aspects of amino acid analysis are sample preparation, including hydrolysis and pre-column derivatization (does the method used to produce amino acids from the peptide or protein under study accurately represent the composition?); preparation of the solvents used for the chromatographic fractionation; and the reproducibility of the analytical system. Reproducibility of the analytical system includes reproducibility of the spectrophotometric analytical system, maintenance of an extremely reproducible flow rate, and the accuracy of the data analysis system. Implicit in this last area is the accuracy of the amino acid standard used for calibration of the system. With ninhydrin-based detection systems, consideration needs to be given to the possibility of reagent decomposition.

The preparation of the sample will be discussed first. The strategy is somewhat different depending upon the qualitative and quantitative demands on the information desired. If the objective is to rigorously determine the amino acid composition of newly isolated protein, it is essential to know the *chemical* purity of the preparation. Here it would be assumed that a single macromolecular component (protein) is present. In the absence of other analytical information regarding the nature of the protein preparation (i.e., carbohydrate content, nucleic acid content, etc.), it is of critical importance to know precisely how much material (mass) is present at the start so that the investigation can accurately determine what portion (percentage) of the sample is recovered as amino acid. The careful investigator would also know what portion of the sample is moisture or inorganic components detected as % ash. Impurities can also contribute to artifactual results other than inaccurate estimates of the amount of protein present. The presence of nucleic acid in a sample can result in the formation of artifactual amounts of glycine.[1] Inorganic constituents can pose a number of difficulties. A systematic study has explored the effect of cupric ions and ferric ions[2] on the recovery of amino acids after acid hydrolysis (6 *N* HCl, 20 to 22 h at 110°C). The presence of these metal ions results in the conversion of cysteine/cystine to cysteic acid and methionine to methionine sulfoxide.

In general, adequate results for the amino acid composition of a protein (excluding tryptophan content) can be obtained from careful hydrolysis in 6 *N* HCl. Effective exclusion of oxygen during the hydrolysis is essential for the accurate determination of composition. This is easily achieved by the careful deaeration of the sample prior to placement in a hydrolysis oven or block. Accurate values for serine, threonine, and tyrosine require samples hydrolyzed for several periods of time (i.e., 22, 48, and 72 h) such that extrapolation back to zero time can be accomplished.

Hydrolysis in 6 *N* HCl precludes the determination of tryptophan since this amino acid is destroyed under these conditions. The work of Hugli and Moore[3] provided a reliable method for base hydrolysis and subsequent analysis. The conditions developed by these investigators result in the accurate determination of tryptophan, and the presence of carbohydrate does not interfere with the analysis. Although alkaline hydrolysis and spectroscopy are still of value (the use of base hydrolysis, for example, is now not essential for the determination of tryptophan but is essential for the determination of γ-carboxyglutamic acid),[4–6] the work of Liu and Chang[7] has largely supplanted their use in the quantitative determination of tryptophan in proteins which do not contain substantial carbohydrate. This group reported the development of the use of tosyl acid (*p*-toluenesulfonic acid) for the hydrolysis of proteins to determine tryptophan.[7] This procedure was effective despite having several disadvantages. The major problem is that tosyl acid is a solid which generally must be recrystallized from HCl prior to use and it is difficult to obtain completely free of HCl. The inclusion of indole or 3-(2-aminoethyl) indole was observed to improve the yield of tryptophan and, therefore, the reproducibility of analyses. The use of 4.0 *N* methanesulfonic acid as outlined in a subsequent publication from Liu's laboratory has proved to be more effective than tosyl acid.[8] The condition for the hydrolysis of proteins are 4.0 *N* methane-sulfonic acid (containing 0.2% 3-[2-aminoethyl] indole) at 115°C for 22 h (it may be necessary to include other time periods of hydrolysis such as 48 and 72 h for the reasons given above for 6 *N* HCl as well as to assure the cleavage of all peptide bonds). Sartin and co-workers[9] have reported that the inclusion of mercaptoacetic acid, phenol, and 3-(2-aminoethyl) indole in 6.0 *N* HCl (100-μl sample in H_2O, 100 μl concentrated HCl, 1 μl mercaptoacetic acid, 10 μl of 5% phenol, and 5 μl of 2% 3-(2-aminoethyl) indole) resulted in the effective recovery of tryptophan for subsequent analysis. Chiou and Wang[10] have modified this technique by using 4 *M* methanesulfonic acid at 160°C for 45 min. Complete recovery of cysteine and tryptophan is obtained under these conditions. The presence of carbohydrate (greater than 5%) results in the excessive loss of tryptophan using any of these procedures such that alkaline hydrolysis or some alternative analytical approach (i.e., 2-hydroxy-5-nitrobenzyl bromide, *N*-bromosuccinimide) may be preferable. There have been significant advances in hydrolysis technology. The use of a mixture of hydrochloric acid and trifluoroacetic acid (TFA) (HCl:TFA; 2:1) has been recommended for the hydrolysis of hydrophobic proteins.[11] The technique is rapid (25 min at 166°C), and TFA is suggested to be necessary for the hydrolysis of peptide bonds involving hydrophobic amino acids. Tryptophan is destroyed under these conditions. Chiou and Wang[12] have described the use of microwave irradiation for the acid hydrolysis (6 *N* HCl) of peptides and proteins. Microwave irradiation (0.96 kW) for 8 min yielded hydrolyzates comparable to those obtained with 6 *N* HCl (24 h at 110°C). Conditions for the acid hydrolysis of proteins blotted onto polyvinyl difluoride membranes have also been reported.[13,14] A gas-phase hydrolysis technique has been described by Meltzer et al.,[15] which uses conventional HCl hydrolysis conditions (24 h at 110°C). Although there is no time advantage with this technique, it does permit the simultaneous processing of a large number of samples.

Most analytical systems are based on the detection of the respective amino acids in the effluent from the analytical columns by reaction with ninhydrin. The ninhydrin systems in use are based either on a methyl cellosolve solvent or dimethyl sulfoxide. We use the dimethyl sulfoxide-based system as we find this reagent more stable on storage; the dimethyl sulfoxide does not have the severe toxic properties of methyl cellosolve; and the color yields for most amino acids are superior. The source of ninhydrin is critical and we have found that only Pierce Chemical Company, Rockford, IL, has continually provided material of excellence. We have, on occasion, used material from other sources which has had equivalent performance, but there was extensive batch-to-batch variation.

Figure 1 shows the chromatographic separation of amino acids via conventional ion-exchange technology following by post-column detection with the ninhydrin reagent. This data was obtained using Beckman Instruments System Gold®. Figure 2 shows the analysis of physiological fluids using the same technology.

The use of fluorogenic reagents to detect amino acids has been increasing during the past decade. Fluorescamine (4-phenylspiro[furan-2-(3H)-1′-phthalan]-3,3′-dione) was introduced by Udenfriend and co-workers[16] for the detection of amino acids. At the time, the reagent provided greater sensitivity for most amino acids than did the ninhydrin reagent, but did not detect amino acids such as proline or hydroxyproline. Felix and Turkelsen[17] reported the use of post-column derivatization of proline with *N*-chlorosuccinimide which permitted effective reaction with fluorescamine. Further studies[18] on the chemistry of the reaction of fluorescamine with amino acids have created a solid base for the use of this reagent for amino acid analysis using colorimetric rather than fluorescence detection. Further developments of note in the use of this reagent include the use of alkaline sodium hypochlorite for the post-column derivatization of amino acids.

Another fluorescence-based system based on the use of *o*-phthalaldehyde has been developed.[19,20] Our laboratories do not have any practical experience with either fluorescamine or *o*-phthalaldehyde, but it is our impression that *o*-phthalaldehyde has been more useful. It is water soluble which makes this compound somewhat easier to use from a technical viewpoint. Both reagents require the postcolumn derivatization of amino acids which requires a sometimes tedious plumbing arrangement. It is also our impression that solvent purity is a more critical factor with either of these reagents than with ninhydrin.

Amino acid analysis by high-performance liquid chromatography (HPLC) following precolumn derivatization with either dansyl chloride[21] or phenylisothiocyanate[22] has proved to be extremely useful. Sensitivity can be

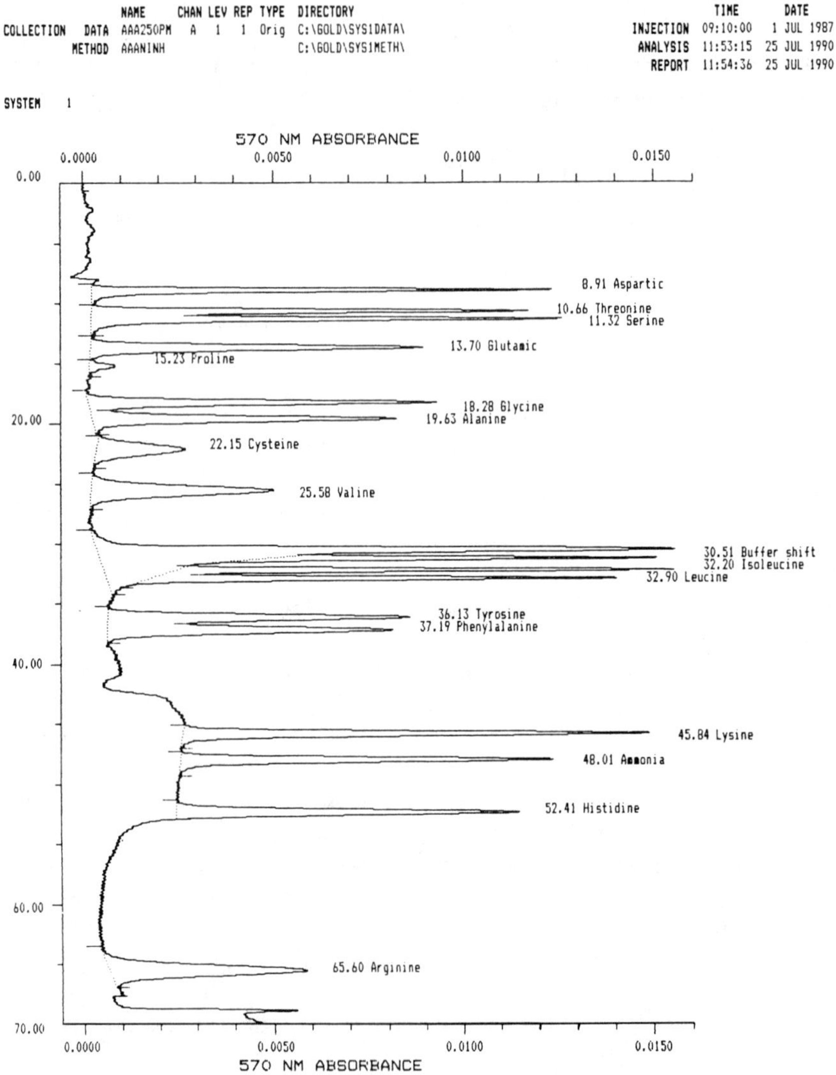

FIGURE 1. Amino acid analysis of protein hydrolyzate with post-column derivatization with ninhydrin. Note buffer change artifacts at injection, 28 min, and 44 min. Response was monitored at 570 and 440 nm (proline). Column temperature was 50°C and coil temperature was 30°C. Obtained using System Gold® technology (Figure courtesy of Beckman Instruments.)

easily achieved at the nanomole level using either technique. The separation of PTC-amino acids using HPLC analysis on an ISCO Instruments, Lincoln, NE system is shown in Figure 3.

In conclusion, the major advances in the amino acid analysis of proteins in the past decade has not been in the actual chemistry of the analysis, but rather in improving hydrolysis techniques, chromatographic separations, and sensitivity improvements. Recent publications are included in the references.[23–40]

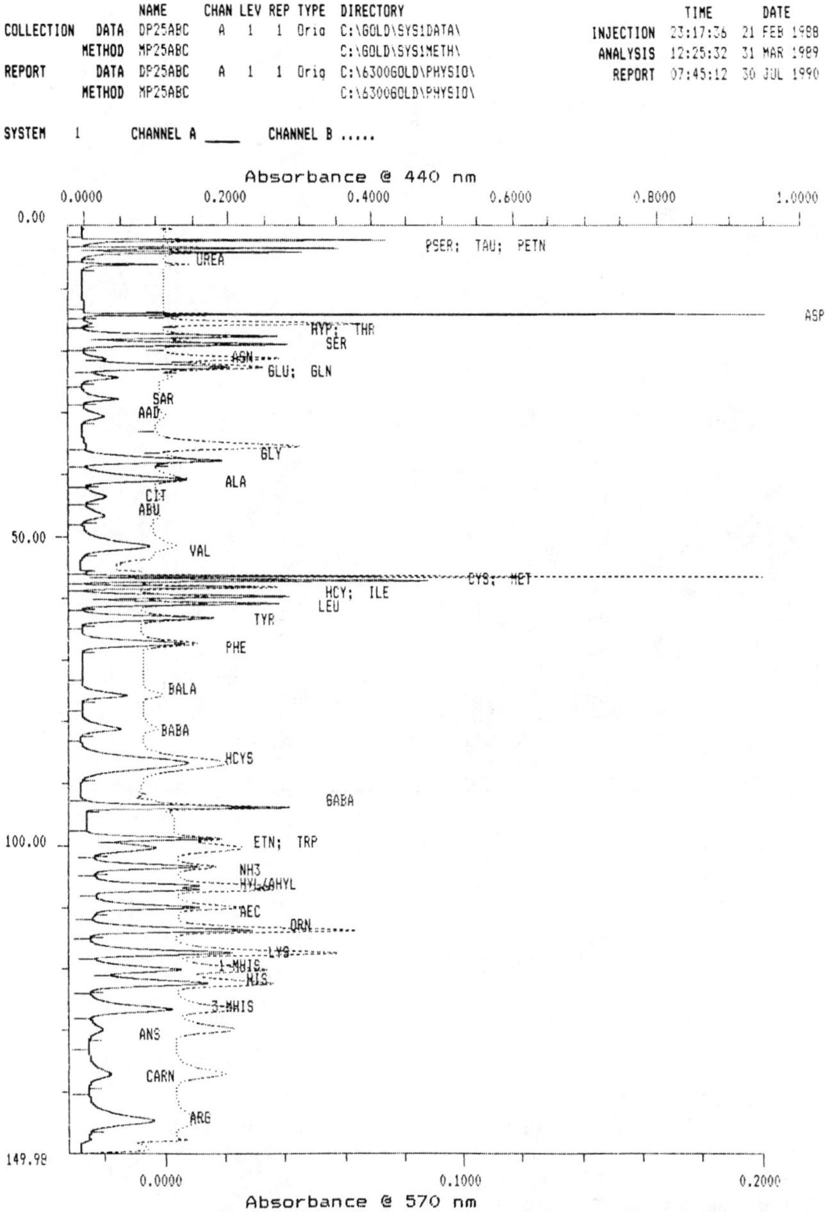

FIGURE 2. Amino acid analysis of physiological fluids with post-column derivatization with ninhydrin. Response was monitored at both 570 and 440 nm. Column temperature was 50°C and coil temperature was 30°C. Obtained using System Gold® technology. (Figure courtesy of Beckman Instruments.)

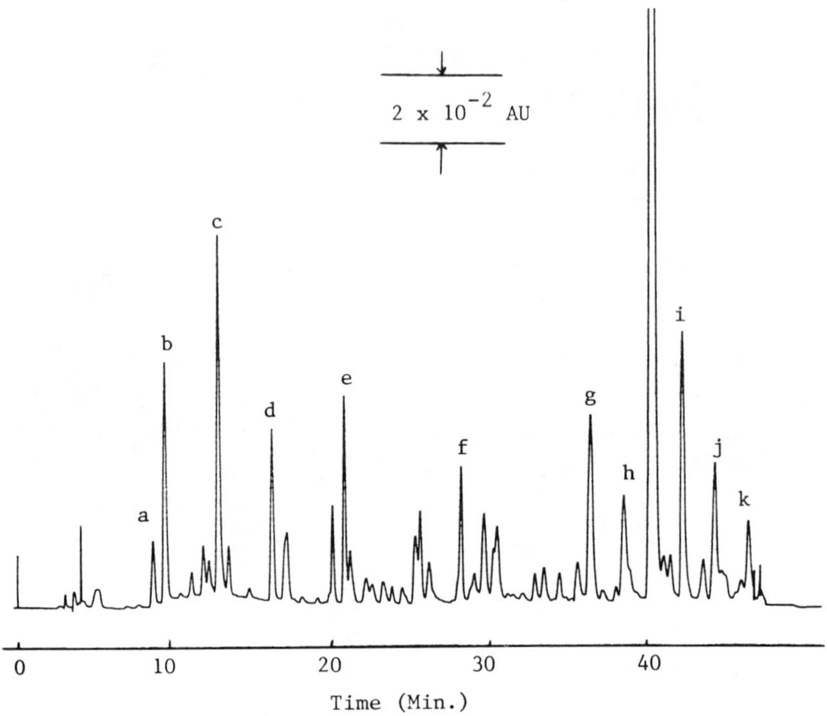

FIGURE 3. Amino acid analysis using pre-column derivatization with phenylisothiocyanate. The column was a C^{18}-reverse phase with 50 m*M* sodium acetate, pH 6.5 as the initial solvent. Elution was accomplished with an acetonitrile/methanol gradient. The effluent was monitored at 269 nm using an ISCO HPLC System. a, asx; b, glx; c, gly; d, his; e, ala; f, thr; g, tyr; h, val; i, leu; j, phe, k, lys. (Courtesy of ISCO Scientific.)

REFERENCES

1. **Paddock, G. V., Wilson, G. B., and Wang, A.-C.,** Contribution of hydrolyzed nucleic acids and their constituents to the apparent amino acid composition of biological components, *Biochem. Biophys. Res. Commun.,* 87, 946, 1979.
2. **Brummel, M., Gerbeck, C. M., and Montgomery, R.,** Effects of metals on analytical procedures for amino acids and carbohydrates, *Anal. Biochem.,* 31, 331, 1969.
3. **Hugli, T. E. and Moore, S.,** Determination of the tryptophan content of proteins by ion-exchange chromatography of alkaline hydrolysates, *J. Biol. Chem.,* 247, 2828, 1972.
4. **Hauschka, P. V.,** Quantitative determination of γ-carboxyglutamic acid in proteins, *Anal. Biochem.,* 80, 212, 1977.
5. **Kuwada, M. and Katayama, K.,** A high-performance liquid chromatographic method for the simultaneous determination of γ-carboxyglutamic acid and glutamic acid in proteins, bone, and urine, *Anal. Biochem.,* 117, 259, 1981.
6. **Price, P. A.,** Analysis for gamma-carboxyglutamic acid, *Meth. Enzymol.,* 91, 13, 1983.
7. **Liu, T.-Y. and Chang, W. H.,** Hydrolysis of proteins with *p*-toluenesulfonic acid. Determination of tryptophan, *J. Biol. Chem.,* 246, 2842, 1971.
8. **Simpson, R. J., Neuberger, M. R., and Liu, T.-Y.,** Complete amino acid analysis of proteins from a single hydrolysate, *J. Biol. Chem.,* 251, 1936, 1976.
9. **Sartin, J. L., Hugli, T. E., and Liao, T.-H.,** Reactivity of the tryptophan residues in bovine pancreatic deoxyribonuclease with *N*-bromosuccinimide, *J. Biol. Chem.,* 255, 8633, 1980.
10. **Chiou, S.-H. and Wang, K.-T.,** Simplified protein hydrolysis with methanesulfonic acid at elevated temperature for the complete amino acid analysis of proteins, *J. Chromatogr.,* 448, 404, 1988.
11. **Tsugita, A. and Scheffler, J.-J.,** A rapid method for acid hydrolysis of protein with a mixture of trifluoroacetic acid and hydrochloric acid, *Eur. J. Biochem.,* 124, 585, 1982.
12. **Chiou, S.-H. and Wang, K.-T.,** Peptide and protein hydrolysis by microwave irradiation, *J. Chromatogr.,* 491, 424, 1989.
13. **Tous, G. I., Fausnaugh, J. L., Akinyosoye, O., Lackland, H., Winter-Cash, P., Victorica, F. J., and Stein, S.,** Amino acid analysis on polyvinyl difluoride membranes, *Anal. Biochem.,* 179, 50, 1989.
14. **Nakagawa, S. and Fukuda, T.,** Direct amino acid analysis of proteins electroblotted onto polyvinylidene difluoride membrane from sodium dodecyl sulfate-polyacrylamide gel, *Anal. Biochem.,* 181, 75, 1989.
15. **Meltzer, N. M., Tous, G. I., Gruber, S., and Stein, S.,** Gas-phase hydrolysis of proteins and peptides, *Anal. Biochem.,* 160, 356, 1987.
16. **Udenfriend, S., Stein, S., Bohlen, P., and Dairman, W.,** A new fluorometric procedure for assay of amino acids, peptides and proteins in the picomole range, in *Chemistry and Biology of Peptides,* Meienhofer, J., Ed., Ann Arbor Science, Ann Arbor, MI, 1972, 655.
17. **Felix, A. M. and Turkelsen, G.,** Total fluorometric amino acid analysis using fluorescamine, *Arch. Biochem. Biophys.,* 157, 177, 1973.
18. **Felix, A. M., Toome, V., DeBernardo, S., and Weigele, M.,** Colorimetric amino acid analysis using fluorescamine, *Arch. Biochem. Biophys.,* 168, 601, 1975.
19. **Cronin, J. R. and Hare, P. E.,** Chromatographic analysis of amino acids and primary amines with *o*-phthalaldehyde detection, *Anal. Biochem.,* 81, 151, 1977.
20. **Lund, E., Thomsen, J., and Brunfeldt, K.,** The use of *o*-phthalaldehyde for fluorescence detection in conventional amino acid analyzers. Sub-nanomole sensitivity in the analysis of phenylthiohydantoin-amino acids, *J. Chromatogr.,* 130, 51, 1977.
21. **Stocchi, V., Piccoli, G., Magnani, M., Palma, F., Biagiarelli, B., and Cucchiarini, L.,** Reversed-phase high-performance liquid chromatography separation of dimethylaminoazobenzene sulfonyl- and dimethylaminoazobenzene thiohydantoin-amino acid derivatives for amino acid analysis and microsequencing studies at the picomole level, *Anal. Biochem.,* 178, 107, 1989.

22. **Cohen, S. A. and Strydom, D. J.,** Amino acid analysis utilizing phenylisothiocyanate derivatives, *Anal. Biochem.,* 174, 1, 1988.
23. **Muramoto, K. and Kamiya, H.,** Recovery of tryptophan in peptides and proteins by high-temperature and short-term acid hydrolysis in the presence of phenol, *Anal. Biochem.,* 189, 223, 1990.
24. **Engelhart, W. G.,** Microwave hydrolysis of peptides and proteins for amino acid analysis, *Am. Biotechnol. Lab.,* November, 30, 1990.
25. **Nakazawa, M. and Manabe, K.,** The direct hydrolysis of proteins containing tryptophan on polyvinylidene difluoride membranes by mercaptoethanesulfonic acid in the vapor phase, *Anal. Biochem.,* 206, 105, 1992.
26. **Cheng, Y.-F. and Dovichi, N. J.,** Subattomole amino acid analysis by capillary zone electrophoresis and laser-induced fluorescence, *Science,* 242, 562, 1988.
27. **Kamps, M. P. and Sefton, B. M.,** Acid and base hydrolysis of phosphoproteins bound to immobilon facilitates analysis of phosphoamino acids in gel-fractionated proteins, *Anal. Biochem.,* 176, 22, 1989.
28. **Barkholt, V. and Jensen, A. L.,** Amino acid analysis: determination of cysteine plus half-cystine in proteins after hydrochloric acid hydrolysis with a disulfide compound as additive, *Anal. Biochem.,* 177, 318, 1989.
29. **García Alvarez-Coque, M. C., Medina Hernández, M. J., Villanueva Camañas, R. M., and Mongay Fernández, C.,** Formation and instability of *o*-phthalaldehyde derivatives of amino acids, *Anal. Biochem.,* 178, 1, 1989.
30. **Stocchi, V., Piccoli, G., Magnani, M., Palma, F., Biagiarelli, B., and Cucchiarini, L.,** Reversed-phase high-performance liquid chromatography separation of dimethylaminoazobenzene sulfonyl- and dimethylaminoazobenzene thiohydantoin-amino acid derivatives for amino acid analysis and microsequencing studies at the picomole level, *Anal. Biochem.,* 178, 107, 1989.
31. **Blankenship, D. T., Krivanek, M. A., Ackermann, B. L., and Cardin, A. D.,** High-sensitivity amino acid analysis by derivatization with *O*-phthalaldehyde and 9-fluorenylmethyl chloroformate using fluorescence detection: applications in protein structure determination, *Anal. Biochem.,* 178, 227, 1989.
32. **Tous, G. I., Fausnaugh, J. L., Akinyosoye, O., Lackland, H., Winter-Cash, P., Victorica, F. J., and Stein, S.,** Amino acid analysis on polyvinylidene difluoride membranes, *Anal. Biochem.,* 179, 50, 1989.
33. **Chiou, S.-H. and Wang, K.-T.,** Peptide and protein hydrolysis by microwave irradiation, *J. Chromatogr.,* 491, 424, 1989.
34. **Nakagawa, S. and Fukuda, T.,** Direct amino acid analysis of proteins electroblotted onto polyvinylidene difluoride membrane from sodium dodecyl sulfate-polyacrylamide gel, *Anal. Biochem.,* 181, 75, 1989.
35. **May, M. E. and Brown, L. L.,** Instability of orthophthalaldehyde reagent for amino acid analysis, *Anal. Biochem.,* 181, 135, 1989.
36. **Liu, T.-Y. and Boykins, R. A.,** Hydrolysis of proteins and peptides in a hermetically sealed microcapillary tube: high recovery of labile amino acids, *Anal. Biochem.,* 182, 383, 1989.
37. **D'Aniello, A., Petrucelli, L., Gardner, C., and Fisher, G.,** Improved method for hydrolyzing proteins and peptides with inducing racemization and for determining their true D-amino acid content, *Anal. Biochem.,* 213, 290, 1993.
38. **Jue, R. A. and Hale, J. E.,** Identification of cysteine residues alkylated with 3-bromopropylamine by protein sequence analysis, *Anal. Biochem.,* 210, 39, 1993.
39. **Yang, C.-Y., Yang, T., and Yeh, B. K.,** Liquid chromatographic analysis of amino acids: using dimethylamino-azobenzenesulfonyl chloride and hypersil ODS column to analyze the composition of apo B peptides, *J. Prot. Chem.,* 12, 11, 1993.
40. **Waxman, E., Rusinova, E., Hasselbacher, C. A., Schwartz, G. P., Laws, W. R., and Ross, J. B. A.,** Determination of the tryptophan:tyrosine ratio in proteins, *Anal. Biochem.,* 210, 425, 1993.

Chapter 3

PEPTIDE SEPARATION BY REVERSE-PHASE HIGH-PERFORMANCE LIQUID CHROMATOGRAPHY

High-performance liquid chromatography (HPLC) is an extremely useful method for the characterization of proteins with putative site-specific chemical modification. The variety of approaches available for generating reproducible peptides from complex proteins has continued to increase. The most meaningful advance has been the application of on-line mass spectrometry to the analysis of peptides. The reader is directed to several reviews and a large number of papers in this area.[1–22] Capillary electrophoresis is another technique which can be used for the analysis of small quantities of proteins and peptides.[23–29] The assay of fractions generated in capillary electrophoresis can also be done by mass spectrometric methods.[30–32]

While the past 5 years has seen a marked increase in the sophistication of matrices available, the basic technique of HPLC has not changed. There has been an increase in the use of affinity matrices[33–41] for the separation of peptides. Affinity capillary electrophoresis has also been described.[42]

The original expectation in development of HPLC was that existing thin-layer and column chromatographic separation methods would be transferred to columns packed with small, uniform-sized, rigid particles. This implied a stationary phase of unmodified silica. However, the retention properties of a silica surface are very sensitive to moisture levels. Elution times may be difficult to reproduce, and long column equilibration times may be required following solvent changes. Attempts to improve reproducibility of retention properties by masking the surface silanol groups led to the development of packing materials having covalently bonded surface phases. The most successful of these have been the so-called C-18 or ODS (octadecylsilyl) packings, which have 18-carbon hydrocarbon chains bonded to the surface silanols. The use of these and other hydrophobic bonded-phase packings is known as "reverse-phase" chromatography because the stationary phase is nonpolar and the mobile phase polar, which happens to be the reverse of most older methods of partition liquid chromatography.

The bonded reverse-phase supports are versatile and provide reproducible retention, rapid equilibration to new solvent conditions, and good column life when used within the pH range of 2 to 7.5. Mobile phases are based on water or aqueous buffers; sample retention is controlled by the proportion of a water-miscible organic solvent, often acetonitrile or an alcohol, in the mobile phase. The higher the volume fraction of organic solvent, the "stronger" the mobile phase, that is, the more hydrophobic and better able it is to elute sample components from the nonpolar stationary phase. Simple mixtures of chemically similar small molecules can be adequately separated in a reasonable time with a single mobile phase composition ("isocratic" elution). More complex mixtures or those having a wider range of polarities require a solvent compositional gradient for efficient analysis. The retention of peptides and proteins on reverse-phase columns is highly dependent on solvent strength[43] and on their amino acid compositions; gradient elution is necessary for most polypeptide mixtures.

The application of reverse-phase HPLC to proteins and peptides lagged behind its use for smaller molecules. The size and chemical complexity of polypeptides cause their retention characteristics to be affected by a number of factors. These include the pH of the mobile phase, the buffer system used, the organic mobile phase component, the nature of the bonded stationary phase surface, and the pore size of the stationary phase. Usable elution systems have now been worked out and high-resolution separations can be achieved, particularly of peptides up to about two dozen amino acids in length.

For small peptides, such as most of those produced by tryptic digestion, ordinary C-18 columns can be used. We have obtained good results with the system recommended by Schroeder and co-workers for the mapping of hemoglobin tryptic peptides.[44] A Beckman Ultrasphere-ODS column, 0.46×25 cm, is eluted at 1 ml/min with a phosphate-acetonitrile gradient at ambient temperature. The aqueous component is approximately 50 m*M* phosphate, pH 2.95. The gradient rate is a 0.25% increase in acetonitrile per minute up to 28% acetonitrile, followed by 1.1% increase per minute to a final acetonitrile concentration of 62%. A typical chromatogram of a tryptic digest is shown in Figure 1.

The phosphate-acetonitrile eluent gives good resolution of peptides and allows absorbance monitoring at 210 nm for high-sensitivity detection. Its main defect is a lack of volatility. This level of phosphate salt will interfere in sequence analysis. Schroeder et al.[44] recommend rechromatography of individual peaks using 10 m*M* ammonium acetate adjusted to pH 6.07 with dilute acetic acid as the aqueous component of the mobile phase. We have generally used 5 m*M* phosphate, pH 6.5, for rechromatography, still with an acetonitrile gradient. Up to 1 ml of this eluent can be used for sequence analysis without further desalting. In either case, the change in pH causes changes in relative peptide mobility so that any peaks which coeluted initially are separated during rechromatography.

Larger peptides give broader peaks[45] and are more sensitive to the nature of both the column packing and the mobile phase. Schroeder et al.[44] note poor elution of hemoglobin "core" peptides in the phosphate-acetonitrile system; they form sharper peaks if oxidized, and also are eluted better from a Zorbax TMS (trimethylsilyl) column than from the Ultrasphere-ODS. Similarly, the broad peak late in the chromatogram shown in Figure 1 is a 35-amino acid glycopeptide (the "activation peptide" of human blood coagulation factor IX). To separate it from the phosphate and the coeluting peaks, we chromatographed it on a Vydac TP C-8 column with a gradient from 0.1% trifluoroacetic acid

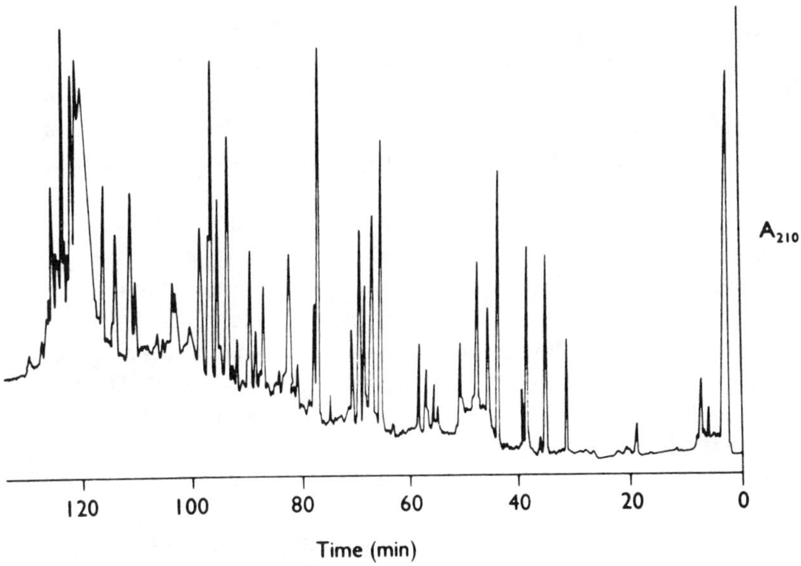

FIGURE 1. Tryptic digest of reduced and carboxymethylated human blood coagulation factor IX. Column, Ultrasphere-ODS, 0.46 × 25 cm; flow rate 1 ml/min; temperature 25°C. Solvent A = 0.05 *M* sodium phosphate, pH 2.95; solvent B = acetonitrile. Gradient program: 0 to 28% B in 114 min, then to 62% B in 30 min; both segments linear.

(TFA) to 2-propanol at a flow rate of 1.0 ml/min, a gradient rate of 0.5%/min, and a temperature of 45°C.

The Vydac TP packing material has 330-Å diameter pores. Several laboratories have shown that supports having 300 to 500-Å pores give better results with large peptides and proteins than the more commonly used supports which have pores of about 100 Å diameter.[46–49]

Either 1-propanol or 2-propanol is a stronger solvent for polypeptides than acetonitrile[50,51] and may be required for elution of some samples not eluted by the latter. However, propanol systems have a higher viscosity than acetonitrile systems. This causes higher column back-pressure and lower efficiency (broader peaks).[45]

Phosphate or phosphoric acid has been found to give particularly good results in the HPLC of peptides.[50–55] It is presumed that ion pairing between peptide $R\text{-}NH_3^+$ groups and the hydrophilic anion $H_2PO_4^-$ increase the polarity of the migrating species, thus decreasing their retention. Some workers add perchlorate as a chaotropic agent to phosphate mobile phases.[56–58] Trifluoroacetic acid also works well;[59] larger perfluoridated acids have been used to provide more hydrophobic counterions and thus increase retention of hydrophilic peptides.[47]

Other mobile phase buffers used in reverse-phase HPLC of polypeptides include triethylamine phosphate[60–65] and, as noted previously, ammonium acetate.[66–68] All the mobile phase systems mentioned to this point are used with absorbance detection of peptides at 205 to 230 nm (and at 254 or 280 nm or other wavelengths if desired). Winkler et al.[69,70] have proposed using absorbance at 215 nm instead of 210 or 205 nm to detect effluent peptide material when using an acetonitrile gradient in TFA to avoid baseline problems concomitant with increasing acetonitrile concentrations. We have found this to be quite useful. Fluorescence detection following post-column derivatization of peptides allows the use of pyridine formate or acetate buffers and is very sensitive but destructive of the portion of the sample diverted to the detector. The fluorescence detector produces some peak broadening from postcolumn mixing of reagents, although this affects only the recorded trace and not the actual resolution of peptides in the portion of the eluent not passed through the detector. They found ultraviolet (UV) absorbance detection more susceptible to impurity and baseline problems.

Chemically modified peptides may be detected by their change in mobility relative to peaks of an unmodified digest, by specific absorbance or fluorescence characteristics of the modifying group, and/or by radiolabel incorporated with the modifying group.

Cyanopropyl columns are less retentive than C-18 or C-8 columns and therefore useful for more hydrophobic peptides. Alkylphenyl columns have been used for peptide separation.[71,72]

Column performance for polypeptide separations is not necessarily related to performance for separation of small molecules. In the latter situation, the sample is considered to undergo a series of equilibration steps between the mobile and stationary phases as it passes through the column.[73] Retention and separation are thus dependent on the column length (number of "theoretical plates" or equilibration steps). With polypeptides this is not the case as 5-cm columns separate as well as 25-cm columns,[74] only loading capacity is affected. Pearson et al.[74] attribute this phenomenon to the operation of an adsorption-desorption mechanism instead of the multistep partitioning which occurs with smaller molecules.

Several groups have determined sets of coefficients for predicting retention of a peptide from its amino acid composition.[75,76] Each set is valid only for a specific chromatographic system (mobile phase, stationary phase, pH, temperature, flow rate, gradient rate, etc.).

Figure 2 shows the separation of the tryptic peptides from lysozyme on a C-18 column using a gradient of acetonitrile in TFA. Figure 3 depicts the separation of the cyanogen bromide peptides from sperm whale myoglobin using a C-4 column with an acetonitrile gradient in TFA. Figure 4 shows the separation of the tryptic peptides from sperm whale myoglobin on a C-18 column using an acetonitrile gradient in TFA.

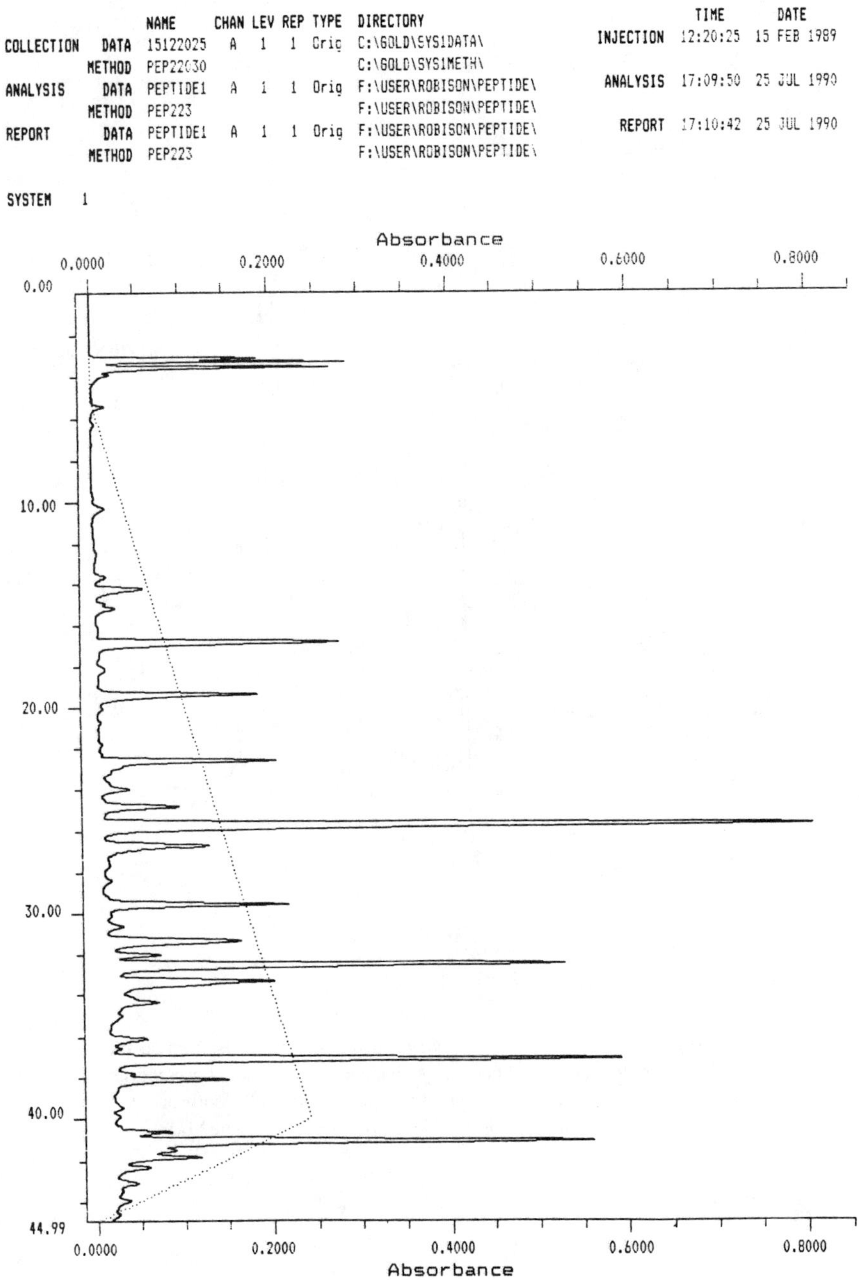

FIGURE 2. HPLC separation of the tryptic digest of lysozyme on an Ultrasphere™ C-18 column. Initial solvent (solvent A) was 5% acetonitrile in 0.1% TFA, and solvent B is 75% acetonitrile in 0.1% TFA. A gradient from 0 to 30% B (shown by dotted line) was used at a flow rate of 1 ml/min. (This figure is courtesy of Beckman Instruments.)

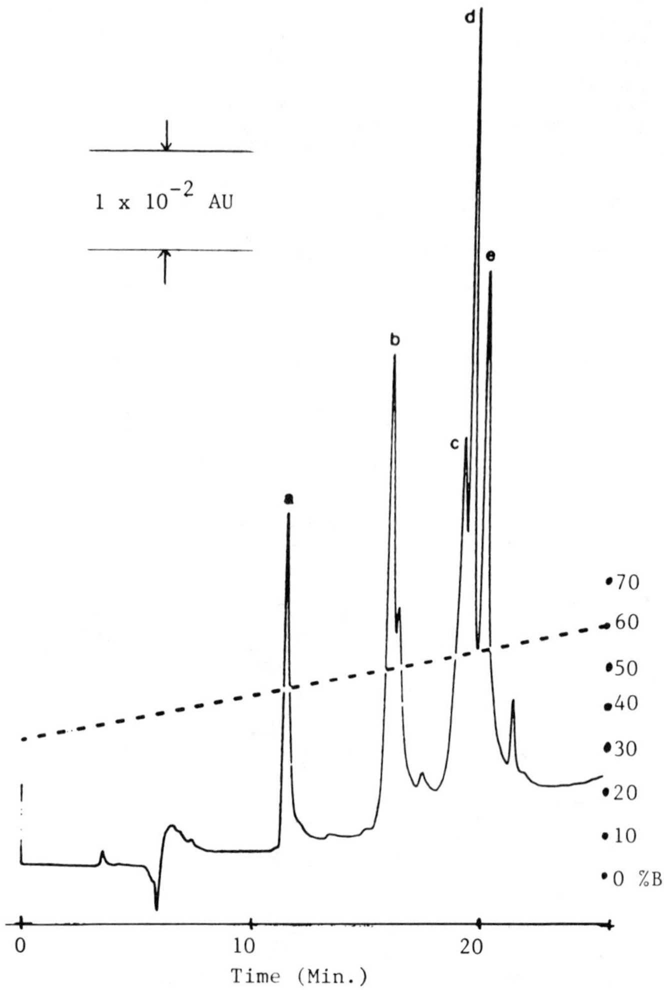

FIGURE 3. The separation of cyanogen bromide peptides from sperm whale myoglobin on a C-4 column. A gradient (dotted line) of acetonitrile in 0.1% TFA was used to develop the column. Effluent absorbance was monitored at 230 nm. (Courtesy of ISCO Scientific, Lincoln, NE.)

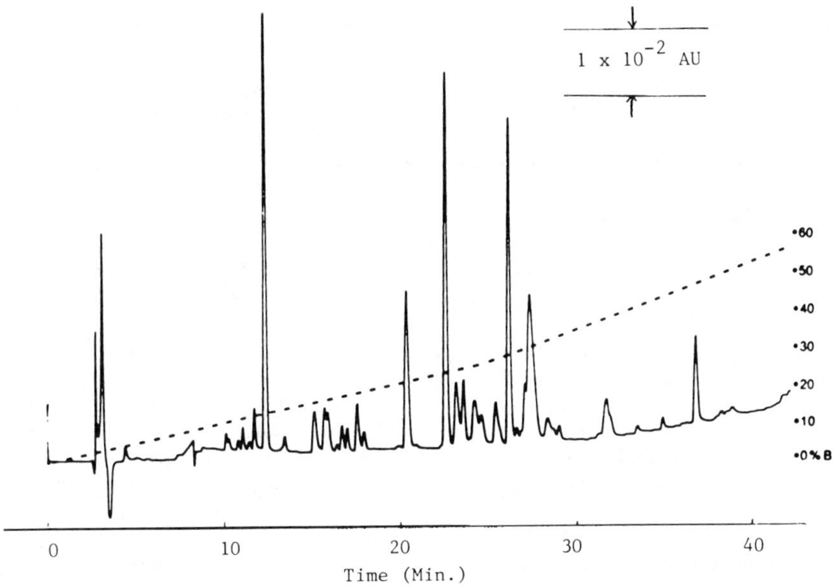

FIGURE 4. The separation of tryptic peptides from sperm whale myoglobin on a C-18 column. A gradient (dotted line) of acetonitrile in 0.1% TFA was used to develop the column. Effluent absorbance was monitored at 230 nm. (Courtesy of ISCO Scientific, Lincoln, NE.)

REFERENCES

1. **Goodlett, D. R., Armstrong, F. B., Creech, R. J., and Van Breemen, R. B.,** Formylated peptides from cyanogen bromide digests identified by fast atom bombardment mass spectrometry, *Anal. Biochem.,* 186, 116, 1990.
2. **Caesar, J. P., Jr., Sheeley, D. M., and Reinhold, V. N.,** Femtomole oligosaccharide detection using a reducing-end derivative and chemical ionization mass spectrometry, *Anal. Biochem.,* 191, 247, 1990.
3. **Nielsen, P. F., Landis, B., Svoboda, M., Schneider, K., and Przybylski, M.,** N-terminal sequence determination of polypeptides and peptide mixtures by Edman degradation combined with Californium-252 plasma desorption mass spectrometry, *Anal. Biochem.,* 191, 302, 1990.
4. **Barr, J. R., Anumula, K. R., Vettese, M. B., Taylor, P. B., and Carr, S. A.,** Structural classification of carbohydrates in glycoproteins by mass spectrometry and high-performance anion-exchange chromatography, *Anal. Biochem.,* 192, 181, 1991.
5. **Loo, J. A., Loo, R. R. O., Light, K. J., Edmonds, C. G., and Smith, R. G.,** Multiply charged negative ions by electrospray ionization of polypeptides and proteins, *Anal. Chem.,* 64, 81, 1992.
6. **Krutzsch, H. C.,** Determination of polypeptide amino acid sequences from the carboxyl terminus using angiotensin I converting enzyme, *Biochemistry,* 19, 5290, 1980.

7. **Barber, M., Wolstenholme, W. A., Guinand, M., Michel, G., Das, B. C., and Lederer, E.,** Determination of amino acid sequences in oligopeptides by mass spectrometry. II. The structure of peptidolipin NA, *Tetranhedron Lett.,* 18, 1331, 1965.
8. **Thorgersen, H. C., Petersen, T. E., Sottrup-Jensen, L., Magnusson, S., and Morris, H. R.,** The N-terminal sequences of blood coagulation factor X_1 and X_2 light chains. Mass-spectrophotometric identification of twelve residues of gamma-carboxyglutamic acid in their vitamin K-dependent domains, *Biochem. J.,* 175, 613, 1978.
9. **Smith, J. B., Thevenon-Emeric, G., Smith, D. L., and Green, B.,** Elucidation of the primary structures of proteins by mass spectrometry, *Anal. Biochem.,* 193, 118, 1991.
10. **Tuineman, A. A., Thomas, D. A., Cook, K. D., Zue, C.-B., Naider, F., and Becker, J. M.,** Mass spectrometric signature of S-prenylated cysteine peptides, *Anal. Biochem.,* 193, 173, 1991.
11. **Carr, S. A., Hemling, M. E., Bean, M. F., and Roberts, G. D.,** Integration of mass spectrometry in analytical biochemistry, *Anal. Chem.,* 63, 2802, 1991.
12. **Hillenkamp, F., Karas, M., Beavis, R. C., and Chait, B. T.,** Matrix-assisted laser desorption/ionization mass spectrometry of biopolymers, *Anal. Chem.,* 83, 1193A, 1991.
13. **van Breeman, R. B. and Davis, R. G.,** Rates of peptide proteolysis measured using liquid chromatography and continuous-flow fast atom bombardment mass spectrometry, *Anal. Chem.,* 64, 2233, 1992.
14. **Carr, S. A., Roberts, G. D., Jurewicz, A., and Frederick, B.,** Structural fingerprinting of Asn-linked carbohydrates from specific attachment sites in glycoproteins by mass spectrometry: application to tissue plasminogen activator, *Biochimie,* 70, 1445, 1988.
15. **Furuya, M., Akashi, S., and Hirayama, K.,** The primary structure of human EGF produced by genetic engineering, studied by high-performance tandem mass spectrometry, *Biochem. Biophys. Res. Commun.,* 163, 1100, 1989.
16. **Piot, J.-M., Guillochon, D., Zhao, Q., Ricart, G., Fournet, B., and Thomas, D.,** Identification of peptides, from a peptic haemoglobin hydrolysate produced at pilot-plant scale, by high-performance liquid chromatography and mass spectrometry, *J. Chromatogr.,* 481, 221, 1989.
17. **Roepstorff, P.,** Plasma desorption mass spectrometry of peptides and proteins, *Acc. Chem. Res.,* 22, 421, 1989.
18. **Nexo, E., Jorgensen, P. E., Thim, L., and Roepstorff, P.,** Purification and characterization of a low and a high molecular weight form of epidermal growth factor from rat urine, *Biochim. Biophys. Acta,* 1037, 388, 1990.
19. **Chowdhury, S. K., Katta, V., and Chait, B. T.,** Electrospray ionization mass spectrometric peptide mapping: a rapid, sensitive technique for protein structure analysis, *Biochem. Biophys. Res. Commun.,* 167, 686, 1990.
20. **Henzel, W. J., Bourell, J. H., and Stults, J. T.,** Analysis of protein digests by capillary high-performance liquid chromatography and on-line fast atom bombardment mass spectrometry, *Anal. Biochem.,* 187, 228, 1990.
21. **Andrews, P. C. and Dixon, J. E.,** Application of fast atom bombardment mass spectrometry to posttranslational modifications of neuropeptides, *Meth. Enzymol.,* 168, 72, 1989.
22. **Huddleston, M. J., Bean, M. F., and Carr, S. A.,** Collisional fragmentation of glycopeptides by electrospray ionization LC/MS and LC/MS/MS: methods for selective detection of glycopeptides in protein digests, *Anal. Chem.,* 65, 877, 1993.
23. **Olechno, J. D., Tso, J. M. Y., Thayer, J., and Wainright, A.,** Capillary electrophoresis: a multifaceted technique for analytical chemistry. II. Separations and injection, *Am. Lab.,* December, 30, 1990.
24. **Pickering, M. V.,** Capillary electrophoresis: proteins, problems, and promises, *LC/GC,* 7, 752, 1989.
25. **Cobb, K. A. and Novotny, M. V.,** Peptide mapping of complex proteins at the low-picomole level with capillary electrophoretic separations, *Anal. Chem.,* 64, 879, 1992.

26. **Ganzler, K., Greve, K. S., Cohen, A. S., Karger, B. L., Guttman, A., and Cooke, N. C.,** High-performance capillary electrophoresis of SDS-protein complexes using UV-transparent polymer networks, *Anal. Chem.*, 64, 2665, 1992.
27. **Heegaard, N. H. H. and Robey, F. A.,** Use of capillary zone electrophoresis to evaluate the binding of anionic carbohydrates to synthetic peptides derived from human serum amyloid P component, *Anal. Chem.*, 64, 2479, 1992.
28. **Rush, R. S., Derby, P. L., Strickland, T. W., and Rohde, M. F.,** Peptide mapping and evaluation of glycopeptide microheterogeneity derived from endoproteinase digestion of erythropoietin by affinity high-performance capillary electrophoresis, *Anal. Chem.*, 65, 1834, 1993.
29. **Karger, B. L., Cohen, A. S., and Guttman, A.,** High-performance capillary electrophoresis in the biological sciences, *J. Chromatogr.*, 492, 585, 1989.
29a. **Swedberg, S. A.,** Characterization of protein behavior in high-performance capillary electrophoresis using a novel capillary system, *Anal. Biochem.*, 185, 51, 1990.
30. **Henzel, W. J., Bourell, J. H., and Stults, J. T.,** Analysis of protein digests by capillary high-performance liquid chromatography and on-line fast atom bombardment mass spectrometry, *Anal. Biochem.*, 187, 228, 1990.
31. **Colburn, J. C.,** Capillary electrophoresis separations of peptides: practical aspects and applications, in *Capillary Electrophoresis. Theory and Practice*, Grossman, P. D. and Colburn, J. C., Eds., Academic Press, San Diego, CA, 1991, 237.
32. **Smith, R. D., Wahl, J. H., Goodlett, D. R., and Hofstadler, S. A.,** Capillary electrophoresis/mass spectrometry, *Anal. Chem.*, 65, 574A, 1993.
33. **Jacquot-Dourges, M. A., Muller, D., Barritault, D., and Jozefonvicz, J.,** High-performance affinity chromatography of basic fibroblast growth factor on polystyrene sulphonate resins modified with serine, *J. Chromatog.*, 510, 141, 1990.
33a. **Ohlson, S., Hansson, L., Glad, M., Mosbach, K., and Larsson, P.-O.,** High performance liquid affinity chromatography: a new tool in biotechnology, *Trends Biotechnol.*, 7, 179, 1989.
34. **Fassina, G., Swaisgood, H. E., and Chaiken, I. M.,** Quantitative high-performance affinity chromatography: evaluation of use for analyzing peptide and protein interactions, *J. Chromatog.*, 376, 87, 1986.
35. **Clarke, A. J. and Svensson, B.,** Identification of an essential tryptophanyl residue in the primary structure of glucoamylase G2 from Aspergillus niger, *Carlsberg Res. Commun.*, 49, 599, 1984.
36. **Regnier, F. E.,** The role of protein structure in chromatographic behavior, *Science*, 238, 319, 1987.
37. **Ladisch, M. R. and Kohlmann, K. L.,** Recombinant human insulin, *Biotechnol. Progr.*, 8, 469, 1992.
38. **Torres, J. L., Guzman, R., Carbonell, R. G., and Kilpatrick, P. K.,** Affinity surfactants as reversibly bound ligands for high-performance affinity chromatography, *Anal. Biochem.*, 171, 411, 1988.
39. **Green, E. D. and Baenziger, J. U.,** Characterization of oligosaccharides by lectin affinity high-performance liquid chromatography, *TIBS*, 14, 168, 1989.
40. **Smith, J. A. and O'Hare, M.,** Effect of stationary and mobile phases on hydrophobic interaction chromatography of proteins and peptides, *J. Chromatogr.*, 496, 71, 1989.
41. **Kaul, R., Olsson, U., and Mattiasson, B.,** Affinity chromatography on a hydrophobic matrix using a heterobifunctional ligand, *J. Chromatogr.*, 438, 339, 1988.
42. **Rush, R. S., Derby, P. L., Strickland, T. W., and Rohde, M. F.,** Peptide mapping and evaluation of glycopeptide microheterogeneity derived from endoproteinase digestion of erythropoietin by affinity high-performance capillary electrophoresis, *Anal. Chem.*, 65, 1834, 1993.
43. **Wehr, C. T., Correia, L., and Abbott, S. R.,** Evaluation of stationary and mobile phases for reversed phased high performance liquid chromatography of peptides, *J. Chromatogr Sci.*, 20, 114, 1982.

44. **Schroeder, W. A., Shelton, J. B., and Shelton, J. R.,** High performance liquid chromatography in the identification of human hemoglobin variants, in *Advances in Hemoglobin Analysis,* Hanash, S. M. and Brewer, G. J., Eds., Alan R. Liss, New York, 1981, 1.
45. **Meek, J. L. and Rossetti, Z. L.,** Factors affecting retention and resolution of peptides in high-performance liquid chromatography, *J. Chromatogr.,* 211, 15, 1981.
46. **Lewis, R. V., Fallon, A., Stein, S., Gibson, K. D., and Udenfriend, S.,** Supports for reverse-phase high-performance liquid chromatography of large proteins, *Anal. Biochem.,* 104, 153, 1980.
47. **van der Rest, M., Bennett, H. P. J., Solomon, S., and Glorieux, F. H.,** Separation of collagen cyanogen bromide-derived peptides by reversed-phase high-performance liquid chromatography, *Biochem. J.,* 191, 253, 1980.
48. **Pearson, J. D., Mahoney, W. C., Hermodson, M. A., and Regnier, F. E.,** Reversed-phase supports for the resolution of large denatured protein fragments, *J. Chromatogr.,* 207, 325, 1981.
49. **Wilson, K. J., Van Wieringen, E., Klauser, S., Berchtold, M. W., and Hughes, G. J.,** Comparison of the high-performance liquid chromatography of peptides and proteins on 100- and 300-Å reversed-phase supports, *J. Chromatogr.,* 237, 407, 1982.
50. **Mahoney, W. C. and Hermodson, M. A.,** Separation of large denatured peptides by reverse phase high performance liquid chromatography. Trifluoroacetic acid as a peptide solvent, *J. Biol. Chem.,* 255, 11199, 1980.
51. **Wilson, K. J., Honegger, A., Stötzel, R. R., and Hughes, G. J.,** The behaviour of peptides on reverse-phase supports during high-pressure liquid chromatography, *Biochem. J.,* 199, 31, 1981.
52. **Hancock, W. S., Bishop, C. A., Prestidge, R. L., Harding, D. R. K., and Hearn, M. T. W.,** High-pressure liquid chromatography of peptides and proteins. II. The use of phosphoric acid in the analysis of underivatised peptides by reversed-phase high-pressure liquid chromatography, *J. Chromatogr.,* 153, 391, 1978.
53. **Fullmer, C. S. and Wasserman, R. H.,** Analytical peptide mapping by high performance liquid chromatography. Application to intestinal calcium-binding proteins, *J. Biol. Chem.,* 254, 7208, 1979.
54. **O'Hare, M. J. and Nice, E. C.,** Hydrophobic high-performance liquid chromatography of hormonal polypeptides and proteins on alkylsilane-bonded silica, *J. Chromatogr.,* 171, 209, 1979.
55. **Nice, E. C. and O'Hare, M. J.,** Simultaneous separation of β-lipotrophin, adrenocorticotropic hormone, endorphins and enkephalins by high-performance liquid chromatography, *J. Chromatogr.,* 162, 401, 1979.
56. **Meek, J. L.,** Prediction of peptide retention times in high-pressure liquid chromatography on the basis of amino acid composition, *Proc. Natl. Acad. Sci. U.S.A.,* 77, 1632, 1980.
57. **Wilson, K. J., Honegger, A., and Hughes, G. J.,** Comparison of buffers and detection systems for high-pressure liquid chromatography of peptide mixtures, *Biochem. J.,* 199, 43, 1981.
58. **Strydom, D. J. and Vallee, B. L.,** Characterization of human alcohol dehydrogenase isoenzymes by high-performance liquid chromatographic peptide mapping, *Anal. Biochem.,* 123, 422, 1982.
59. **Bennett, H. J. P., Browne, C. A., and Solomon, S.,** The use of perfluorinated carboxylic acids in the reversed-phase HPLC of peptides, *J. Liquid Chromatogr.,* 3, 1353, 1980.
60. **Browne, C. A., Bennett, H. J. P., and Solomon, S.,** The isolation of peptides by high-performance liquid chromatography using predicted elution positions, *Anal. Biochem.,* 124, 201, 1982.
61. **Rivier, J. E.,** Use of trialkyl ammonium phosphate (TAAP) buffers in reverse phase HPLC for high resolution and high recovery of peptides and proteins, *J. Liquid Chromatogr.,* 1, 343, 1978.

62. **Biedermann, K., Montali, U., Martin, B., Svendsen, I., and Ottesen, M.,** The amino acid sequence of proteinase A inhibitor 3 from baker's yeast, *Carlsberg Res. Commun.*, 45, 225, 1980.
63. **Chaiken, I. M. and Hough, C. J.,** Mapping and isolation of large peptide fragments from bovine neurophysins and biosynthetic neurophysin-containing species by high-performance liquid chromatography, *Anal. Biochem.*, 107, 11, 1980.
64. **Hancock, W. S., Capra, J. D., Bradley, W. A., and Sparrow, J. T.,** The use of reversed-phase high-performance liquid chromatography with radial compression for the analysis of peptide and protein mixtures, *J. Chromatogr.*, 206, 59, 1981.
65. **Hearn, M. T. W. and Grego, B.,** High-performance liquid chromatography of amino acids, peptides and proteins. XXXVI. Organic solvent modifier effects in the separation of unprotected peptides by reversed-phase liquid chromatography, *J. Chromatogr.*, 218, 497, 1981.
66. **Wilson, J. B., Lam, H., Pravatmuang, P., and Huisman, T. H. J.,** Separation of tryptic peptides of normal and abnormal α, β, γ, and δ hemoglobin chains by high-performance liquid chromatography, *J. Chromatogr.*, 179, 271, 1979.
67. **Hancock, W. S. and Sparrow, J. T.,** Use of mixed-mode, high-performance liquid chromatography for the separation of peptide and protein mixtures, *J. Chromatogr.*, 206, 71, 1981.
68. **Oray, B., Jahani, M., and Gracy, R. W.,** High-sensitivity peptide mapping of triosephosphate isomerase: a comparison of high-performance liquid chromatography with two-dimensional thin-layer methods, *Anal. Biochem.*, 125, 131, 1982.
69. **Winkler, G., Wolshann, P., Briza, P., Heinz, F. X., and Kunz, C.,** Spectral properties of trifluoroacetic acid-acetonitrile gradient systems for separation of picomole quantities of peptides by reversed-phase high-performance liquid chromatography, *J. Chromatogr.*, 34, 83, 1985.
70. **Winkler, G.,** Increasing the sensitivity of UV detection in protein and peptide separations when using TFA-acetonitrile gradients, *LC/GC*, 5, 1044, 1987.
71. **Hancock, W. S., Bishop, C. A., Meyer, L. J., Harding, D. R. K., and Hearn, M. T. W.,** High-pressure liquid chromatography of peptides and proteins. VI. Rapid analysis of peptides by high-pressure liquid chromatography with hydrophobic ion-pairing of amino groups, *J. Chromatogr.*, 161, 291, 1978.
72. **Hancock, W. S., Bishop, C. A., Battersby, J. E., Harding, D. R. K., and Hearn, M. T. W.,** High-pressure liquid chromatography of peptides and proteins. XI. The use of cationic reagents for the analysis of peptides by high-pressure liquid chromatography, *J. Chromatogr.*, 168, 377, 1979.
73. **Martin, A. J. P. and Synge, R. L. M.,** A new form of chromatogram employing two liquid phases. I. A theory of chromatography. II. Application to the micro-determination of the higher monoamino-acids in proteins, *Biochem. J.*, 35, 1358, 1941.
74. **Pearson, J. D., Lin, N. T., and Regnier, F. E.,** The importance of silica type for reverse-phase protein separations, *Anal. Biochem.*, 124, 217, 1982.
75. **Su, S.-J., Grego, B., Niven, B., and Hearn, M. T. W.,** Analysis of group retention contributions for peptides separated by reversed-phase-high performance liquid chromatography, *J. Liquid Chromatogr.*, 4, 1745, 1981.
76. **Sasagawa, T., Okuyama, T., and Teller, D. C.,** Prediction of peptide retention times in reversed-phase high-performance liquid chromatography during linear gradient elution, *J. Chromatogr.*, 240, 329, 1982.

Chapter 4

METHODS FOR SEQUENCE DETERMINATION

I. INTRODUCTION

Determination of the amino acid sequence around a site of chemical modification provides unequivocal identification of its location in the proteins, providing the complete sequence of the protein is known. Although analysis of amino acid composition alone may allow identification of the peptide containing the modification, even a few cycles of sequence analysis serve to verify its purity and identity. While the remainder of this chapter will be concerned with a description of classical techniques for the determination of amino acid sequence by solution chemistry, it is emphasized that mass spectrometry is rapidly becoming the technique of choice for sequence analysis.[1–8]

Most methods for the determination of amino acid sequence use the degradation scheme developed by Edman.[9,10] (The procedure was originally described as a micro-method requiring "only" 10 mg of amino acid per cycle.) Phenylisothiocyanate (PITC) reacts with the NH_2-terminal amino acid of a polypeptide at basic pH (see Scheme 1) to form a phenylthiocarbamyl derivative (Reaction 1). Acidification, generally with anhydrous acid, then cleaves off the first amino acid as its 2-anilino-5-thiazolinone derivative and exposes the amino group of the second amino acid (Reaction 2). The derivatized amino acid is removed by extraction, the remaining polypeptide is dried, and the cycle is repeated. The large difference in pH between coupling and cleavage allows the degradation to proceed by discrete steps, unlike exopeptidase digestions, which begin to release the second amino acid before all of the first has been removed.

Procedures for sequence analysis by the Edman degradation fall into two major categories differing in the type of method used for determination of the amino acid at each step. In so-called indirect Edman procedures, a portion of the sample is removed at the beginning and after each cycle and analyzed either quantitatively for total amino acid composition or qualitatively for the identity of the NH_2-terminal. In direct methods, the thiazolinone produced at each step is extracted, converted to the more stable thiohydantoin (Reaction 3), and identified. Manual procedures of both types are used; automated sequenators have generally used direct identification.

The following points apply to all Edman degradation techniques.

1. All procedures and identification methods should be practiced first on known peptides.
2. Ideally, known examples of derivatives from all amino acids from actual degradations should be observed in the identification system to be used.

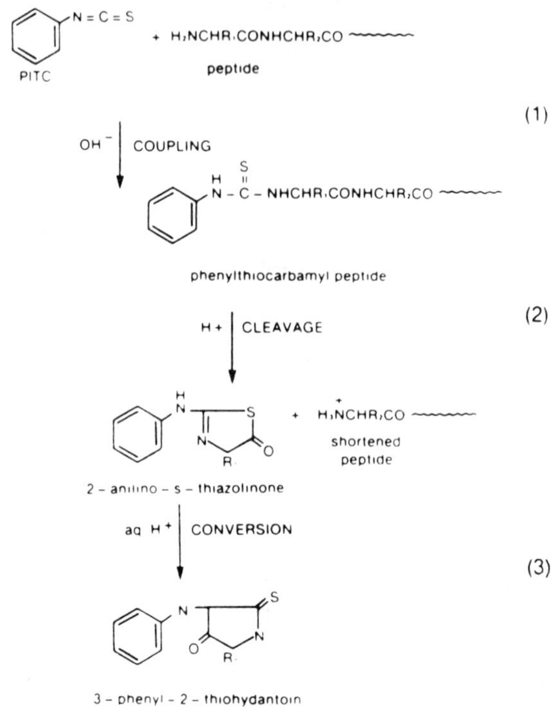

SCHEME 1.

3. Purity of reagents and solvents, particularly freedom from aldehydes and oxidants, is important.
4. Degradations are performed in an atmosphere of nitrogen or argon to exclude oxygen and air pollutants that can cause side reactions and block further degradation.[11,12] Manual degradation may be carried out in a glove box[13] or closed reaction vessel[14] to facilitate maintenance of an inert atmosphere.

II. INDIRECT EDMAN DEGRADATION

A. Subtractive

An aliquot of the peptide is removed after each cycle and subjected to hydrolysis and quantitative amino acid analysis. A typical procedure is outlined by Konigsberg.[15]

The medium for coupling can be (1) 50% aqueous pyridine containing 2% triethylamine, (2) *N*-ethylmorpholine per acetic acid per 95% ethanol per water (60:1.5:500:438), or (3) 50% pyridine containing 5% dimethylallylamine. The peptide is dissolved in coupling buffer and treated with a 50-fold excess of PITC at 37°C for 2 h. (Exclusion of oxygen and contaminants during coupling is particularly important in the subtractive method, since partial blockage of peptide will cause nonintegral loss of amino acid.)

After coupling, the sample is evaporated nearly to dryness, extracted three times with 1 to 2 ml benzene, and then dried. Cleavage is carried out in anhydrous trifluoroacetic acid (TFA) at 25°C for 1 h or at 40°C for 15 min. The TFA is evaporated; the residue is dissolved in 0.2 *M* acetic acid and heated at 40°C for 10 min. It is then extracted three times with benzene. An appropriate aliquot of the acetic acid solution is removed for hydrolysis and amino acid analysis (see Chapter 2). The remaining sample is dried, and the cycle is repeated.

Subtractive degradation is applicable only to peptides small enough that loss of a single amino acid is clearly distinguishable in the amino acid analysis. The amount of peptide needed depends on the sensitivity of the analyzer and the number of Edman cycles to be carried out. Glu/Gln and Asp/Asn are not differentiated by the usual acid hydrolysis procedure, since Gln and Asn are deamidated.

B. Dansyl-Edman

Intensely fluorescent derivatives result from the reaction of 1-dimethylaminonaphthalene-5-sulfonyl (dansyl) chloride with free amino groups. Dansyl amino acid derivatives are stable to acid hydrolysis and exhibit yellow fluorescence.[16] The NH_2-terminal of a peptide therefore can be identified by reacting it with dansyl chloride, hydrolyzing the peptide, and identifying the fluorescent derivative by electrophoresis or chromatography.[16–19] For sequence determination, an aliquot is removed at each cycle of degradation and analyzed in this way. Since 10 pmol of dansyl amino acid is readily detected,[19] the amount removed can be less than in subtractive Edman degradation. Also, the sample need only be washed after cleavage, so extractive losses of peptide may be less than with other procedures. The method requires no major equipment and is considered simple and easily learned.[20]

Instructions for dansyl-Edman degradation of about 1 nmol of peptide are given by Bruton and Hartley.[19] The sample is dissolved in 20 μl of water, a 10-pmol aliquot is removed for dansylation, and 20 μl of 5% (v/v) PITC in pyridine is added to the remainder. The tube is flushed with nitrogen and heated at 45°C for 1 h, then dried in a vacuum desiccator at 60°C for 30 min. TFA (20 μl) is added, and the sample is incubated at 45°C for 30 min and dried *in vacuo* over NaOH. The residue is dissolved in 25 μl of water and extracted with four 150-μl portions of *n*-butyl acetate. The aqueous solution is dried *in vacuo* and redissolved in 20 μl of water. Another 10-pmol aliquot is removed for dansylation, and the cycle is repeated on the remainder.

The aliquots removed at each cycle are dried in a vacuum desiccator over P_2O_5. Two successive 2-μl portions of 0.1 *M* $NaHCO_3$ are added and dried to remove ammonia. Dansyl chloride solution (1 μl of a 1:1 mixture of water and a solution of 2.5 mg dansyl chloride per milliliter acetone) is added, and the sample is incubated at 37°C for 1 h. After drying, 5 µl of 6 *N* HCl is added. The tube is sealed, and the peptide is hydrolyzed at 105°C for 16 h. The dansyl amino acids are identified by two-dimensional thin-layer chromatography on 5 × 5 cm polyamide sheets.[18] A fine capillary is used for sample application to keep the spot as small as possible. A mixture of standards is run on the reverse side of the sheet. The solvent for the first dimension is 1.5% (v/v) formic acid. In the second dimension, benzene/acetic acid 9:1 (or less toxic toluene/acetic acid 10:1) is run, followed in the same direction by ethyl acetate/methanol/ acetic acid 20:1:1. The plate is dried and examined under UV light after each step. Resolution of certain spots may require a fourth solvent, 0.05 *M* trisodium phosphate/ethanol 3:1, also run in the second direction. As in the subtractive method, glutamine and asparagine are deamidated during hydrolysis. The thiazolinones may be recovered from the organic solvent, converted to the phenylthiohydantoins, and identified by one of the methods described later.

Gray and Smith developed a rapid dansyl-Edman procedure for five or six cycles on small peptides.[21] The sample is divided at the start into portions corresponding to the number of cycles to be carried out, and one tube is set aside before each cycle. The desired number of cycles is performed, with drying at 70°C *in vacuo* after each coupling and each cleavage, but with no solvent wash until the end. At that time all tubes are washed with water-saturated ethyl acetate to remove phenylthiourea and diphenylthiourea. Dansylation, hydrolysis, and identification are performed as described above.

Detailed instructions for a modified dansyl-Edman procedure suitable for analysis of proteins (which tend to become insoluble in the solvents used for peptides) are given by Weiner et al.[22]

III. DIRECT EDMAN DEGRADATION

A. Manual PITC

An extensive discussion of the factors affecting the Edman chemistry and of possible reagent, solvent, and temperature options for manual degradation is given by Tarr.[14] The manual Edman procedure described by Peterson et al.[23] has been widely used.

Levy recommends the following technique for 1 to 10 nmol of peptide.[24] Acid-washed polypropylene microcentrifuge tubes are used, and all manipulations of liquids are done with polypropylene micropipet tips. Argon is preferred for flushing, since it is heavier than nitrogen and thus stays in the tubes better. The buffer for coupling consists of 15 ml pyridine, 1.18 ml dimethylallylamine, and 10 ml water; the pH is adjusted to 9.5 with TFA. Norleucine, 25 to 50 nmol, is added at the beginning of each cycle as a carrier and internal standard.

Coupling takes place with 40 µl buffer plus 3 µl PITC for 30 min at 50°C under argon. The sample is washed twice with heptane/ethyl acetate 10:1 and once with heptane/ethyl acetate 2:1 with vigorous mixing followed by centrifugation to separate the phases. The sample (aqueous phase) is dried under vacuum. Cleavage follows in 20 µl TFA for 20 min at 50°C. After vacuum drying, the sample is dissolved in 40 μl of 30% pyridine, extracted with three 150-μl portions of benzene/ethyl acetate (1:2), and the cycle is repeated.

The benzene/ethyl acetate extract containing the thiazolinone is dried. Conversion to the phenylthiohydantoin (PTH) is done in 40 μl 1 *N* HCl for 10 min at 80°C, followed by extraction with three 50-μl portions of ethyl acetate. All PTHs (phenylthiohydantoin derivatives) except PTH-Arg, -His, and -$CysSO_3H$ are extracted by the ethyl acetate. The PTHs are identified by HPLC or one of the other methods described in Section IV of this chapter.

There have been significant advances in manual techniques for sequence analysis of peptides and proteins. Shively and co-workers[25,26] have developed an attractive apparatus for performing manual gas-phase Edman degradation reactions. Brandt and Frank[27] have also reported a manual technique for the gas-phase Edman degradation.

B. Manual DABITC/PITC

The Edman degradation can be carried out with the compound dimethylaminoazobenzene-4′-isothiocyanate (DABITC) in place of PITC.[28,29] The thiohydantoin amino acid derivatives (DABTHs) of this compound are red (ε_M, 436 nm ≅ 34,000) so they can be detected visually on thin layer sheets, with the further advantage that derivatives of non-amino acid impurities appear blue or are invisible. However, quantitative coupling with DABITC requires a temperature of 75°C. Consequently, a double coupling is performed instead at 52°C; the second coupling is with PITC to drive the reaction to completion.

In the method developed by Chang et al.,[29] 2 to 8 nmol of peptide or protein are dried in an acid-washed tube and dissolved in 80 μl of 50% pyridine. Freshly prepared DABITC solution (2.82 mg/ml pyridine), 40 μl, is added. The tube is flushed with nitrogen and heated at 52°C for 50 min. PITC, 10 μl, is added, followed by another 30-min incubation at 52°C. The mixture is extracted two or three times by 0.5-ml portions of heptane/ethyl acetate 2:1, with vortexing followed by centrifugation. The aqueous phase is dried under high vacuum. The sample is dissolved in 50 μl anhydrous TFA, flushed with nitrogen, sealed with a glass stopper, and heated at 52°C for 15 min. After drying in a vacuum desiccator, it is dissolved in 50 μl of water and extracted with 200 μl of butyl acetate. The sample is dried, and the cycle is repeated.

For conversion, the butyl acetate extract is evaporated; 20 μl of water and 40 μl of acetic acid saturated with HCl are added. Conversion is at 52°C for 50 min. The sample is dried, redissolved in ethanol, and applied to polyamide sheets for thin-layer chromatography. Acetic acid/water 1:2 is run in the first dimension and toluene/*n*-hexane/acetic acid 2:1:1 in the second. DABTH-Leu and -Ile are resolved on silica gel plates in chloroform/ethanol 100:3. The plates are dried and exposed to HCl vapor; DABTH spots are red. Sensitivity of detection is 5 to 25 pmol on polyamide and 25 to 50 pmol on silica.

Hughes, Wilson, and co-workers[30,31] have adapted the DABITC/PITC degradation scheme for use in automated sequenators. They prefer 1 *N* HCl or 20% TFA for the conversion step. DABITC is light sensitive and not very stable in solution, so appropriate precautions must be taken. Fischer and Howden[32] have described a manual sequence technique for picomole quantities of peptide using the 4-*N,N'*-dimethylaminoazobenzene 4′-thiohydantoin derivatives.

C. Spinning-Cup Sequenator

In 1967, Edman and Begg[33] published a description of a machine for automatically performing the Edman degradation of proteins. Essentially a "robot chemist", the sequenator or sequencer carries out the coupling, wash, cleavage, and extraction steps of the cycle. Conversion of thiazolinones to thiohydantoins is still often performed manually, although automated converters have been devised.[34,35]

The heart of the sequenator is a spinning glass cup in a temperature-controlled reaction chamber. The protein or peptide sample is deposited in a thin film in the lower half of the cup. Reagents for coupling or cleavage are added to cover the film. Solvents for extraction are run into the cup for several minutes so that the liquid flows up over the precipitated sample, out through a tube at the top of the cup, and is directed by a valve to waste or a fraction collector. The sample can be dried by a stream of nitrogen followed by progressive degrees of vacuum. An inert atmosphere is maintained in the reaction chamber.

In addition to the convenience of unattended operation, the sequenator gives better results than manual degradation in extended sequence analyses of intact proteins. These tend to become insoluble precipitates after a few cycles of manual degradation, with concomitant poor yields and out-of-phase results. The spinning cup of the sequenator keeps the sample spread out in a thin film, thus maintaining efficiency of reaction and extraction.

For small peptides, on the other hand, the sequenator does not have as clear a performance advantage over manual techniques. With these samples, it is more difficult to achieve adequate extraction of reagents, byproducts, and thiazolinones without also washing out the sample. Manual procedures offer more flexibility and better control. A great deal of effort has gone into improving sequenator retention of small peptides, including modifying them to decrease their solubility in organic solvents[36,37] and use of nondegradable or artificial carriers.[38,39] The most successful carrier has been the polymer 1,5-dimethyl-1,5-diazaundecamethylene polymethobromide, commonly known as Polybrene.[40,41] Addition of 1 to 3 mg of Polybrene to the cup with the sample allows complete sequence analysis of even subnanomolar amounts of many peptides. Le Caer and Rossier[42] have reported that polyethyleneimine is superior to Polybrene for sequencing peptides and proteins since the precycling used to condition Polybrene-coated glass fiber filters was no longer necessary. Extremely hydrophobic peptides may still require manual degradation, but in general, automated sequence analysis is now as feasible for peptides as for proteins.

As with all sequence analysis techniques, the amount of sample required for automated Edman degradation is determined by the sensitivity of the identification method used, the repetitive yield on successive cycles of degradation, and the number of cycles to be run. Identification depends on both absolute sensitivity and levels of background contaminants. Besides whatever background is obtained from reagent impurities and side reactions, an increasing level of amino acid background arises during a sequence analysis because of small amounts of nonspecific cleavage of peptide bonds.[33] This background is proportional to the size of the polypeptide and runs on intact proteins, commonly limited in length by the decreasing signal-to-noise ratio arising from decreasing yield and increasing background from nonspecific cleavage and cycle-to-cycle overlap. Peptides up to a few dozen residues in length, on the other hand, produce very little cumulative background and can be followed to the limits imposed by the sensitivity of the detection method used and/or the background from the degradation procedure.

We are able to achieve 93% repetitive yield on 250 pmol of a peptide using an unmodified Beckman 890C sequenator, Polybrene carrier, and Beckman chemicals. Identification of PTHs by HPLC with this system is possible down to 10 to 50 pmol. The first amino acids to be unidentifiable tend to be serine and proline.

We used the 0.1 *M* Quadrol sequenator program of Brauer et al.[40] with the following modifications. After coupling, the buffer is flushed from the delivery line with a short delivery of ethyl acetate. Delivery of heptafluorobutyric acid (HFBA) for cleavage is modified to use a minimum amount of acid, to avoid blowing acid vapors through the effluent valve, and to provide thorough flushing of acid from the delivery line. HFBA is delivered for 3 s, followed by 20 s of restricted vacuum to draw it into the cup. After cleavage and a preliminary drying step, *n*-chlorobutane is delivered briefly through the same line and dried. Extraction with *n*-chlorobutane follows. Three mg of Polybrene is added to the cup and put through one complete cycle before adding the sample. (Some authors find more extensive precycling of Polybrene necessary,[41–43] so there may be lot-to-lot variations in its purity.)

Some laboratories have devised extensive modifications of the spinning-cup sequenator.[41,44,45] These changes in conjunction with meticulous purification reagents provide decreased background levels and improved repetitive yields, thus allowing extended degradation on subnanomolar quantities of sample. Applications related to the topic of this book should not generally require such measures.

Automated devices for the conversion of thiazolinones to PTHs have described.[35,36] Although they increase convenience and possibly recovery levels of unstable derivatives, their use places more exacting demands on system cleanliness, reagent purity, and the resolution of the identification method and may in turn necessitate the further sequenator and reagent improvements described below. With an automatic converter the conversion mixture is dried down in a single phase. Thus, all the PTHs must be distinguished from each other and from all byproducts during identification, whereas manual conversion divides the products between an organic and an aqueous phase, which are analyzed separately.

D. Solid-Phase Sequenator

One approach to the problem of sample loss during washing is covalent attachment of the peptide to a solid support.[46] The support beads are then placed in a temperature-controlled column or reaction chamber, and reagents and solvents are passed through.

Methods of covalent attachment have been summarized by Laursen.[47] Peptides can be attached to appropriate supports through the COOH-terminal homoserine produced by cyanogen bromide digestion, through lysine ε-NH_2 groups, or through carboxyl groups. In the latter method, aspartic and glutamic acid side chains tend to become bound in addition to the COOH-terminal carboxyl, causing gaps at those points in the sequence. Support materials are derivatized glass or polystyrene beads.[46,48]

The necessary repertoire of materials and attachment methods is cumbersome, but they may be used to achieve separation of peptide mixtures by selective coupling and sequences of COOH-terminal portions of larger peptides by trypsin digestion of bound material.[47]

Commercially available equipment for automated solid-phase Edman degradation is less expensive than liquid-phase (spinning-cup) machines. It may be possible to use less expensive chemicals, since more extensive washing is possible. Disadvantages are the extra time and work required for sample attachment to the solid support and sample losses due to incomplete coupling.

Powers[49] describes solid-phase methods for spinning-cup sequenators. These include both placing beads with peptides attached in the sequenator cup (and hoping they will not wash out and plug the effluent valve) and attaching peptides directly to the glass wall of the cup. These techniques have been mostly superseded by the simpler use of Polybrene carrier, but may be useful for particular problems.

Chang[50] has performed manual sequence analysis by the DABITC/PITC procedure on peptides attached to glass beads. No advantage was found over the corresponding liquid-phase manual method, except for small peptides where extractive loss would be a problem without attachment.

E. Gas-Liquid Solid-Phase Sequenator

The gas-phase automatic sequenator devised by Hewick et al.[51] is now the instrument of choice for performing the Edman degradation. The reader is directed to excellent reviews by Shively and co-workers,[52,53] Hunkapiller et al.,[54] and Allen[55] for a complete description of this technology. This technology has been quite useful in the direct sequence analysis of proteins blotting from polyacrylamide gel electrophoretograms.[56–58] Bhown et al.[59] have described a method for the conversion of a spinning-cup sequenator to a gas-phase instrument.

IV. IDENTIFICATION METHODS FOR DIRECT EDMAN DEGRADATION

A. High-Performance Liquid Chromatography

HPLC offers high resolution, quantitative analysis, and high sensitivity; it is currently the method of choice for identification of PTH amino acids in sequence analysis.

The methods most commonly used today are similar in general principle to that of Zimmerman et al.,[60] i.e., a 25-cm C-18 reverse-phase column is used, with isocratic or gradient elution by a mobile phase of buffered aqueous sodium acetate and acetonitrile at elevated temperature. Other bonded-phase column packings such as cyanopropyl[61] or phenylalkyl[62] types are also used, as are other solvents and buffer salts. Even C-18 columns from different manufacturers differ in retention characteristics, so experimental conditions optimized for one column may have to be adjusted if a column from another source is used.

The desired result is adequate separation of all PTH amino acids from each other and from any byproducts of the Edman degradation. If the conversion procedure used includes extraction from an acidic aqueous phase by ethyl acetate, then PTH-Arg, -His, and hydrophilic byproducts remain in the aqueous phase and need only be separated from each other, which is not difficult. Retention of the PTHs extracted by ethyl acetate is adjusted by changing the temperature and the proportion of organic solvent in the mobile phase.

We find it helpful in establishing optimal conditions to measure the retention of the individual PTHs over a range of solvent strengths and temperatures. If retention times are plotted on a logarithmic scale as a function of solvent strength or temperature, separation between lines for the individual PTHs is essentially proportional to resolution.[63]

All the PTHs except PTH-dehydrothreonine (derived from PTH-Thr) are easily detected at 254 nm with the common mercury lamp absorbance monitor. PTH-dehydrothreonine is detectable at 313 or 323 nm. In the case of serine, little or no PTH-Ser is seen in samples from our sequencing system (manual conversion in 1 *N* HCl) nor is there any peak visible at 323 nm; the only peak seen elutes after all the other PTHs and absorbs at 254 nm. It may be a dehydrated and polymerized product, as described by Chang for the DABITC degradation.[64] Other laboratories do report detection of serine as PTH-dehydroserine absorbing at 313 to 323 nm and eluting at an intermediate position among the PTHs[45,65] or (in the presence of dithiothreitol) as a deriva-

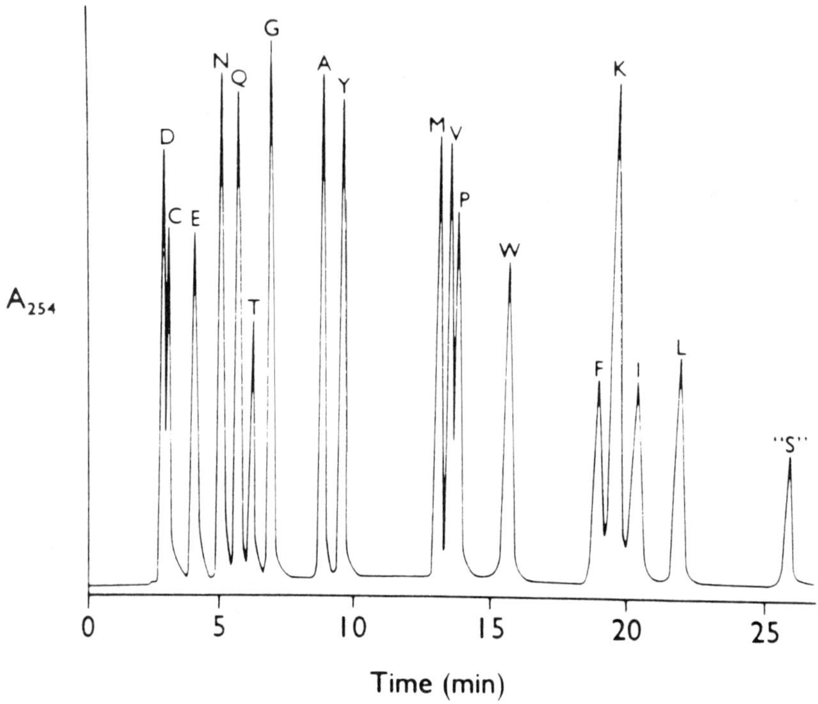

FIGURE 1. Separation of PTH amino acids on Ultrasphere-ODS column, 0.46 × 25 cm. Flow rate 1 ml/min, temperature 55°C. Solvent A = 0.01 *M* sodium acetate, pH 4.9; solvent B = acetonitrile. Gradient program: percent B increased linearly from 27 to 46.9% in 11 min, then dropped immediately to 30% and held there 1 min, and finally increased to 49% in 10 min. PTHs: D, aspartic acid; C, carboxymethylcysteine; E, glutamic acid; N, asparagine; Q, glutamine; T, threonine; G, glycine; A, alanine; Y, tyrosine; M, methionine; V, valine; P, proline; W, tryptophan; F, phenylalanine; K, lysine; I, isoleucine; L, leucine; and "S", serine product.

tive eluting near PTH-Ala and absorbing at 254 nm.[41,66] It is, therefore, essential that elution and absorbance characteristics be determined for the product(s) of Edman degradation of known serine residues in the particular sequence determination procedure to be used.

Gradient elution is more rapid than isocratic, but it may not save time in repetitive analyses because of the time required for column reequilibration. It also causes baseline fluctuations at high detector sensitivity. The principal advantage of gradient elution in PTH analysis may be increased sensitivity of detection for the late-eluting serine derivative mentioned above. Reproducibility and flexibility are important in a solvent gradient system for PTH analysis. We use a complex gradient profile which includes a nominal temporary drop in solvent strength at one point (although the actual change is damped by the mixing chamber and is more likely only a plateau, see Figure 1). Total analysis time including reequilibration is under 35 min. A faster analysis is possible in which PTH-Lys elutes between PTH-Trp and -Phe, but background peaks (e.g., diphenylthiourea) from the sequenator then co-elute with PTH-Trp.

Detection of PTHs is possible at the picomole level with the UV monitor. As discussed earlier (Section III.C), sensitivity in actual sequence analysis depends partly on levels of background from the sample and the degradation system.

B. Thin-Layer Chromatography

Thin-layer methods are inexpensive and have the particular advantage of speed, since derivatives from a number of cycles of degradation can be processed simultaneously. Lack of quantitation is less of a problem in sequence analysis of pure samples of peptides, since they do not develop the degree of overlap and background from nonspecific cleavage that long runs on proteins produce. PTHs are detected under 254 nm UV light as dark spots on a fluorescent background.

PTH identification procedures using silica gel plates include those of Jeppsson and Sjöquist,[67] Solal and Bernard,[68] and Inagami and Murakami.[69] We have found that a change in the binder used in Eastman plates in the late 1970s made them incompatible with some of the organic solvents recommended in these papers. However, the solvent xylene/95% ethanol/acetic acid, 50:50:0.5, recommended by Inagami,[70] for PTH-His and -Arg is compatible with the new binder, and the same mixture with twice the proportion of xylene works well for most of the other PTHs. Inagami and Murakami[69] recommend marking the positions of spots under UV light, then spraying the plate with 0.5% ninhydrin in *n*-butanol, drying, and heating at 95 to 110°C for 10 to 15 min. Many of the PTHs develop characteristic colors useful for their identification. Sensitivity is about 1 to 5 nmol.

Kulbe[71] recommends solvent systems for one- and two-dimensional TLC on polyamide sheets. Solvent I is toluene/*n*-pentane/acetic acid, 60:30:16, and Solvent II is 25% aqueous acetic acid. Solvent II is run at right angles to Solvent I for maximum resolution, or in the same direction if several sequence cycles are to be analyzed on one sheet. Sensitivity is 0.05 to 0.2 nmol.

C. Hydrolysis and Amino Acid Analysis

This procedure is slower than others, a disadvantage for its use as the primary means of identification in extended sequence analysis. However, it does provide quantitative results and is valuable as a supplementary or confirmatory means of identification.

Smithies et al.[72] recommend hydrolysis of the thiazolinones or PTHs in 57% hydriodic acid (HI) at 127°C for 20 h. PTH-alanine, -serine, -carboxymethylcysteine, or -cysteine all hydrolyze to alanine. Threonine is identified as α-aminobutyric acid. PTH-tryptophan gives glycine plus alanine, and methionine is destroyed. Alkaline hydrolysis in 0.2 *M* NaOH plus 0.1 *M* sodium dithionite allows recovery of methionine and tryptophan and differentiation of alanine from serine or cysteine.

Mendez and Lai[73] prefer hydrolysis in 5.7 *N* HCl containing 0.1% $SnCl_2$ for 4 h at 150°C.

V. COOH-TERMINAL DEGRADATION

Methods for sequential degradation from the carboxyl terminals of peptides have not yet achieved the efficiency of the NH_2-terminal methods just described. Most such procedures involve reaction with thiocyanate or thiocyanic acid to form the 2-thiohydantoin derivative of the COOH-terminal amino acid.[74] Meuth et al.[75] describe advances in technique, stating that repetitive yield is about 90%. Attachment of the sample to a derivatized glass support aids in separating peptide from reagents, but seems to lower recoveries of some amino acid thiohydantoins. In 1987, Hawke et al.[76] reported a significant advance in the isothiocyanate-based chemistry for the stepwise degradation of peptides and proteins from the carboxyl terminus using trimethylsilylisothiocyanate. Bailey and Shively[77] have extended this study and, by using aqueous triethylamine for cleavage, have developed a useful technique for the stepwise degradation of peptides from the carboxyl terminus. There have been several more recent articles concerned with the development of this approach to primary structure analysis.[78–80]

Finally, there has been a significant advance in using blotted proteins and protein fragments for direct sequence determination.[81–85]

REFERENCES

1. **Chait, B. T., Wang, R., Beavis, R. C., and Kent, S. B. H.,** Protein ladder sequencing, *Science,* 262, 89, 1993.
2. **Nielsen, P. F., Landis, B., Svoboda, M., Schneider, K., and Przybylski, M.,** N-terminal sequence determination of polypeptides and peptide mixtures by Edman degradation combined with Californium-252 plasma desorption mass spectrometry, *Anal. Biochem.,* 191, 302, 1990.
3. **Smith, J. B., Thevenon-Emeric, G., Smith, D. L., and Green, B.,** Elucidation of the primary structures of proteins by mass spectrometry, *Anal. Biochem.,* 193, 118, 1991.
4. **Nexo, E., Jorgensen, P. E., Thim, L., and Roepstorff, P.,** Purification and characterization of a low and a high molecular weight form of epidermal growth factor from rat urine, *Biochim. Biophys. Acta,* 1037, 388, 1990.
5. **Yu, W., Vath, J. E., Huberty, M. C., and Martin, S. A.,** Identification of the facile gas-phase cleavage of the Asp-Pro and Asp-Xxx peptide bonds in matrix-assisted laser desorption time-of-flight mass spectrometry, *Anal. Chem.,* 65, 3015, 1993.
6. **Gesquiere, J.-C. and Tartar, A.,** "Mac Mass," a software dedicated to mass spectrometry of synthetic peptides, *Anal. Biochem.,* 215, 299, 1993.

7. **Carr, S. A., Hemling, M. E., Bean, M. F., and Roberts, G. D.,** Integration of mass spectrometry in analytical biotechnology, *Anal. Chem.*, 63, 2802, 1991.
8. **Hauer, C. R., Staudenmann, W., Kuster, T., Neuheiser, F., Hughes, G. J., Seto-Ohsima, A., Tanokura, M., and Heizmann, C. W.,** Protein sequence determination by ESI-MS and LSI-MS tandem mass spectrometry: parvalbumin primary structures from cat, gerbil and monkey skeletal muscle, *Biochim. Biophys. Acta,* 1160, 1, 1992.
9. **Edman, P.,** A method for the determination of the amino acid sequence in peptides, *Arch. Biochem.*, 22, 475, 1949.
10. **Edman, P.,** Method for determination of the amino acid sequence in peptides, *Acta Chem. Scand.*, 4, 283, 1950.
11. **Ilse, D. and Edman, P.,** The formation of 3-phenyl-2-thiohydantoins from phenylthiohydantoin amino acids, *Aust. J. Chem.*, 16, 411, 1963.
12. **Schroeder, W. A.,** Degradation of peptides by the Edman method with direct identification of the PTH-amino acid, *Meth. Enzymol.*, 11, 445, 1967.
13. **Meagher, R. B.,** Rapid manual sequencing of multiple peptide samples in a nitrogen chamber, *Anal. Biochem.*, 67, 404, 1975.
14. **Tarr, G.,** Improved manual sequencing methods, *Meth. Enzymol.*, 47, 335, 1977.
15. **Konigsberg, W.,** Subtractive Edman degradation, *Meth. Enzymol.*, 11, 461, 1967.
16. **Gray, W. R. and Hartley, B. S.,** A fluorescent end-group reagent for proteins and peptides, *Biochem. J.*, 89, 59P, 1963.
17. **Gray, W. R.,** Sequential degradation plus dansylation, *Meth. Enzymol.*, 11, 469, 1967.
18. **Hartley, B. S.,** Strategy and tactics in protein chemistry, *Biochem. J.*, 119, 805, 1970.
19. **Bruton, C. J. and Hartley, B. S.,** Chemical studies on methionyl-tRNA synthetase from *Escherichia coli, J. Mol. Biol.*, 52, 165, 1970.
20. **Croft, L. R.,** *Introduction to Protein Sequence Analysis,* John Wiley & Sons, New York, 1980, 110.
21. **Gray, W. R. and Smith, J. R.,** Rapid sequence analysis of small peptides, *Anal. Biochem.*, 33, 36, 1970.
22. **Weiner, A. M., Platt, T., and Weber, K.,** Amino-terminal sequence analysis of proteins purified on a nanomole scale by gel electrophoresis, *J. Biol. Chem.*, 247, 3242, 1972.
23. **Peterson, J. D., Nehrlich, S., Oyer, P. E., and Steiner, D. F.,** Determination of the amino acid sequence of the monkey, sheep and dog pro-insulin C-peptides by a semi-micro Edman degradation procedure, *J. Biol. Chem.*, 247, 4866, 1972.
24. **Levy, W.,** Manual Edman sequencing techniques for proteins and peptides at the nanomole level, *Meth. Enzymol.*, 79, 27, 1981.
25. **Shively, J. E., Miller, P., and Ronk, M.,** Microsequence analysis of peptides and proteins. VI. A continuous flow reactor for sample concentration and sequence analysis, *Anal. Biochem.*, 163, 517, 1987.
26. **Haniu, M. and Shively, J. E.,** Microsequence analysis of peptides and proteins. IX. Manual gas-phase microsequencing of multiple samples, *Anal. Biochem.*, 173, 296, 1988.
27. **Brandt, W. F. and Frank, G.,** Manual gas-phase isothiocyanate degradation, *Anal. Biochem.*, 168, 314, 1988.
28. **Chang, J. Y., Creaser, E. H., and Bentley, K. W.,** 4-*N,N*-Dimethylamino-azobenzene 4′-isothiocyanate, a new chromophoric reagent for protein sequence analysis, *Biochem. J.*, 153, 607, 1976.
29. **Chang, J. Y., Brauer, D., and Wittmann-Liebold, B.,** Micro-sequence analysis of peptides and proteins using 4-*N,N*-dimethylaminoazobenzene 4′-isothiocyanate/phenylisothiocyanate double coupling method, *FEBS Lett.*, 93, 205, 1978.
30. **Hughes, G. J., Winterhalter, K. H., Lutz, H., and Wilson, K. J.,** Microsequence analysis. III. Automatic solid-phase sequencing using DABITC, *FEBS Lett.*, 108, 92, 1979.
31. **Wilson, K. J., Hunziker, P., and Hughes, G. J.,** Microsequence analysis. IV. Automatic liquid-phase sequencing using DABITC, *FEBS Lett.*, 108, 98, 1979.

32. **Fischer, P. M. and Howden, M. E. H.,** Analysis of 4-*N,N*′-dimethylaminoazobenzene 4-thiohydantoin amino acids at sub-picomole levels by high-performance liquid chromatography: simultaneous manual sequencing of picomole quantities of several polypeptides, *Anal. Biochem.,* 177, 46, 1989.
33. **Edman, P. and Begg, G.,** A protein sequenator, *Eur. J. Biochem.,* 1, 80, 1967.
34. **Wittmann-Liebold, B., Graffunder, H., and Kohls, H.,** A device coupled to a modified sequenator for the automated conversion of anilinothiazolinones into PTH amino acids, *Anal. Biochem.,* 75, 621, 1976.
35. **Horn, M. J. and Bonner, A. G.,** Automatic conversion, microsequencing and other advances in solid-phase sequence analysis, in *Solid Phase Methods in Protein Sequence Analysis,* Previero, A. and Coletti-Previero, M.-A., Eds., Elsevier/North Holland, Amsterdam, 1977, 163.
36. **Rochat, H., Bechis, G., Kopeyan, C., Gregoire, J., and Van Rietschoten, J.,** Use of parvalbumin as a protecting protein in the sequenator: an easy and efficient way for sequencing small amounts of peptides, *FEBS Lett.,* 64, 404, 1976.
37. **Silver, J. and Hood, L.,** Automated microsequence analysis in the presence of a synthetic carrier, *Anal. Biochem.,* 60, 285, 1974.
38. **Tarr, G. E., Beecher, J. F., Bell, M., and McKean, D. J.,** Polyquaternary amines prevent peptide loss from sequenators, *Anal. Biochem.,* 84, 622, 1978.
39. **Klapper, D. G., Wilde, C. E., III, and Capra, J. D.,** Automated amino sequence analysis of small peptides utilizing Polybrene, *Anal. Biochem.,* 85, 126, 1978.
40. **Brauer, A. W., Margolies, M. N., and Haber, E.,** The application of 0.1 *M* Quadrol to the microsequence of proteins and the sequence of tryptic peptides, *Biochemistry,* 14, 3029, 1975.
41. **Hunkapiller, M. W. and Hood, L. E.,** Direct microsequence analysis of polypeptides using an improved sequenator, a nonprotein carrier (Polybrene), and high pressure liquid chromatography, *Biochemistry,* 17, 2124, 1978.
42. **Le Caer, J. P. and Rossier, J.,** On the use of polyethyleneimine as a carrier for protein sequencing: comparison with Polybrene, *Anal. Biochem.,* 169, 246, 1988.
43. **Henschen-Edman, A. and Lottspeich, F.,** Aspects on automated and microscale sequencing, in *Methods in Peptide and Protein Sequence Analysis,* Birr, C., Ed., Elsevier/North-Holland, Amsterdam, 1980, 105.
44. **Wittmann-Liebold, B. and Lehmann, A.,** New approaches to sequencing by micro- and automatic solid phase technique, in *Methods in Peptide and Protein Sequence Analysis,* Birr, C., Ed., Elsevier/North Holland, Amsterdam, 1980, 49.
45. **Shively, J. E.,** Sequence determinations of proteins and peptides at the nanomole and subnanomole level with a modified spinning cup sequenator, *Meth. Enzymol.,* 79, 31, 1981.
46. **Laursen, R. A.,** Solid-phase Edman degradation. An automatic peptide sequencer, *Eur. J. Biochem.,* 20, 89, 1971.
47. **Laursen, R. A.,** Coupling techniques in solid-phase sequencing, *Meth. Enzymol.,* 47, 277, 1977.
48. **Machleidt, W. and Wachter, E.,** New supports in solid-phase sequencing, *Meth. Enzymol.,* 47, 263, 1977.
49. **Powers, D. A.,** Solid-phase sequencing in spinning-cup sequenators, *Meth. Enzymol.,* 47, 299, 1977.
50. **Chang, J. Y.,** Manual solid phase sequence analysis of polypeptides using 4-*N,N*-dimethylaminoazobenzene 4′-isothiocyanate, *Biochim. Biophys. Acta,* 578, 188, 1979.
51. **Hewick, R. M., Hunkapiller, M. W., Hood, L. E., and Dreyer, W. J.,** A gas-liquid solid phase peptide and protein sequenator, *J. Biol. Chem.,* 256, 7990, 1981.
52. **Shively, J. E.,** Comparison of the spinning cup and gas-phase instruments, in *Methods of Protein Microcharacterization,* Shively, J. E., Ed., Humana Press, Clifton, NJ, 1986, 195.
53. **Shively, J. E., Paxton, R. J., and Lee, T. D.,** Highlights of protein structural analysis, *Trends Biochem. Sci.,* 14, 246, 1989.

54. **Hunkapiller, M. W., Granlund-Moyer, K., and Whitely, N. W.,** Gas-phase protein/peptide sequencer, in *Methods of Protein Microcharacterization,* Shively, J. E., Ed., Humana Press, Clifton, NJ, 1986, 223.
55. **Allen, G.,** *Sequencing of Proteins and Peptides,* Elsevier, Amsterdam, 1989.
56. **Xu, Q.-Y. and Shively, J. E.,** Microsequencing analysis of peptides and proteins. VIII. Improved electroblotting of proteins onto membranes and derivatized glass-fiber sheets, *Anal. Biochem.,* 170, 19, 1988.
57. **Hseih, J.-C., Lin, E.-P., and Tam, M. F.,** Electroblotting onto glass-fiber from an analytical isoelectrofocusing gel: a preparative method for isolating proteins for N-terminal microsequencing, *Anal. Biochem.,* 170, 1, 1988.
58. **Yuen, S., Hunkapiller, M. W., Wilson, K. J., and Yuan, P. M.,** Applications of tandem microbore liquid chromatography and sodium dodecyl sulfate-polyacrylamide gel electrophoresis/electroblotting in microsequence analysis, *Anal. Biochem.,* 168, 5, 1987.
59. **Bhown, A. S., Wayland, J. L., Jr., Lynn, J. D., and Bennett, J. C.,** Conversion of the Beckman liquid phase sequencer to a gas-liquid phase sequencer, *Anal. Biochem.,* 175, 39, 1989.
60. **Zimmerman, C. L., Appella, E., and Pisano, J. J.,** Rapid analysis of amino acid phenylthiohydantoins by high-performance liquid chromatography, *Anal. Biochem.,* 77, 569, 1977.
61. **Johnson, N. D., Hunkapiller, M. W., and Hood, L. E.,** Analysis of phenylthiohydantoin amino acids by high-performance liquid chromatography on DuPont Zorbax cyanopropylsilane columns, *Anal. Biochem.,* 100, 335, 1979.
62. **Henderson, L. E., Copeland, T. D., and Oroszlan, S.,** Separation of amino acid phenylthiohydantoins by high-performance liquid chromatography on phenylalkyl support, *Anal. Biochem.,* 102, 1, 1980.
63. **Noyes, C. M.,** Optimization of complex separations in HPLC: application to phenylthiohydantoin amino acids, *J. Chromatogr.,* 226, 451, 1983.
64. **Chang, J. Y.,** The destruction of serine and threonine thiohydantoins during the sequence determination of peptides by 4-*N*,*N*-dimethylaminoazobenzene 4′-isothiocyanate, *Biochim. Biophys. Acta,* 578, 175, 1979.
65. **Wittmann-Liebold, B.,** Application of HPLC-techniques to the separation of PTH-amino acid derivatives and peptides, in *High Performance Chromatography in Protein and Peptide Chemistry,* Lottspeich, F., Henschen, A., and Hupe, K.-P., Eds., Walter de Gruyter, Berlin, 1981, 223.
66. **Shively, J. E., Hawke, D., and Jones, B. N.,** Microsequence analysis of peptides and proteins. III. Artifacts and the effects of impurities on analysis, *Anal. Biochem.,* 120, 312, 1982.
67. **Jeppsson, J.-O. and Sjöquist, J.,** Thin-layer chromatography of PTH amino acids, *Anal. Biochem.,* 18, 264, 1967.
68. **Solal, M. C. and Bernard, J. L.,** Miniature thin-layer chromatography of phenylthiohydantoin amino acids. Application of automatic Edman degradation, *J. Chromatogr.,* 80, 140, 1973.
69. **Inagami, T. and Murakami, K.,** Identification of phenylthiohydantoin amino acids by thin-layer chromatography on a plastic-backed silica-gel plate, *Anal. Biochem.,* 47, 501, 1972.
70. **Inagami, T.,** Simultaneous identification of PTH derivatives of histidine and arginine by thin-layer chromatography, *Anal. Biochem.,* 52, 318, 1973.
71. **Kulbe, K. D.,** Micropolyamide thin-layer chromatography of phenylthiohydantoin amino acids (PTH) at the subnanomolar level. A rapid microtechnique for simultaneous multisample identification after automated Edman degradation, *Anal. Biochem.,* 59, 564, 1974.
72. **Smithies, O., Gibson, D., Fanning, E. M., Goodfliesh, R. M., Gilman, J. G., and Ballantyne, D. L.,** Quantitative procedures for use with the Edman-Begg sequenator. Partial sequences of two unusual immunoglobulin light chains, Rzf and Sac, *Biochemistry,* 10, 4912, 1971.

73. **Mendez, E. and Lai, C. Y.,** Regeneration of amino acids from thiazolinones formed in the Edman degradation, *Anal. Biochem.,* 68, 47, 1975.
74. **Stark, G. S.,** Sequential degradation of peptides from their carboxyl termini with ammonium thiocyanate and acetic anhydride, *Biochemistry,* 7, 1796, 1968.
75. **Meuth, J. L., Harris, D. E., Dwulet, F. E., Crowl-Powers, M. L., and Gurd, F. R. N.,** Stepwise sequence determination from the carboxyl terminus of peptides, *Biochemistry,* 21, 3750, 1982.
76. **Hawke, D. H., Lahm, W.-H., Shively, J. E., and Todd, C. W.,** Microsequence analysis of peptides and proteins: trimethylsilylisothiocyanate as a reagent for COOH-terminal sequence analysis, *Anal. Biochem.,* 166, 298, 1987.
77. **Bailey, J. M. and Shively, J. E.,** Carboxy-terminal sequencing: formation and hydrolysis of C-terminal peptidylthiohydantoins, *Biochemistry,* 29, 3145, 1990.
78. **Inglis, A. S.,** Chemical procedures for C-terminal sequencing of peptides and proteins, *Anal. Biochem.,* 195, 183, 1991.
79. **Boyd, V. L., Bozzini, M., Zon, G., Noble, R. L., and Mattaliano, R. J.,** Sequencing of peptides and proteins from the carboxyl terminus, *Anal. Biochem.,* 206, 344, 1992.
80. **Shenoy, N. R., Shively, J. E., and Bailey, J. M.,** Studies in C-terminal sequencing: new reagents for the synthesis of peptidylthiohydantoins, *J. Prot. Chem.,* 12, 195, 1993.
81. **Mozdzanowski, J. and Speicher, D. W.,** Microsequence analysis of electroblotted proteins, I. Comparison of electroblotting recoveries using different types of PVDF membranes, *Anal. Biochem.,* 207, 11, 1992
82. **Rein, D. F. and Speicher, D. W.,** Microsequence analysis of electroblotted proteins. II. Comparison of sequence performance on different types of PVDF membranes, *Anal. Biochem.,* 207, 19, 1992.
83. **Lombard-Platet, G. and Jalinot, P.,** Funnel-Well SDS-PAGE: A rapid technique for obtaining sufficient quantities of low-abundance proteins for internal sequence analysis, *BioTechniques,* 15, 669, 1993.
84. **Alimi, E., Martinage, A., Sautiere, P., and Chevaillier, P.,** Electroblotting proteins onto carboxymethylcellulose membranes for sequencing, *BioTechniques,* 15, 912, 1993.
85. **Reim, D. F. and Speicher, D. W.,** High-sensitivity gas phase sequence analysis of proteins and peptides on PVDF membranes using short cycle times, *Anal. Biochem.,* 214, 87, 1993.

Chapter 5

CHEMICAL CLEAVAGE OF PEPTIDE BONDS

The elucidation of the covalent structure of a protein requires the development of specific, reproducible methods for cleavage of the protein into fragments of a size amenable to structural analysis. Proteolytic enzymes such as trypsin, chymotrypsin, and pepsin have proved quite useful in the cleavage of specific peptide bonds in proteins. In addition to the use of specific proteases, the nature of certain amino acid residues has permitted the development of nonenzymatic chemical methods for the cleavage of certain peptide bonds. Site-specific chemical cleavage can also be used for the cleavage of fusion proteins[1] and *in situ* in polyacrylamide gels or on protein blots when the use of proteases is impractical.

Partial acid hydrolysis is the oldest of the various chemical approaches to the cleavage of specific peptide bonds.[2–6] The general principle of partial acid hydrolysis is based on the use of dilute acid at a pH just adequate to maintain the β-carboxyl group of aspartic acid in the protonated form. Under these conditions, peptide bonds in which the carboxyl moiety is contributed by aspartic acid are cleaved 100-fold more rapidly than other peptide bonds. Specifically, the use of 0.03 *N* HCl *in vacuo* at 105°C for 20 h has been found to be satisfactory. This cleavage also can occur in matrix-assisted laser desorption mass spectrometry.[7]

Cleavage of methionine-containing peptide bonds with cyanogen bromide[8,9] is certainly the most widely used method for specific chemical cleavage of peptide bonds (Figure 1). The reaction cleaves peptide bonds in which methionine contributes the carboxyl moiety. Methionine is converted into homoserine lactone during this process. The reaction is reasonably quantitative although, as indicated below, variable amounts of cyanogen bromide (CNBr) might be required. Second, the methionine content of most proteins is low[10] enough that a reasonably small number of fragments are obtained, providing a distinct advantage in primary structure analysis. Finally, the knowledge of the primary structure around methionine residues is of particular value in the design of primary DNA probes for the isolation and characterization of cDNA fragments in recombinant DNA research. The chemistry of this reaction is straightforward, involving the nucleophilic attack of the thioether sulfur on the carbon in cyanogen bromide followed by cyclization to form the iminolactone, which is hydrolyzed by water resulting in cleavage of the peptide bond. At acid pH this reaction does not generally, in and by itself, affect any other amino acid with the exception of cysteine, which is converted to cysteic acid. In this regard, it is noted that one would rarely be working with a protein or peptide containing free sulfhydryl groups. The yield of cleavage is measured either by the loss of methionine or by the sum of homoserine and homoserine lactone after acid hydrolysis. This value is probably best determined by allowing

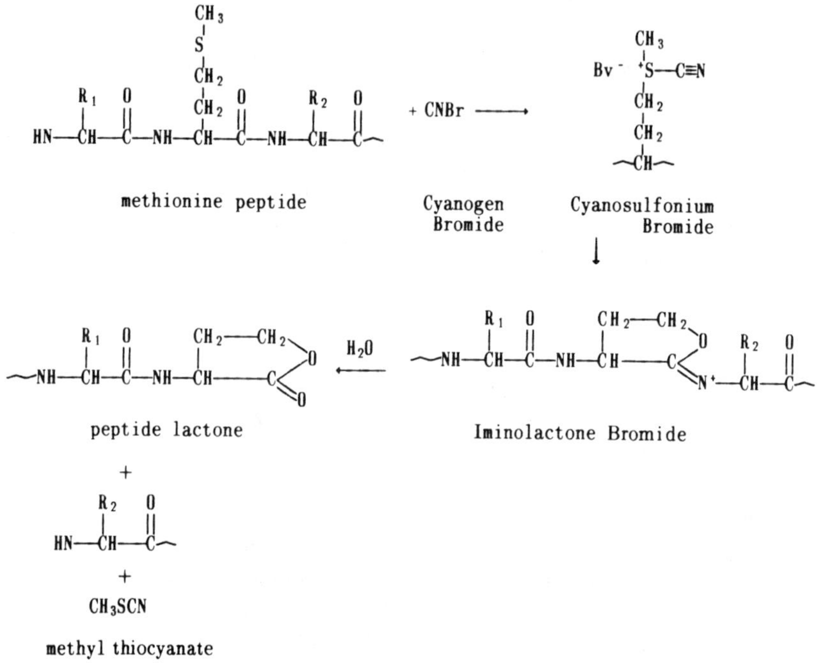

FIGURE 1. A scheme for the cleavage of methionine-containing peptide bonds with cyanogen bromide.

complete conversion to homoserine with base at room temperature. Cleavage of peptide chains at methionine with cyanogen bromide proceeds best with a fully denatured protein in mild acid. Early work with this reaction used 0.1 *M* HCl as the solvent or 0.1 *M* HCl in 6 *M* guanidine hydrochloride. Most recent studies have used 70% formic acid, trifluoroacetic acid,[11] or an equal mixture of formic acid and trifluoroacetic acid.[12] Use of formic acid has, on occasion, resulted in the blocking of amino-terminal residues via reaction with formaldehyde (present as a contaminant in the formic acid).[13] Acetic acid can also be used as solvent for this reaction. In general, the reaction proceeds effectively with a 20- to 100-fold molar excess of cyanogen bromide (added either as a solid to the protein or peptide dissolved in the solvent of choice). Solutions of acetic acid could also be used as solvent for the cyanogen bromide reaction. The molar ratio of cyanogen bromide to methionyl residues needs to be established for each peptide and protein under study. In the work on the structure of the pancreatic deoxyribonuclease, it was necessary to use a 3000-fold molar excess to cleave a particular methionine-serine peptide bond.[14] In this regard it is of interest that methionine-123 in human serum albumin is converted to homoserine lactone by treatment with cyanogen bromide without concomitant peptide bond cleavage.[15] The conversion of methionine to methionine sulfoxide under conditions used for the cyanogen bromide cleavage has been reported.[16] With a tenfold molar excess of cyanogen bromide for 22 h at ambient temperature, 1% conversion to methionine sulfoxide was observed in 70% formic acid; 8% conversion in 0.1 *M* HCl; 64% conversion in 0.1 *M* citrate, pH 3.5; and 97% conversion in 0.1 *M* phosphate, pH 6.5.

Simpson and Nice[17] developed a procedure for the *in situ* cyanogen bromide cleavage of proteins absorbed to the glass-fiber membranes developed for the gas-phase protein-peptide sequenator. The procedure was originally developed to determine internal sequence information from N-terminal blocked proteins. In the procedure described, satisfactory sequence information was obtained from 2.2 nmol (24 μg) cytochrome c. Cyanogen bromide (20 μl containing a 20-fold molar excess of reagent in 70% formic acid) was applied to the membrane. The membrane was then placed in a vacuum desiccator over CNBr/HCOOH for 16 h at ambient temperature. The membrane was air dried and taken to the sequenator.

Xu and Shively[18] have described improvements on the electroblotting of proteins. These investigators reported higher degrees of success with polyvinyldifluoride (PVDF) membranes. Transfer yields were markedly improved upon pretreatment of the membranes with Polybrene.

Scott et al.[19] described an additional approach to the cyanogen bromide cleavage of proteins on PDVF membranes followed by elution of the reaction products from the membranes with 2% sodium dodecyl (SDS) sulfate per 1% Triton X-100 in 50 m*M* Tris, pH 9.2. (Using bovine serum albumin as a model protein, these investigators obtained a 90% recovery of eluted peptides [10 μg protein applied to PVDF membrane].[20]) The heavy and light chains of antibody were separated by electrophoresis and transferred to a PVDF membrane. After staining with Ponceau Red, the bands were excised and placed in a 500-μl Eppendorf tube. Cyanogen bromide (150 μl; 150 μ*M* in 70% formic acid) was added, and the reaction was allowed to proceed for 16 h at ambient temperature. The solvent was removed *in vacuo* and the peptides eluted (90 min at ambient temperature) with 75 μl of the Tris-detergent solvent described above. The peptides were separated by SDS-polyacrylamide gel electrophoresis and the bands subjected to gas-phase sequence analysis. Useful sequence information was obtained from 25 to 50 μg of parent immunoglobulin.

Sokolov et al.[21] describe a modified method for direct cyanogen bromide cleavage directly within the polyacrylamide gel. After identification of the proteins by either staining or autoradiography, the gel is sliced and dried. The dried gel is taken into 30 μl cyanogen bromide (200 mg/ml in 70% HCOOH). The reaction is allowed to proceed for 16 h at ambient temperature or 3 h at 37°C. The gel is then dried, and the reaction products are separated by a second electrophoretic step. The drying of the gel prior to the cleavage step is critical to avoid protein loss during this procedure. An alternative approach has been developed by Jahnen et al.[22] These investigators isolated the fragments from the cyanogen bromide cleavage of proteins on polyacrylamide gel (the gel slices were dried by lyophilization prior to the cyanogen bromide cleavage step) by either a second electrophoretic step or by HPLC after elution. The electrophoretic step is recommended over the HPLC step.

The other methods for chemical cleavage of specific peptide bonds have been used somewhat infrequently. This is, in part, a reflection of the considerable success experienced with the cyanogen bromide reaction as well as the increased ability to determine more primary structure during a single run with the improved automated Edman degradation. There is, however, continued interest because of the possibility of peptide mapping as well as the ability to perform these reactions on a solid support such as a transfer membrane (i.e., nitrocellulose membranes, polyvinyldifluoride membranes).

A number of methods have been proposed for chemical cleavage at cysteine residues. One approach is based on the conversion of cysteine to dehydroalanine[23–25] and subsequent hydrolysis with either acid or base to release pyruvic acid and involves the conversion of cysteine to the dialkyl sulfonic salt with methyl bromide or methyl iodide at pH 6.0 and subsequent β-elimination in dilute bicarbonate with mild heating.[26] The use of 2,4-dinitrofluorobenzene for the modification of cysteine to form the *S*-dinitrophenyl derivatives at pH 5.6 has been reported.[25] The β-elimination of these derivatives was accomplished with sodium methoxide in methanol. Cleavage of the dehydroalanine-containing peptide bond was accomplished by heating (100°C) in dilute acid (0.01 *M* HCl) for 1 h. This reaction mixture was then lyophilized, treated with a volume of 0.1 *M* NaOH equivalent to the original volume of acid and one fifth volume of 30% hydrogen peroxide, and then heated at 37°C for 30 min. The reaction mixture was then neutralized with acetic acid, and excess peroxide was removed with catalase. Alternatively, cleavage can be accomplished with bromide or performic acid.

The cleavage of peptide bonds containing cystine (disulfide groups) has been examined in some detail.[26] In this reaction, cyanide reacts with cysteine to yield a sulfhydryl and a thiocyano group. The thiocyano-containing derivative at pH <8 will cyclize to form an acyliminothiazolidine ring which will then undergo hydrolysis to cleave the peptide bond (Figure 2). The formation of the iminothiazolidine can be followed by absorbance at 235 nm. An application of this approach has been advanced by Jacobson et al.[27] *S*-cyanocysteine is obtained by reaction of cysteine or cystine with 2-nitro-5-thiocyano-benzoic acid. Cleavage of the *S*-cyanocysteine is achieved by incubation in 0.1 *M* sodium borate, 6 *M* guanidine, pH 9.0 at 37°C with the formation of 2-iminothiazolidine-4-carboxyl peptides. Virtually 100% cleavage was achieved for several proteins. This results in the formation of a free carboxyl group and a "blocked" amino-terminal peptide (2-iminothiazolidine-4-carboxyl) derivative. This reaction has been recently used for the identification of reactive sulfydryl groups in phosphoglycerate kinase.[28] In this study, the two highly reactive cysteine residues (there are seven cysteine residues in the native protein) were modified with 5,5′-dithiobis-(2-nitrobenzoate). An excess of potassium cyanide (KCN) was added, resulting in the formation of *S*-cyanocysteine. The excess cyanide was removed by gel filtration. Incubation

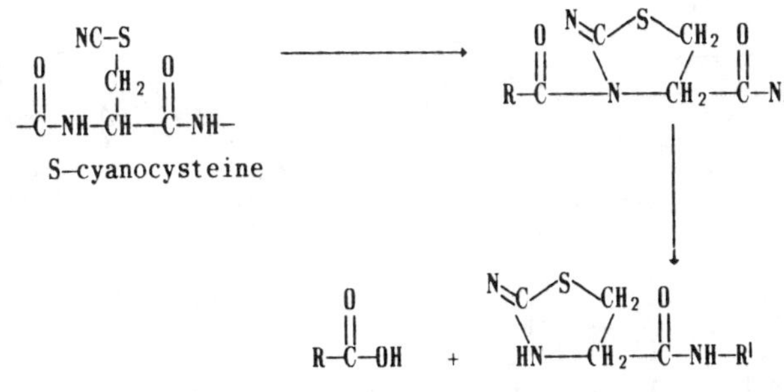

FIGURE 2. A scheme for the cleavage of *S*-cyanocysteine-containing peptide bonds.

of the *S*-cyano protein in 6.0 *M* guanidine hydrochloride, pH 8.0 at 50°C resulted in peptide bond cleavage. Schaffer and Stark[29] have proposed a catalyst prepared from nickel chloride and sodium borohydride for the conversion of the 2-iminothiazolidine-4-carboxylate to alanine. These investigators also noted that cleavage could occur at phenylalanyl-seryl and phenylalanyl-threonyl peptide bonds. Lu and Gracy[30] have employed 2-nitro-5-thiocyanobenzoic acid to convert the cysteinyl residues in human placental glucosephosphate isomerase to *S*-cyanocysteine, followed by cleavage at the modified cysteine residues. Conversion to the *S*-cyanocysteinyl derivative was accomplished with a five- to tenfold molar excess of 2-nitro-5-thiocyanobenzoic acid in 0.2 *M* Tris-acetate, 6 *M* guanidinium chloride, pH 9.0 (protein previously incubated with adequate dithiothreitol — fourfold molar excess over sulfhydryls in the protein) for 5 h at 37°C. The modified protein was dialyzed extensively against 10% acetic acid and lyophilized. Cleavage of *S*-cyanylated protein was achieved by incubation in 0.2 *M* Tris-acetate, 6 *M* guanidium chloride, pH 9.0 at 37°C for 2 h. The average extent of cleavage obtained was approximately 80%.

Specific cleavage at tryptophanyl residues in peptides and proteins has been frequently used to obtain specific fragments. Cleavage of tryptophanyl peptide bonds with *N*-bromosuccinimide can occur as a side reaction of the *N*-bromosuccinimide oxidation of tryptophanyl residues (see Chapter 12), but it generally requires a substantial molar excess of reagent with mild acid.[31] This reaction is generally accomplished in 70% acetic acid. Although cleavage is generally restricted to tryptophanyl residues, cleavage can also occur at tyrosyl and histidinyl residues. Cleavage at tryptophanyl residues under the above conditions generally occurs with an efficiency of 50 to 80% with peptides, but is substantially less with proteins (10 to 50%).

BNPS-skatole (2-(2-nitrophenylsulfenyl)-3-methyl-3-bromoindolenine) has been used for the cleavage of peptide bonds involving tryptophan.[32] The reaction conditions are similar to those utilized for *N*-bromosuccinimide, and the reaction mechanism is similar in terms of the production of an "active bromide". It is reported to be somewhat more selective than *N*-bromosuccinimide, but nonspecific cleavages do occur as does the conversion

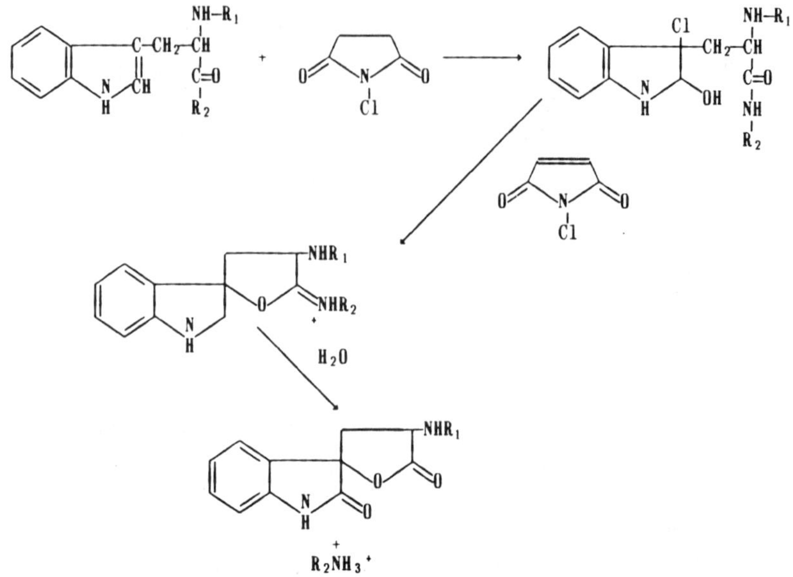

FIGURE 3. A scheme for the *N*-chlorosuccinimide cleavage of peptide bonds containing tryptophan.

of methionine to methionine sulfoxide. The yield of peptide bond cleavage is similar to that reported with *N*-bromosuccinimide.

The specific cleavage of tryptophanyl peptide bonds with *N*-chlorosuccinimide[33] has been reported. The peptide bond cleavages (Figure 3) obtained with *N*-chlorosuccinimide (NCS) are much more specific than those achieved with either *N*-bromosuccinimide or BPNS-skatole. The cleavage of tryptophanyl peptide bonds requires a twofold excess of *N*-chlorosuccinimide in 50% acetic acid under ambient conditions. Cleavage of other peptide bonds was not detected under these conditions, but methionine is converted to methionine sulfoxide and cysteine to cystine. Model peptides were cleaved in approximately 40% yield, while with several proteins yields from 19 to 50% were reported. Mechanistically, the reaction proceeds as described above for *N*-bromosuccinimide. The *N*-chlorosuccinimide should be recrystallized from ethyl acetate prior to use. Lischwe and Sung[34] examined the *N*-chlorosuccinimide cleavage of proteins in some detail. Cytochrome c was used as the model protein. The protein was dissolved in water (1 nmol/ml), and the *N*-chlorosuccinimide was dissolved in a buffer composed of 1.0 ml glacial acetic acid, 1 g urea, and 1 ml water. Four volumes of the protein solution were mixed with ten volumes of the *N*-chlorosuccinimide in this buffer (the resulting solution in 4.68 *M* with respect to urea in 27.5% acetic acid). Approximately 50% of the tryptophanyl peptide bonds were cleaved after 30 min of reaction using a tenfold molar excess of reagent. The oxidation of methionine (to methionine sulfoxide) and cysteine (to cysteic acid) occurs as a side reaction. More recently, the specific cleavage of tryptophanyl peptide bonds by *N*-chlorosuccinimide under these reaction conditions has been used to study epitope distribution in α-1-antiprotease inhibitor.[35] The use of 2,4,6-tribromo-4-methylcyclohexadione (TBC) for the cleavage of tryptophanyl peptide bonds in proteins has been advanced by Burstein and Patchornik.[36] The cleavage is fairly specific for tryptophanyl residues, but modification of other amino acid residues was noted (tyrosine, methionine, cysteine, etc). Optimal conditions for the cleavage reaction were a threefold excess of reagent at pH 3.0 at ambient conditions for 15 min. Generally, 60 to 80% acetic acid is used as the solvent and the reaction is allowed to proceed in the dark. Approximately 50% cleavage of tryptophanyl-containing peptide bonds is obtained with synthetic peptides such as *N*-benzyloxy-carbonyl-tryptophanyl-glycine, while 5 to 60% yields are reported with proteins such as lysozyme.

The cleavage of protein at asparaginyl-glycyl peptide bonds with hydroxylamine[37] has proved useful in selected circumstances. The reaction is generally performed in the presence of 6 *M* guanidium chloride, pH 9.0 with 2 *M* NH_2OH. The pH of the solution is maintained either with a pH-stat or with 0.2 *M* potassium carbonate. Generally, as with other means of peptide bond cleavage, optimal results are obtained with the reduced and alkylated protein. The reaction will yield a new amino-terminal amino acid and aspartyl hydroxyamate.

Cleavage at peptide bonds where the carboxyl group is contributed by tryptophan occurs upon reaction with *o*-iodosobenzoic acid. The reaction has been studied in detail by Mahoney and co-workers.[38,39] The reaction can be reasonably specific for tryptophan, although some modification of methionine to form methionine sulfoxide is observed. The reaction is performed in 60 to 80% acetic acid in the presence of a denaturing agent such as guanidine. The occasional modification of tyrosyl residues seen with some preparations of *o*-iodosobenzoic acid has been shown to be a property of *o*-iodoxybenzoic acid contamination of certain *o*-iodosobenzoic acid preparations.[39] Pretreatment of the *o*-iodosobenzoic acid preparations with *p*-cresol obviates cleavage at tyrosyl peptide bonds. The site-specific oxidative cleavage of peptide bonds (Figure 4)[40] by either cupric ions or ferric ions is an exciting recent development in this area.

Schepartz and Cuenoud[41] have reported the specific cleavage of calmodulin with a reagent based on the structure of trifluoroperazine (Figure 5). In these experiments, the trifluoroperazine-ethylenediaminetetraacetic acid reagent (TFE-EDTA), ferric ions, and dithiothreitol were required for effective peptide bond cleavage in 10 m*M* Tris, pH 7.2. In a directly related approach, Hoyer et al.[42] attached EDTA to biotin (Figure 6). The resulting derivative was used to place either a cupric ion or a ferric ion close to the biotin binding site on streptavididin. Relatively specific cleavage was obtained in solvent 50 m*M* borate, pH 7.5 containing 20 μ*M* protein, 20 μ*M* cupric chloride, and 20 μ*M* biotin-EDTA.

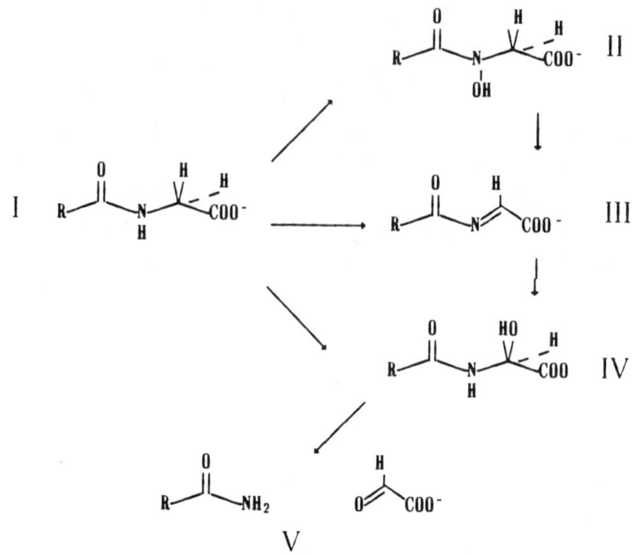

FIGURE 4. A scheme for the oxidative cleavage of peptide bonds. (Adapted from Ramer, S. E., et al., *J. Am. Chem. Soc.*, 112, 8526, 1988.)

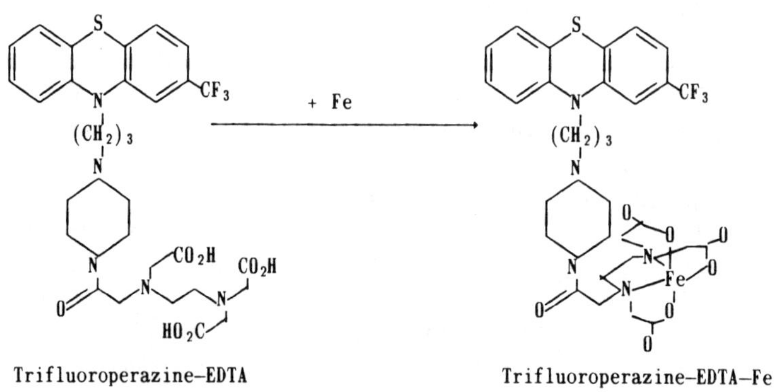

FIGURE 5. The structure of trifluoroperazine-EDTA.

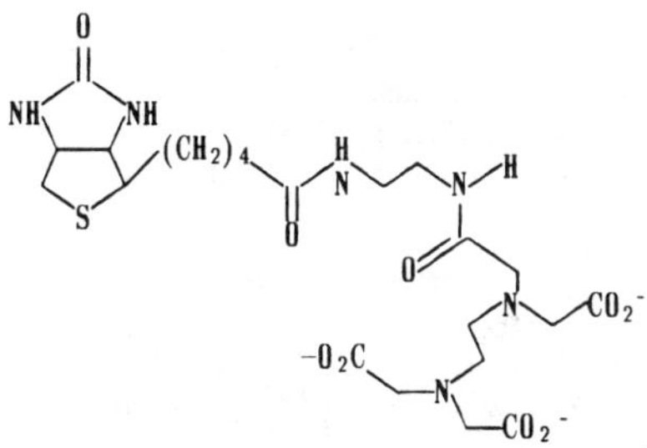

Biotin–EDTA–I

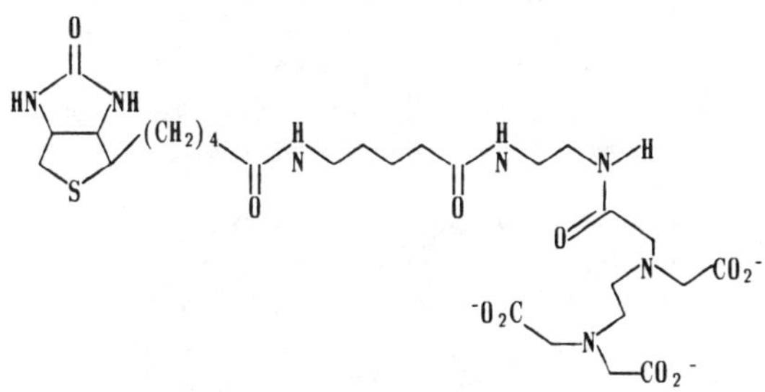

Biotin–EDTA–II

FIGURE 6. The structure of biotin-EDTA.

REFERENCES

1. **Forsberg, G., Baastrup, B., Brobjer, M., Lake, M., Jörnvall, H., and Hartmanis, M.,** Comparison of two chemical cleavage methods for preparation of a truncated form of recombinant human insulin-like growth factor I from a secreted fusion protein, *BioFactors,* 2, 105, 1989.
2. **Sanger, F.,** The terminal peptides of insulin, *Biochem. J.,* 45, 563, 1949.
3. **Sanger, F. and Tuppy, H.,** The amino-acid sequence in the phenylalanyl chain of insulin. I. The identification of lower peptides from partial hydrolysates, *Biochem. J.,* 49, 463, 1951.
4. **Sanger, F. and Thompson, E. O. P.,** The amino acid sequence in the glycyl chain of insulin. I. The identification of lower peptides from partial hydrolysates, *Biochem. J.,* 53, 353, 1953.
5. **Schultz, J.,** Cleavage at aspartic acid, *Meth. Enzymol.,* 11, 255, 1967.
6. **Tsung, C. M. and Fraenkel-Conrat, H.,** Preferential release of apartic acid by dilute acid treatment of tryptic peptides, *Biochemistry,* 4, 793, 1965.
7. **Gross, E.,** The cyanogen bromide reaction, *Meth. Enzymol.,* 11, 238, 1967.
8. **Yu, W., Vath, J. E., Huberty, M. C., and Martin, S. A.,** Identification of the facile gas-phase cleavage of the Asp-Pro and Asp-Xxx peptide bonds in matrix-assisted laser desorption time-of-flight mass spectrometry, *Anal. Chem.,* 65, 3015, 1993.
9. **Spande, T. F. and Witkop, B.,** Cyanogen bromide cleavage of peptides and proteins, *CRC Handbook of Biochemistry-Selected Data for Molecular Biology,* 2nd ed., Sober, H. A. and Harte, R. A., Eds., CRC Press, Boca Raton, FL, 1970, C-137.

10. **Tristram, G. R. and Smith, R. H.,** Amino acid composition of certain proteins, in *The Proteins,* 2nd ed., Neurath, H., Ed., Academic Press, New York, 1963, 45.
11. **Morrison, J. R., Fidge, N. N., and Grego, H.,** Studies on the formation, separation, and characterization of cyanogen bromide fragments of human A1 apolipoprotein, *Anal. Biochem.,* 186, 145, 1990.
12. **Shively, J. E.,** Reverse-phase HPLC isolation and microsequence analysis, in *Methods of Protein Microcharacterization,* Shively, J. E., Ed., Humana Press, Clifton, NJ, 1986, 65.
13. **Shively, J. E., Hawke, D., and Jones, B. N.,** Microsequence analysis of peptides and proteins. III. Artifacts and the effects of impurities on analysis, *Anal. Biochem.,* 120, 312, 1982.
14. **Liao, T.-H., Salnikow, J., Moore, S., and Stein, W. H.,** Bovine pancreatic deoxyribonuclease A. Isolation of cyanogen bromide peptides; complete covalent structure of the polypeptide chain, *J. Biol. Chem.,* 248, 1489, 1973.
15. **Doyen, N. and LaPresle, C.,** Partial non-cleavage by cyanogen bromide of a methionine-cystine bond from human serum albumin and bovine α-lactalbumin, *Biochem. J.,* 177, 251, 1979.
16. **Joppich-Kuhn, R., Corkill, J. A., and Giese, R. W.,** Oxidation of methionine to methionine sulfoxide as a side reaction of cyanogen bromide cleavage, *Anal. Biochem.,* 119, 73, 1982.
17. **Simpson, R. J. and Nice, E. C.,** *In situ* cyanogen bromide cleavage of N-terminally blocked proteins in a gas-phase sequencer, *Biochem. Int.,* 8, 787, 1984.
18. **Xu, Q.-Y. and Shively, J. E.,** Microsequence analysis of peptides and proteins. VIII. Improved electroblotting of proteins onto membranes and derivatized glass-fiber sheets, *Anal. Biochem.,* 170, 19, 1988.
19. **Scott, M. G., Crimmins, D. L., McCourt, D. W., Tarrand, J. J., Eyerman, M. C., and Nahm, M. H.,** A simple *in situ* cyanogen bromide cleavage method to obtain internal amino acid sequence of proteins electroblotted to polyvinyldifluoride membranes, *Biochem. Biophys. Res. Commun.,* 155, 1353, 1988.
20. **Szewczyk, B. and Summers, D. F.,** Preparative elution of proteins blotted to immobilon membranes, *Anal. Biochem.,* 168, 48, 1988.
21. **Sokolov, B. P., Sher, B. M., and Kalinin, V. N.,** Modified method for peptide mapping of collagen chains using cyanogen bromide-cleavage of protein within polyacrylamide gels, *Anal. Biochem.,* 176, 365, 1989.
22. **Jahnen, W., Ward, L. D., Reid, G. E., Moritz, R. L., and Simpson, R. J.,** Internal amino acid sequencing of proteins by *in situ* cyanogen bromide cleavage in polyacrylamide gels, *Biochem. Biophys. Res. Commun.,* 166, 139, 1990.
23. **Witkop, B. and Ramachandran, L. K.,** Progress in non-enzymatic selective modification and cleavage of proteins, *Metabolism,* 13, 1016, 1964.
24. **Patchornik, A. and Sokolovsky, M.,** Nonenzymatic cleavages of peptide chains at the cysteine and serine residues through their conversion into dehydroalanine. I. Hydrolytic and oxidative cleavage of dehydroalanine residues, *J. Am. Chem. Soc.,* 86, 1206, 1964.
25. **Sokolovsky, M., Sadeh, T., and Patchornik, A.,** Nonenzymatic cleavages of peptide chains at the cysteine and serine residues though their conversion to dehydroalanine (DHAL). II. The specific chemical cleavage of cysteinyl peptides, *J. Am. Chem. Soc.,* 86, 1212, 1964.
26. **Catsimpoolas, N. and Wood, J. L.,** Specific cleavage of cystine peptides by cyanide, *J. Biol. Chem.,* 241, 1790, 1966.
27. **Jacobson, G. R., Schaffer, M. H., Stark, G. R., and Vanaman, T. C.,** Specific chemical cleavage in high yield at the amino peptide bonds of cysteine and cystine residues, *J. Biol. Chem.,* 248, 6583, 1973.
28. **Minard, P., Desmadril, M., Ballery, N., Perahia, D., Mouawad, L., Hall, L., and Yon, J. M.,** Study of the fast-reacting cysteines in phosphoglycerate kinase using chemical modification and site-directed mutagenesis, *Eur. J. Biochem.,* 185, 419, 1989.

29. **Schaffer, M. H. and Stark, G. R.,** Ring cleavage of 2-iminothiazolidine-4-carboxylates by catalytic reduction. A potential method for unblocking peptides formed by specific chemical cleavage at half-cysteine residues, *Biochem. Biophys. Res. Commun.*, 71, 1040, 1976.
30. **Lu, H. S. and Gracy, R. W.,** Specific cleavage of glucosephosphate isomerase at cysteinyl residues using 2-nitro-5-thiocyanobenzoic acid: analyses of peptides eluted from polyacrylamide gels and localization of active site histidyl and lysyl residues, *Arch. Biochem. Biophys.*, 212, 347, 1981.
31. **Ramachandran, L. K. and Witkop, B.,** N-Bromosuccinimide cleavage of peptides, *Meth. Enzymol.*, 11, 283, 1967.
32. **Fontana, A.,** Modification of tryptophan with BNPS-skatole (2-(2-nitrophenylsulfenyl)-3-methyl-3-bromoindolenine), *Meth. Enzymol.*, 25, 419, 1972.
33. **Shechter, Y., Patchornik, A., and Burstein, Y.,** Selective chemical cleavage of tryptophanyl peptide bonds by oxidative chlorination with N-chlorosuccinimide, *Biochemistry*, 15, 5071, 1976.
34. **Lischwe, M. A. and Sung, M. T.,** Use of N-chlorosuccinimide/urea for selective cleavage of tryptophanyl peptide bonds in proteins, *J. Biol. Chem.*, 252, 4976, 1977.
35. **Zhu, X.-J. and Chan, S. K.,** The use of monoclonal antibodies to distinguish several chemically modified forms of human alpha-1-proteinase inhibitor, *Biochem. J.*, 246, 19, 1987.
36. **Burstein, Y. and Patchornik, A.,** Selective chemical cleavage of tryptophanyl peptide bonds in peptides and proteins, *Biochemistry*, 11, 4641, 1972.
37. **Bornstein, P. and Balian, G.,** Cleavage at Asn-Gly bonds with hydroxylamine, *Meth. Enzymol.*, 47, 132, 1977.
38. **Mahoney, W. C. and Hermodson, M. A.,** High yield cleavage of tryptophanyl peptide bonds by *o*-iodosobenzoic acid, *Biochemistry*, 18, 3810, 1979.
39. **Mahoney, W. C., Smith, P. K., and Hermodson, M. A.,** Fragmentation of proteins with *o*-iodosobenzoic acid: chemical mechanism and identification of *o*-iodoxybenzoic acid as a reactive contaminant that modifies tyrosyl residues, *Biochemistry*, 20, 443, 1981.
40. **Ramer, S. E., Cheng, H., Palcic, M. M., and Vederas, J. C.,** Formation of peptide amides by peptideylglycine alpha-amidating monooxygenase: a new assay and stereochemistry of hydrogen loss, *J. Am. Chem. Soc.*, 112, 8526, 1988.
41. **Schepartz, A. and Cuenoud, B.,** Site-specific cleavage of the protein calmodulin using a trifluoroperazine-based affinity reagent, *J. Am. Chem. Soc.*, 112, 3247, 1990.
42. **Hoyer, D., Cho, H., and Schultz, P. G.,** A new strategy for selective protein cleavage, *J. Am. Chem. Soc.*, 112, 3249, 1990.

Chapter 6

THE MODIFICATION OF CYSTEINE

Cysteine is potentially the most powerful nucleophile in a protein and, as a result, is frequently the easiest to modify with a variety of reagents. The chemistry of cysteine has been reviewed, and the reader is recommended to these reviews for further detail.[1,2] The unique reactivity of cysteine has prompted investigators to use site-specific mutagenesis to place cysteine at particular points in a protein for the subsequent attachment of structural probes.[3,4] Review of the recent literature would suggest that dithiobis(2-nitrobenzoic acid) (Ellman's Reagent) has been the most frequently used reagent as the extent of reaction can be easily determined by spectral measurement.[5–7]

In addition to their use in the direct site-specific modification of proteins, certain base compounds listed below have been used as means for attaching structural probes. Primary examples are the α-haloalkanoic acids and *N*-alkylmaleimides.[8–11]

Table 1 presents three reagents which have been frequently used for the modification of cysteine in proteins.

Cysteine is relatively sensitive to oxidation, but there is little selectivity in these reactions. Mild oxidizing conditions can result in the formation of disulfide bonds with appropriately aligned cysteinyl residues. Formation of sulfenic acid is generally readily reversible unless stabilized by local conditions[12] and more highly oxidized forms such as cysteine-sulfonic acid are more frequently observed. More rigorous conditions such as treatment with performic acid result in the formation of cysteic acid.

Modification of cysteine with sodium tetrathionate (Figure 1)[13] is similar to oxidation. This reaction has the advantage of ready reversibility by mild reducing agents. Reaction with sodium tetrathionate has been used to study cysteine in chalcone isomerase.[14] The native enzyme was only slowly inactivated by sodium tetrathionate (pH 5.2 [50 m*M* MES], $k^2 < 0.005$ M^{-1} min^{-1}; pH 7.5 [50 m*M* HEPES], $k^2 = 0.009$ M^{-1} min^{-1}; pH 9.4 [50 m*M* CHES], $k^2 = 0.093$ M^{-1} min^{-1}). In the presence of 6.0 *M* urea, the enzyme was rapidly inactivated (pH 7.5, $k^2 > 690$ M^{-1} min^{-1}).

Methyl methanethiosulfonate was introduced by Smith et al.[15] as reagent for the reversible reaction of sulfydryl groups (Figure 2). This concept was extended by Bruice and Kenyon as a mechanism for introducing novel substituents[16] (see Figure 3) into proteins via modification of cysteine sulfydryl groups. These reagents have subsequently been used for the modification of site-specific mutants of carboxypeptidase Y.[17]

Haloacetates, the corresponding amides, and derivatives have been extremely useful reagents for the specific modification of cysteinyl residues. These reagents react with cysteine via a S_N2 reaction mechanism to give the corresponding carboxymethyl or carboxamidomethyl derivatives (see Figure 4). When a rapid

Table 1
COMMONLY USED REAGENTS FOR THE SITE-SPECIFIC MODIFICATION OF CYSTEINE IN PROTEINS

Reagent	Other AA modified	Ref.
Iodoacetate[a]	Histidine, lysine, methionine	1–3
N-Ethylmaleimide	Lysine	4–7
5,5′-Dithiobis-(2-nitrobenzoic acid)		8–13
p-Hydroxymercuribenzoate[b]		14

[a] This includes related α-haloketo compounds such as bromoalkanoic acids, chloroalkanoic acids, and related amides.

[b] Included are related organic mercurial derivatives including mercurinitrophenol derivatives and mercuriphenylsulfonate derivatives.

References for Table 1

1. **Thompson, S. A. and Fiddes, J. C.,** Chemical characterization of the cysteines of basic fibroblast growth factor, *Ann. N.Y. Acad. Sci.,* 638, 78, 1991.
2. **Clark, S. and Konstantopoulos, N.,** Sulphydryl agents modulate insulin- and epidermal growth factor (EGF)-receptor kinase via reaction with intracellular receptor domains: differential effects on basal versus activated receptors, *Biochem. J.,* 292, 217, 1993.
3. **Kaslow, H. R., Schlotterbeck, J. D., Mar, V. L., and Burnette, W. N.,** Alkylation of cysteine 41, but not cysteine 200, decreases the ADP-ribosyltransferase activity of the S1 subunit of pertussis toxin, *J. Biol. Chem.,* 264, 6386, 1989.
4. **Reyes, A. M., Bravo, N., Ludwig, H., Iriarte, A., and Slebe, J. C.,** Modification of Cys-128 of pig kidney fructose 1,6-bisphosphatase with different thiol reagents: size dependent effect on the substrate and fructose-2,6-bisphosphate interaction, *J. Prot. Chem.,* 12, 159, 1993.
5. **May, J. M.,** Differential labeling of the erythrocyte hexose carrier by *N*-ethylmaleimide: correlation of transport inhibition with reactive carrier sulfhydryl groups, *Biochim. Biophys. Acta,* 986, 207, 1989.
6. **Ali, M. S., Roche, T. E., and Patel, M. S.,** Identification of the essential cysteine residue in the active site of bovine pyruvate dehydrogenase, *J. Biol. Chem.,* 268, 22553, 1993.
7. **Siksnys, V. and Pleckaityte, M.,** Role of the reactive cysteine residue in restriction endonuclease C24*fr*9I, *Biochim. Biophys. Acta,* 1160, 199, 1992.
8. **Mattsson, P., Pohjalainen, T., and Korpela, T.,** Chemical modification of cyclomaltodextrin glucanotransferase from Bacillus circulans var. alkalophilus, *Biochim. Biophys. Acta,* 1122, 33, 1992.
9. **Chang, G.-G., Satterlee, J., and Hsu, R. Y.,** Essential sulfydryl group of malic enzyme from *Escherichia coli, J. Prot. Chem.,* 12, 7, 1993.
10. **Nakayama, T., Tanabe, H., Deyashiki, Y., Shinoda, M., Hara, A., and Sawada, H.,** Chemical modification of cysteinyl, lysyl and histidyl residues of mouse liver 17β-hydroxysteroid dehydrogenase, *Biochim. Biophys. Acta,* 1120, 144, 1992.
11. **Okonjo, K. O. and Adejoro, I. A.,** Hemoglobins with multiple reactive sulfydryl groups: the reaction of dog hemoglobin with 5,5′-dithiobis(2-nitrobenzoate), *J. Prot. Chem.,* 12, 33, 1993.
12. **Karlstrom, A. R., Shames, B. D., and Levine, R. L.,** Reactivity of cysteine residues in the protease from human immunodeficiency virus: identification of a surface-exposed region which affects enzyme function, *Arch. Biochem. Biophys.,* 304, 163, 1993.

References for Table 1 (continued)

13. **Sheikh, S. and Katiyar, S. S.,** Chemical modification of octopine dehydrogenase by thiol-specific reagents: evidence for the presence of an essential cysteine at the catalytic site, *Biochim. Biophys. Acta,* 1202, 251, 1993.

14. **Strange, R., Morante, S., Stefanini, S., Chiancone, E., and Desideri, A.,** Nucleation of the iron core occurs at the three-fold channels of horse spleen apoferritin: an EXAFS study on the native and chemically-modified protein, *Biochim. Biophys. Acta,* 1164, 331, 1993.

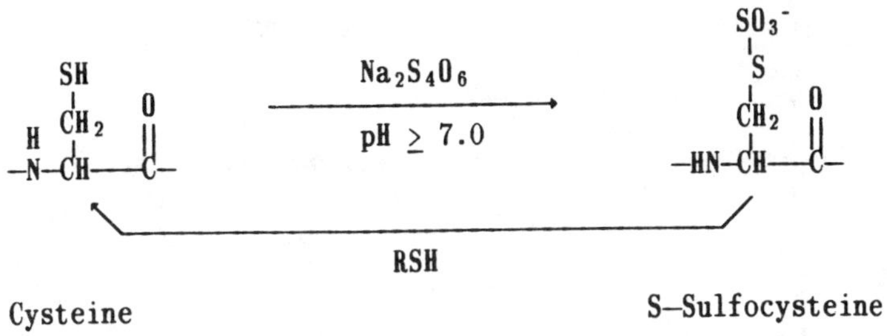

FIGURE 1. Conversion of cysteine to *S*-sulfocysteine by reaction with sodium tetrathionate and reversal by exogenous thiols.

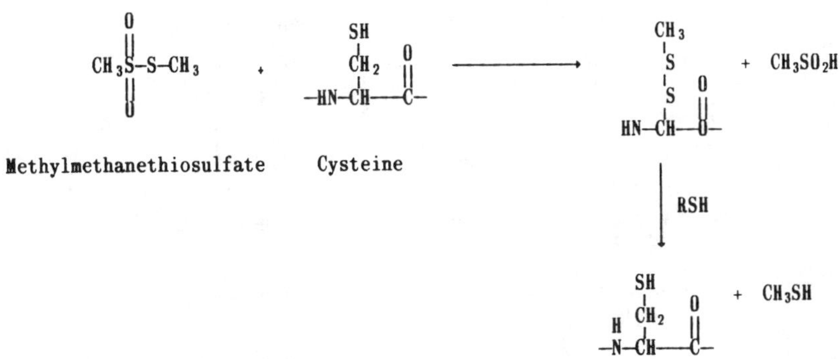

FIGURE 2. The reaction of methyl methanethiosulfate with cysteine and reversal with exogenous thiols.

reaction is desired, the iodine-containing compounds are used. For example, the reaction of iodoacetate with cysteine is approximately twice as fast as the reaction of bromoacetate and 20 to 100 times as rapid as chloroacetate. There are situations in which fast reaction rates are not necessarily desirable, such as the studies of Gerwin on streptococcal proteinase.[18] This particular study was of considerable importance since it emphasized the importance of microenvironmental effects on the reaction of cysteine with α-halo acids and α-halo amides. Chloroacetic acid was far less effective than chloroacetamide. The sulfhydryl group at the active site of streptococcal proteinase has enhanced reactivity in that

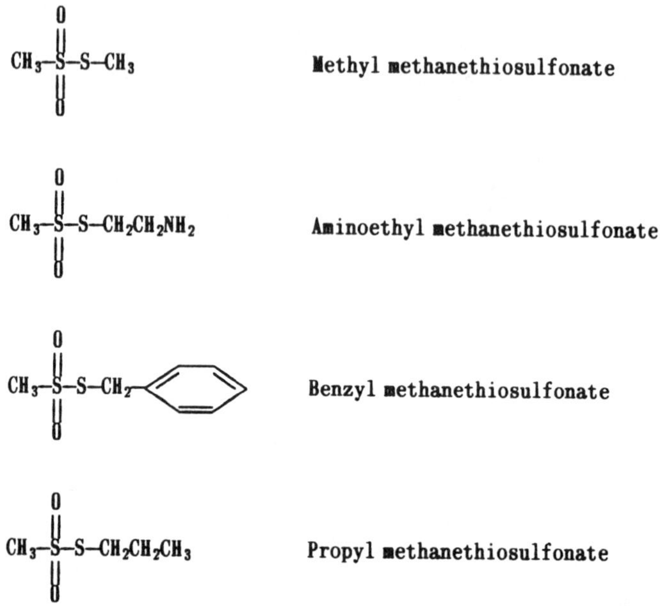

FIGURE 3. Structures of selected alkyl and aryl methanethiosulfates.

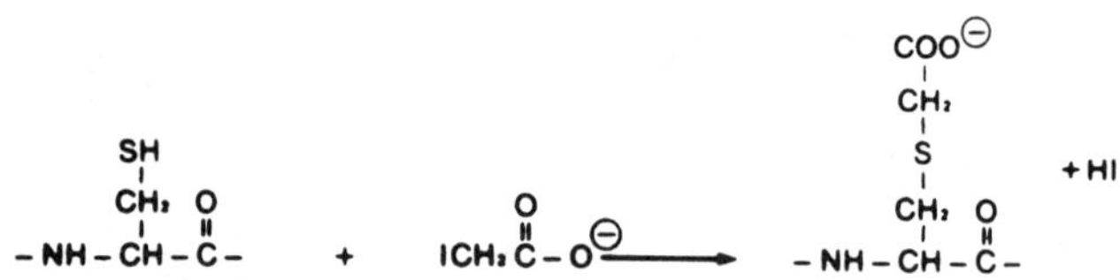

FIGURE 4. The modification of cysteine with iodoacetic acid to form *S*-carboxymethylcysteine.

modification with iodoacetate readily occurred in the presence of 100- to 1000-fold excess of β-mercaptoethanol or cysteine. The enhanced reactivity of the active-site cysteine is also apparent from a comparison of the relative rates of modification of streptococcal proteinase and reduced glutathione. The rate of modification of streptococcal proteinase is 50 to 100 times more rapid than that of glutathione. The unique properties of this cysteine residue can be explained in part by the presence of an adjacent histidyl residue which was demonstrated by an elegant series of studies by Liu.[19] Although histidine residues will react with α-halo acids and amides, the presence of an adjacent cysteine residue precluded the use of this class of reagents to demonstrate the presence of a histidyl residue at the active site of streptococcal proteinase. Liu took advantage of the reversible modification of cysteinyl residues with sodium tetrathionite to modify the active-site histidine.

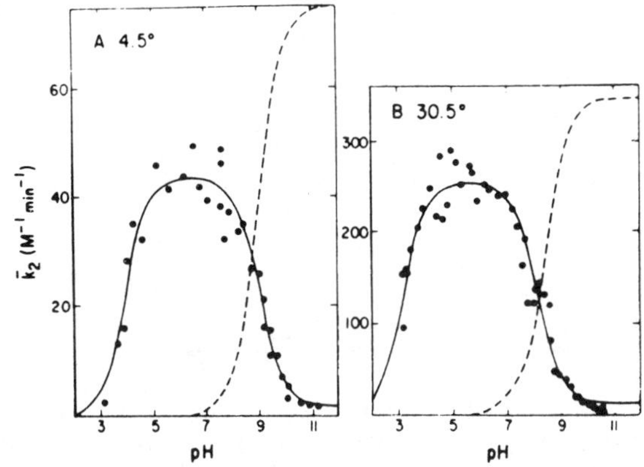

FIGURE 5. The effect of pH on the second-order rate constant for the inactivation of papain by chloroacetic acid at low ionic strength (0.07). The broken lines are the theoretical curves for the reaction of papain with chloroacetamide under the same reaction conditions. (From Chaiken, I. M. and Smith, E. L., *J. Biol. Chem.*, 244, 5095, 1969. With permission.)

The reaction of chloroacetic acid and chloroacetamide with papain has also yielded interesting results.[20,21] In studies with chloroacetamide, the active-site sulfhydryl group of papain reacts at a rate more than tenfold faster than free cysteine (5.78 M^{-1} s^{-1} vs. 0.429 M^{-1} s^{-1}).[20] As was the situation with streptococcal proteinase, there are dramatic differences in the rate of reaction of papain with chloroacetic acid and chloroacetamide. The reaction with chloroacetic acid has a pH optimum of approximately 7, while the optimum for reaction with chloroacetamide is at a pH >9. A comparison of the effect of pH on the reaction of papain with chloroacetic acid and chloroacetamide is shown in Figure 5. This investigation notes the influence of the neighboring histidyl residue as has been discussed for streptococcal proteinase. These data further emphasize the importance of neighboring functional group effects on cysteinyl reactivity in proteins as well as the importance of rigorous evaluation of the effect of pH on the rate of the modification reaction.

The α-halo acids decompose in water, with the rate being far more rapid at alkaline pH. In the case of iodoacetic acid, the products are iodide and glycolic acid. We recrystallize the commercially obtained reagents and store over P_2O_5. The compounds are readily soluble in water. In the case of the free acid, it is useful to dissolve the compound in base prior to addition to the reaction mixture. In the case of α-haloacetyl derivatives, the resultant *S*-carboxymethylcysteine is easily quantitated by amino acid analysis.

Jörnvall and co-workers[22] have used reaction with iodoacetate to probe differences in structure in wild-type β-galactosidase and various mutant forms of the enzyme. The modification reactions were performed in 0.1 *M* Tris, pH 8.1 under nitrogen in the dark. (This condition is of considerable importance

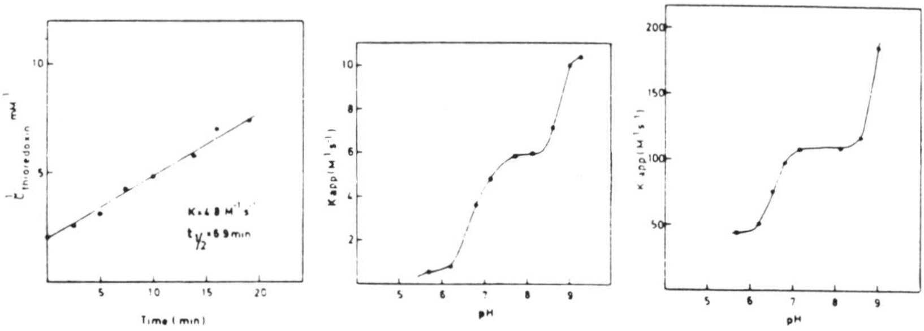

FIGURE 6. The left figure shows a time course for the reaction of thioredoxin and iodoacetic acid at pH 7.2. Analysis of this data yields a single second-order rate constant of 4.8 M^{-1} s^{-1} and a halftime of 6.9 min. The center figure shows the effect of pH on the second-order rate constant for the reaction between iodoacetic acid and thioredoxin. The figure on the right shows the effect of pH on the second-order rate constant for the reaction between iodoacetamide and thioredoxin. (From Kallis, G.-B. and Holmgren, A., *J. Biol. Chem.*, 255, 10261, 1980. With permission.)

since the α-halo acids are photolabile.) The reaction was terminated by the addition of excess β-mercaptoethanol. Kalimi and Love[23] have examined the reaction of the hepatic glucorticoid-receptor with iodoacetamide in 0.010 *M* Tris 0.25 *M* sucrose. Again, this reaction was performed in the dark. Kallis and Holmgren[24] have examined the differences in reactivity of two sulfhydryl groups present at the active site of thioredoxin. The pH dependence of the reaction with iodoacetate suggested that one group had a pKa value of 6.7, while the second was 9.0. Iodoacetamide showed the same pH dependence, but the rate of reaction was approximately 20-fold greater than with iodoacetate. For example, at pH 7.2, the second-order rate constant for reaction with iodoacetate was 5.2 M^{-1} s^{-1}, while it was 107.8 M^{-1} s^{-1} for iodoacetamide. The results from this study are shown in Figure 6. The low pKa of one of the sulfhydryl groups was suggested to be a reflection of the presence of an adjacent lysine residue. Mikami and co-workers have examined the inactivation of soybean β-amylase with iodoacetamide and iodoacetate.[25] Inactivation with iodoacetamide occurred approximately 60 times more rapidly than with iodoacetate at pH 8.6. Hempel and Pietruszko[26] have shown that human liver alcohol dehydrogenase is inactivated by iodoacetamide, but not by iodoacetic acid. These experiments were performed in 0.030 *M* sodium phosphate, pH 7.0 containing 0.001 *M* EDTA.

The reaction of sulfhydryl groups with iodoacetate[27] is still extensively used in the preparation of proteins for primary structure analysis, although pyridylethylation[28,29] is also useful.

Dahl and McKinley-McKee[30] have made a rather detailed study of the reaction of alkyl halides with thiols. It is emphasized that reactivity of alkyl halides not only depends on the halogen, but also on the nature of the alkyl groups. These investigators emphasized that the reactivity of an alkyl halide such as iodoacetate depends not only on the leaving potential of the halide

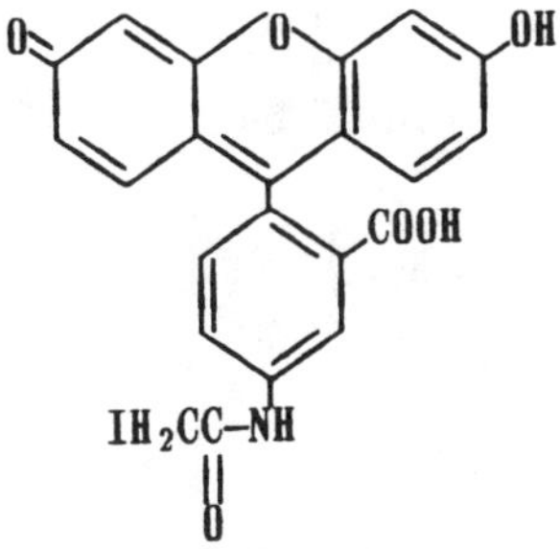

FIGURE 7. Structure of 5-iodoacetamidofluorescein.

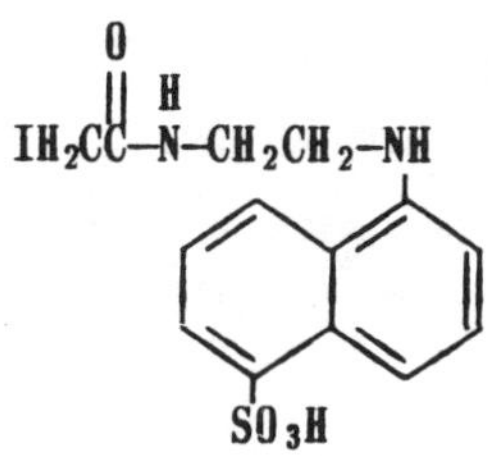

FIGURE 8. Structure of 5-[2-((iodoacetyl)amino-ethyl]naphthalene-1-sulfonic acid (1,5-IAE-DANS).

substituent (I > Br >>> Cl; 130:90:1), but also on the nature of the alkyl group. The rate of reaction of 2-bromoethanol with the sulfydryl group of L-cysteine (pH 9.0) is approximately 1000 times less than that observed with bromoacetic acid. The reactions are extremely pH dependent, emphasizing the importance of the thiolate anion in the reaction.

While haloacetates and haloacetamides continue to be useful,[31–36] there has been far greater interest in the use of this chemistry as a mechanism for introducing a larger molecule which can serve as structural probe. Examples include 5-iodoacetamido-fluorescein[37] (Figure 7), 5-[2-((iodoacetyl)amino)-ethyl]naphthalene-1-sulfonic acid (1,5-IAEDANS)[38–41] (Figure 8), and 4-(2-iodoacetamido)-TEMPO[33] (Figure 9). Iodoacetate has also been used for introducing a biotin probe (Figure 10) in proteins.[32]

N-Ethylmaleimide reacts with sulfhydryl groups (Figure 11) in proteins with considerable specificity.[42–44] This reaction can be followed spectrophotometrically by the decrease in absorbance at 300 nm, the absorbance maximum of *N*-ethylmaleimide. The extinction coefficient of *N*-ethylmaleimide is 620 M^{-1} cm^{-1} at 302 nm.[42] This reaction product yields *S*-succinyl cysteine on acid hydrolysis. Although the reagent is reasonably specific for cysteine, reaction with other nucleophiles must be considered.[45] A "diagonal" procedure for the isolation of cysteine-containing peptides modified with *N*-ethylmaleimide has been reported (Figure 12).[46] This procedure is based

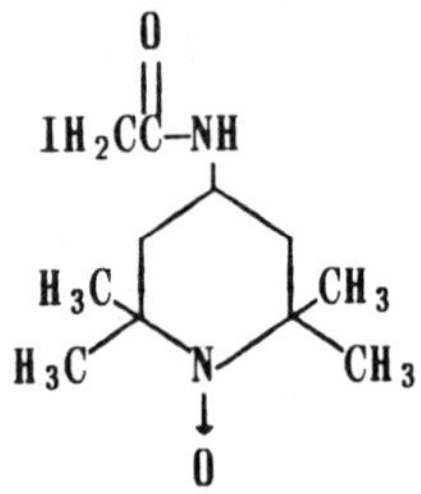

FIGURE 9. Structure of iodoacetamido-TEMPO.

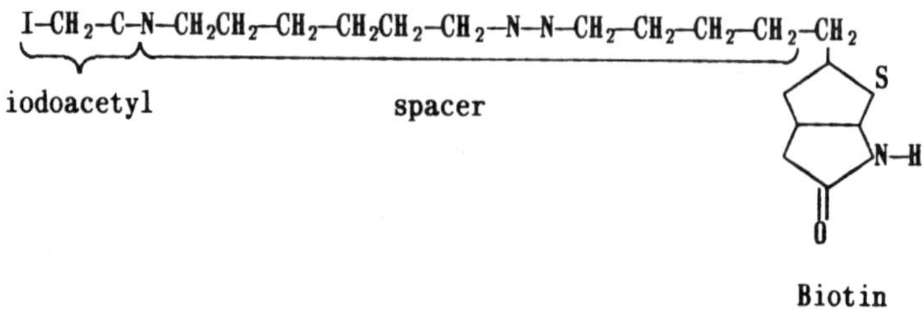

FIGURE 10. Iodoacetyl derivative of biotin for coupling with protein sulfhydryl groups.

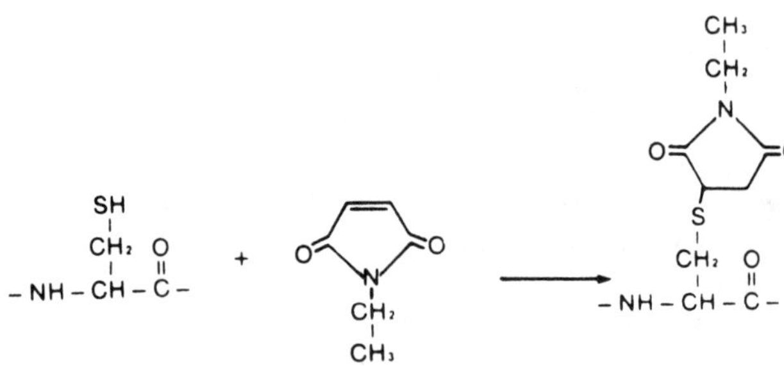

FIGURE 11. The reaction of *N*-ethylmaleimide with cysteine.

on the hydrolysis of the reaction product of cysteine and *N*-ethylmaleimide to cysteine-*S*-*N*-ethyl succinamic acid, generating a new negative charge. Various derivatives of maleimide provide the basis for the design of cross-linking reagents. Brown and Matthews[47,48] have studied the reaction of lactose repressor protein with *N*-ethylmaleimide, two spin-label derivatives of *N*-ethylmaleimides, and a fluorophore derivative. The spin-labeled compounds showed the same pattern of reaction with the three cysteinyl residues as seen with *N*-ethylmaleimide. The fluorophore-derivative (*N*-(3-pyrene)maleimide) shows a slightly different reaction pattern. Other probes include 4-maleimido-2,2,6,6-tetramethylpiperidine-1-oxyl[49] (Figure 13),

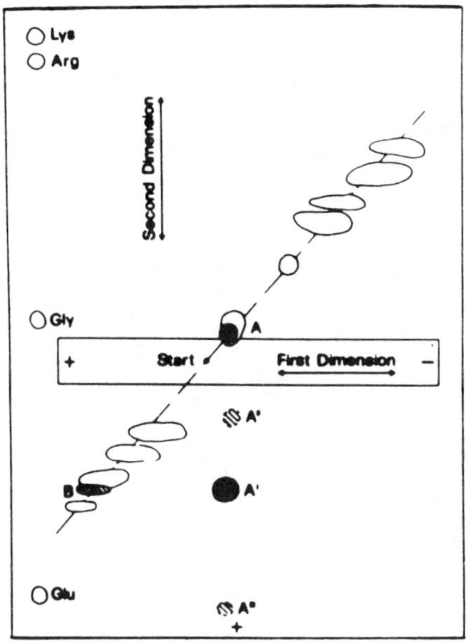

FIGURE 12. A "diagonal" method for the identification of cysteine peptides. A mixture of chymotryptic peptides containing a peptide with a cysteinyl residue alkylated with radiolabeled *N*-ethylmaleimide was subjected to diagonal electrophoresis with intervening exposure to ammonia for 5.5 h at 35°C. The treatment with ammonia results in the hydrolysis of the *N*-ethylsuccinimide to *N*-ethylsuccinamic acid, thus generating a new negative charge permitting the resolution of the labeled peptides from the unlabeled peptides during the second electrophoresis. (From Gehring, H. and Christen, P., *Anal. Biochem.*, 107, 358, 1980. With permission.)

maleimidotetra-methylrhodamine,[39] *N*-(1-pyrenyl)maleimide[50] (Figure 14), 2-(4-maleimidoanilino)naphthalene-6-sulfonic acid,[51] 2,5-dimethoxy-4-stilbenylmaleimide,[51] rhodamine maleimide,[51] and eosin-5-maleimide[52] (Figure 15). Le-Quoc and colleagues have examined the effect of the nature of the *N*-substituent groups on the rate of sulfhydryl group modification in succinate dehydrogenase.[53] The derivatives used were *N*-ethylmaleimide, *N*-butylmaleimide, and *N*-benzylmaleimide. The most reactive thiol groups in succinate dehydrogenase are probably located in an apolar environment since the benzyl derivative reacted twice as fast as the ethyl derivative, as shown in Figure 16 (these modification reactions were performed in 0.050 *M* sodium phosphate, pH 7.6, at 37°C).

Hydrophobic derivatives of *N*-ethylmaleimide (Figure 17) can be used as probes of the environment surrounding a sulfydryl group in membrane anion channels.[54] Reaction with *N*-ethylmaleimide, *N*-benzylmaleimide, and *N,N′*-1,2-phenylenedimaleimide was evaluated and the reaction rate increased with increasing hydrophobicity. *N*-Phenylmaleimide has been used for the modification of a sulfydryl group in the acetylcholine receptor.[55,56] Detergent was

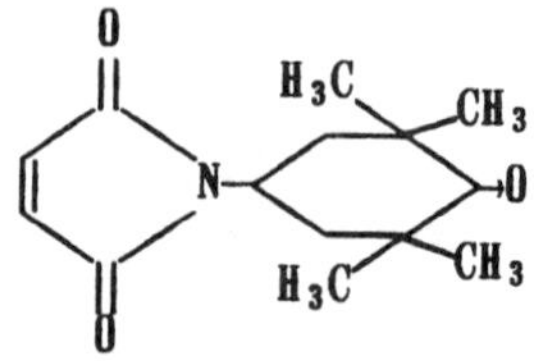

FIGURE 13. The structure of 4-maleimido-2,2,6,6-tetramethylpiperidino-1-oxyl, a spin-label maleimide derivative.

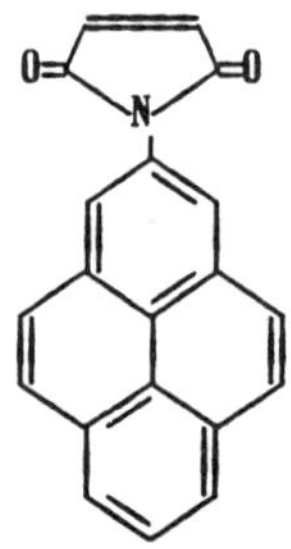

FIGURE 14. The structure of *N*-(1-pyrenyl)maleimide, a hydrophobic fluorescent maleimide-based structural probe.

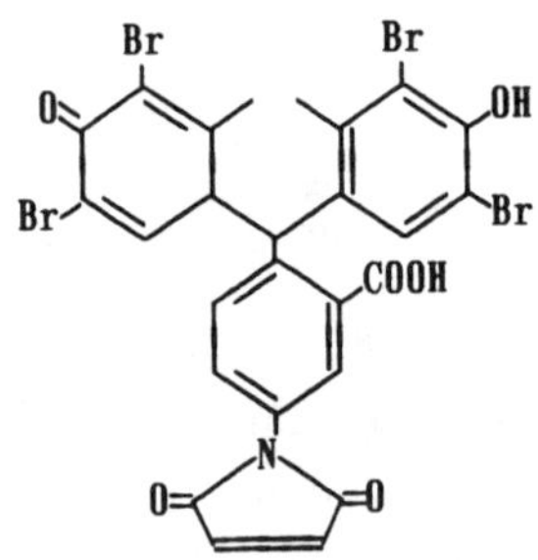

FIGURE 15. The structure of eosin-5-maleimide.

required for the modification reaction (10 m*M* MOPS to 100 m*M* NaCl to 0.1 m*M* EDTA with 0.02% sodium azide and 1% cholate). These studies identified cysteine residues potentially important in membrane function. Subsequent studies using site-specific mutagenesis have supported the importance of these cysteinyl residues.[56]

Localization of sulfydryl groups within membranes has been achieved through the comparison of the reaction with membrane-permeant and membrane-impermeant derivatives (Figure 18).[57–59] The use of these reagents can be traced back to the original observations of Abbott and Schachter in 1976.[60] The

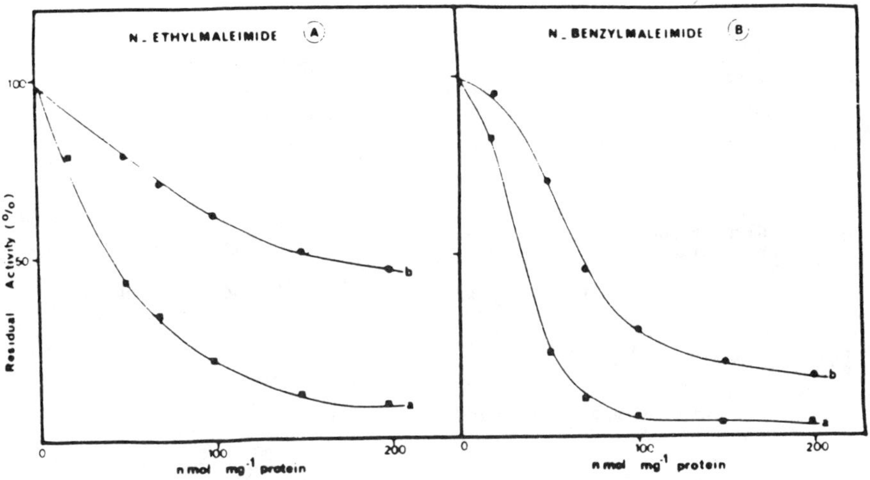

FIGURE 16. The inactivation of succinate dehydrogenase with *N*-ethylmaleimide (panel A) or *N*-benzylmaleimide (panel B) as a function of reagent concentration. Reaction A was performed with enzyme preparations preincubated with 50 m*M* succinate, 3 m*M* thenoyltrifluoroacetone, and 10 μg rotenone. Reaction B was performed with no additions other than the maleimide derivatives in 0.05 *M* sodium phosphate, pH 7.6. (From Le-Quoc, K., Le-Quoc, D., and Guademer, Y., *Biochemistry,* 20, 1705, 1981. With permission.)

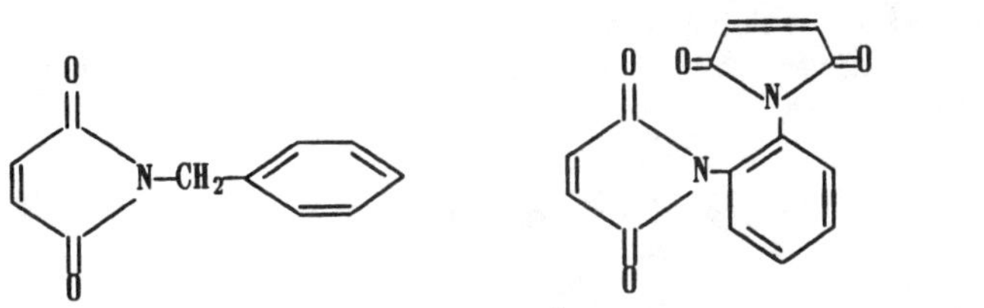

FIGURE 17. The structures of some hydrophobic *N*-alkylmaleimide derivatives, *N*-benzylmaleimide and *N,N′*-1,2,-diphenylenedimaleimide.

basic concept is to provide either a polar derivative or a derivative with steric considerations which preclude passage through or into the membranes (c.f. dextran-maleimide in Figure 18). Maleimide derivatives of glucosamine have been synthesized as affinity labels for the human erythrocyte hexose transport protein.[61]

A particularly novel approach to this problem has been used by Falke's laboratory to analyze aspartate receptor structure.[62] In this study, site-specific mutagenesis was used to place cysteinyl residues at six positions in the peptide chain. A new membrane-impermeant reagent (Figure 19) was used to study the reactivity of the individual sulfydryl residues. From these studies, it was possible to "map" the domain structure of the receptor protein. Cysteinyl

Dextran Maleimide

Glutathione Maleimide

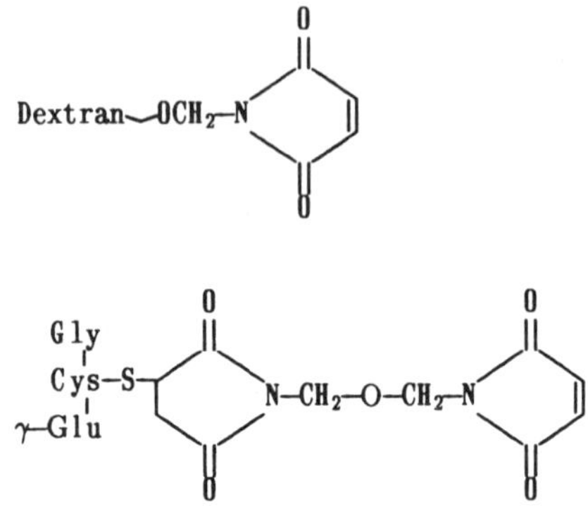

FIGURE 18. Some membrane-impermeant *N*-alkylmaleimide derivatives.

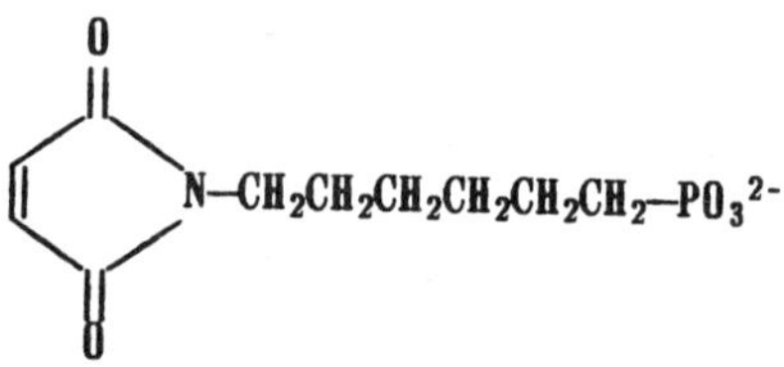

FIGURE 19. The structure of *N*-(6-phosphonyl-*n*-hexyl)-maleimide (PHM), a membrane-impermeant *N*-alkylmaleimide. (From Falke, J. J., et al., *J. Biol. Chem.*, 263, 14850, 1988. With permission.)

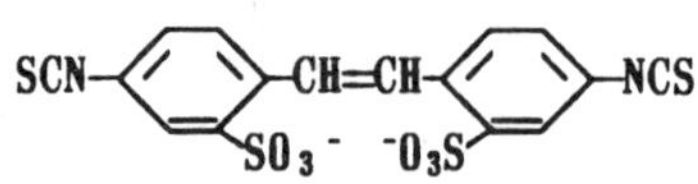

FIGURE 20. The structure of 4,4′-diisothiocyanostilbene-2,2′-disulfonic acid.

residues placed in the surface area could be modified by aqueous reagents, while transmembrane areas could be excluded by lack of reaction with membrane-impermeant reagents. Finally, spatial proximity of multiple cysteinyl residues could be evaluated by disulfide bond formation.

4,4′-Diisothiocyanostilbene-2,2′-disulfonic acid (Figure 20) has been used to study the importance of specific sulfydryl groups in anion transport by membrane proteins.[63–65] Bimanes have proven to be useful structural probes for proteins. The structures of monobromobimane (I) and monobromotrimethylammonium bimane (II) are shown in Figure 21. These are two examples of various derivatives available with monobromobimane being considered a non-polar or hydrophobic probe and monobromotrimethylammonium bimane being considered a polar probe. The reader is directed toward two studies on the use of these reagents for the study of sulfydryl group chemistry in proteins.[41,52]

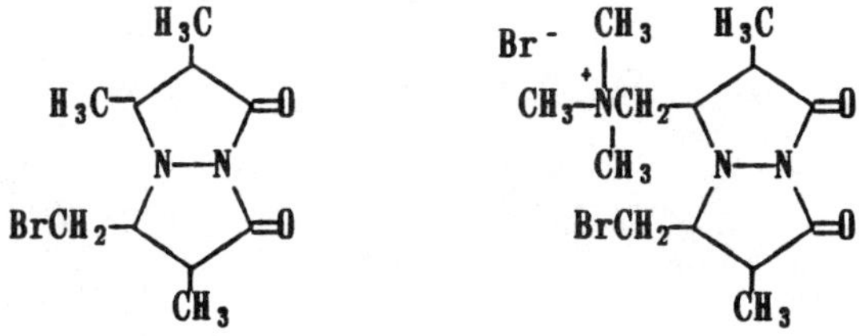

I II

FIGURE 21. Structure of two bimane derivatives which have proved useful in the study of protein sulfydryl groups: (I) monobromobimane, a relatively hydrophobic derivative; and (II) monobromotrimethylammonium bimane, a relatively hydrophilic derivative.

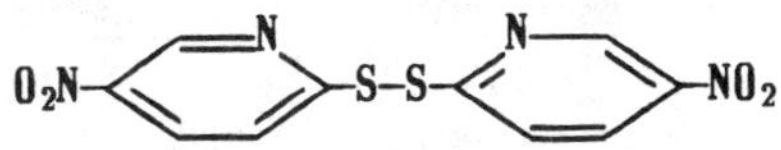

2,2ˡ -dithiobis-(S-nitropyridine)

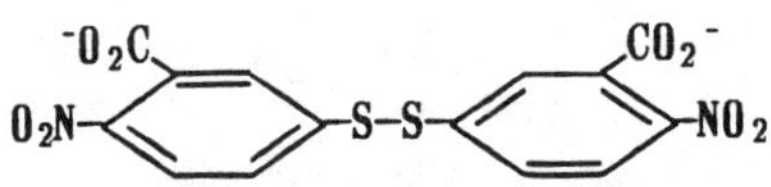

S,Sˡ -dithiobis-(2-nitrobenzoic acid)

FIGURE 22. The structure of 2,2′-dithiobis-(5-nitropyridine) which is relatively hydrophobic compared to 5,5′-dithiobis-(2-nitrobenzoic acid).

The formation of 2-mercapto-5-nitrobenzoic acid, which occurs with the reaction of 2-nitrothiocyanobenzoic acid with thiols to form *S*-cyano derivatives, can be used for the quantitative determination of sulfhydryl groups. 2-Mercapto-5-nitrobenzoic acid has an absorbance maximum at 412 nm with a molar extinction coefficient of 13,600 M^{-1} cm^{-1}.[66] Pecci and co-workers[67] have characterized the reaction of rhodanese with 2-nitrothiocyanobenzoic acid. These investigators used a 1.3 molar excess of reagent in 0.050 *M* phosphate buffer, pH 8.0 at 18°C. The reaction was followed spectrophotometrically by the release of 2-mercapto-5-nitrobenzoic acid and was complete after 6 h.

Cleavage at *S*-cyano-cysteinyl residues was first studied by Vanaman and Stark.[68] *S*-Cyanocysteine was first generated from the mixed disulfide of cysteine and 5-thio-2-nitrobenzoic acid by reaction with KCN (0.05 *M*) at pH 8.2 (0.2 *M* Tris-acetate with 20 m*M* EDTA).

One of the most popular reagents for the modification and determination of the sulfhydryl group has evolved from the early studies of Ellman[66] on 5,5′-dithiobis-(2-nitrobenzoic acid) (DTNB; Ellman's Reagent) (Figure 22). Reac-

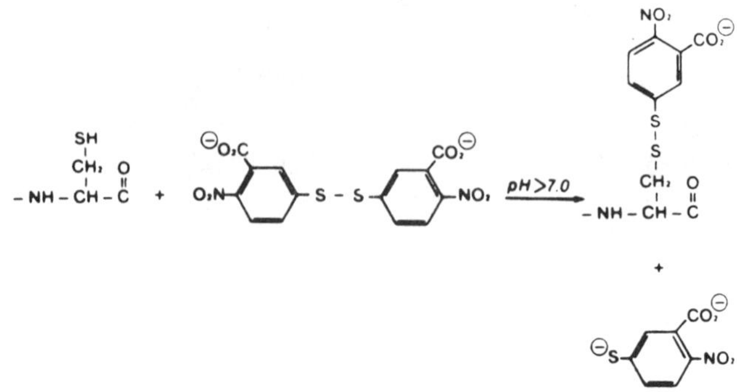

FIGURE 23. The reaction of 5,5′-dithiobis-(2-nitrobenzoic acid) with cysteinyl residues in proteins.

tion with sulfhydryl groups in proteins results in the release of 2-nitro-5-mercaptobenzoic acid (Figure 23), which has a molar extinction coefficient of 13,600 M^{-1} cm^{-1} at 410 nm.[69] Examples of the use of this reagent have included studies on *Escherichia coli* citrate synthase[70] and D-amino acid transaminase[71] (0.1 *M* Tris, 0.002 *M* EDTA, pH 7.5). 4,4′-Dithiodipyridine is similar to 5,5′-dithiobis-(5-nitrobenzoic acid) in that a mixed disulfide is formed between a cysteinyl residue in the protein and the reagent with the concomitant release of pyridine-4-thione.[70] The reaction of 4,4′-dithiodipyridine with protein sulfhydryl groups can be followed by spectroscopy ($\varepsilon_{324\ nm}$ = 19,800 M^{-1} cm^{-1}). The reaction is readily reversed by the addition of a reducing agent such as dithiothreitol. Reaction with 5,5′-dithiobis-(2-nitrobenzoic acid) does result in the loss of activity, and this loss of activity appears to be correlated with the modification of one of the more slowly reacting cysteinyl residues. Other proteins which have been studied with this reagent include rat brain nicotinic-like acetylcholine receptors[72] (calcium-containing Ringers solution, pH 7.4), lipophilin from human myelin[73] (0.001 *M* glycylglycine — 0.0001 *M* EDTA, pH 8.0), and human hemoglobin.[74] The latter study followed the changes in absorbance at 450 nm to monitor the release of the 2-nitro-5-mercaptobenzoic acid. The molar extinction coefficients obtained at 450 nm were 5550 M^{-1} cm^{-1} (pH 6.0); 6510 M^{-1} cm^{-1} (pH 7.0); 6810 M^{-1} cm^{-1} (pH 8.0); 6940 M^{-1} cm^{-1} (pH 9.0); and 7010 M^{-1} cm^{-1} (pH 9.5). The synthesis of a selenium analog of this class of reagents, 6,6-diselenobis(3-nitrobenzoic acid), has been reported.[75] The selenium-containing reagent has the same reaction characteristics as the sulfur-containing compound in terms of specificity of reaction with cysteinyl residues in proteins. The reaction is monitored by spectroscopy following the release of 6-seleno-3-nitrobenzoate which has a maximum at 432 nm (Figure 24). The extinction coefficient for the 6-seleno-3-nitrobenzoate anion varies slightly from 9532 (with excess reagent) to 10,200 M^{-1} cm^{-1} (with either excess cysteine or excess β-mercaptoethanol).

Other studies on the use of DTNB to modify sulfydryl groups in proteins have included the modification of oncomodulin,[76] glutathione synthetase,[77]

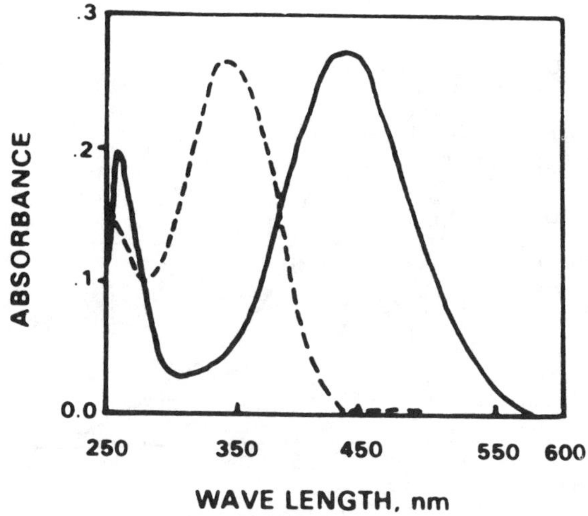

FIGURE 24. The UV absorption spectrum of 6-selenobis(3-nitrobenzoic acid) (SNB^{2-}) (25.9 μ*M*) and 6,6-diselenobis(3-nitrobenzoic acid) (DSNB) (12.9 μ*M*) in 0.2 *M* Tris-HCl, pH 8.2, 1 m*M* EDTA. (From Luthra, M. P., Dunlap, R. B., and Odom, J. D., *Anal. Biochem.*, 117, 94, 1981. With permission.)

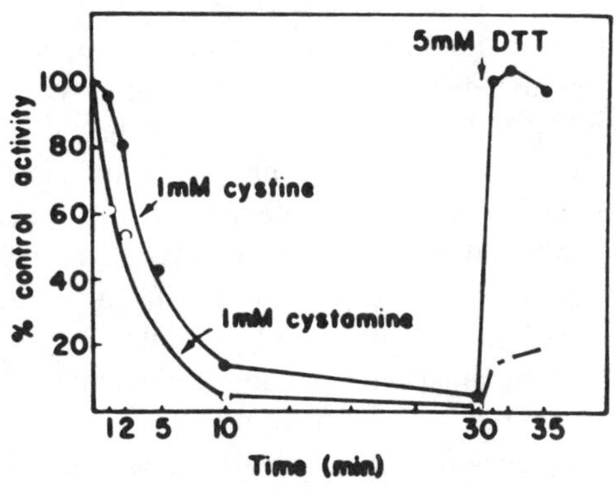

FIGURE 25. The effect of mixed disulfide formation on guanylate cyclase activity. Guanylate cylase was incubated with either 1 m*M* cystine or 1 m*M* cystamine in 0.02 *M* Tris-HCl, pH 7.6, 1 m*M* dithiothreitol, 10% sucrose. After 30 min of incubation, dithiothreitol was added to a final concentration of 5 m*M*. (From Brandwein, J., Lewicki, J., and Murad, F., *J. Biol. Chem.*, 256, 2958, 1981. With permission.)

fibronectin,[78] and streptococcal NADH peroxidase.[79] In studies on the modification of ATP sulfurylase,[80] it was observed that DTNB was less potent than the more hydrophobic dithionitropyridine derivative (see Figure 22).

Cystine or cystamine have proved effective in the modification of guanylate cyclase[80] as shown in Figure 25. Note the ready reversibility of the modifica-

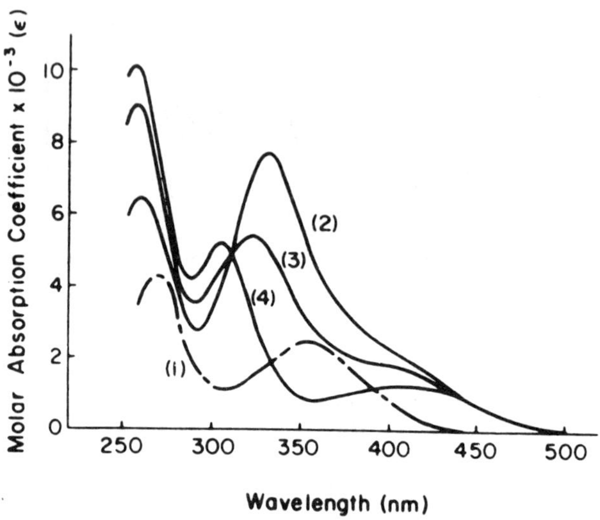

FIGURE 26. The UV absorption spectra of methyl-3-nitro-2-pyridyl disulfide (NPySSMe) and 3-nitro-2-pyridone (NPySH) in 0.050 *M* sodium phosphate: NPySSMe at pH 4.8 (Curve 1), NPySH at pH 4.0, 4.6 (Curve 2), pH 6.4 (Curve 3), and pH 8.4, 8.8 (Curve 4). (From Kimura, T., et al., *Anal. Biochem.*, 122, 274, 1982. With permission.)

tion on the addition of dithiothreitol. Kimura and co-workers[81] introduced methyl 3-nitro-2-pyridyl disulfide and methyl 2-pyridyl disulfide. Both of these reagents modify sulfhydryl groups forming the thiomethyl derivative. Figure 26 shows the spectra of methyl-3-nitro-2-pyridyl disulfide (NPySSMe) and 3-nitro-2-pyridone (NPySH) with determinations of the latter compared at several different conditions of pH. The spectrum of 3-nitro-2-pyridone is pH dependent. There is an isosbestic point at 310.4 nm which can be used to determine the extent of the reaction of methyl-3-nitro-2-pyridyl disulfide with sulfhydryl groups. Similar spectral studies for methyl-2-pyridyl disulfide (PySSMe) and 2-thiopyridone (PySH) are shown in Figure 27. The difference in spectrum obtained does not show the pH dependence of the nitropyridyl derivative (see Figure 26). At 343 nm, the change in extinction coefficient is 7060 M^{-1} cm^{-1}. The extinction coefficient (7600 M^{-1} cm^{-1}) of the 2-thiopyridinone at 343 nm is relatively stable from pH 3 to 8.0.[81] Above pH 8.0, there is a marked decrease reflecting the loss of a proton. Reaction with the sulfhydryl group in the protein clearly proceeds more rapidly at alkaline pH.

In a related study, Drewes and Faulstich[82] prepared 2,4-dinitrophenyl-^{14}C-cysteinyl disulfide (Figure 28) via a facile synthetic method as a means for introducing radiolabeled cysteine into proteins via disulfide exchange with free thiols. The reaction can be monitored by following the release of 2,4-dinitrophenol at 408 nm (408 = 12,700 M^{-1} cm^{-1}). The specificity of this reagent corresponded to that obtained with DTNB. Reaction with the sulfhydryl groups

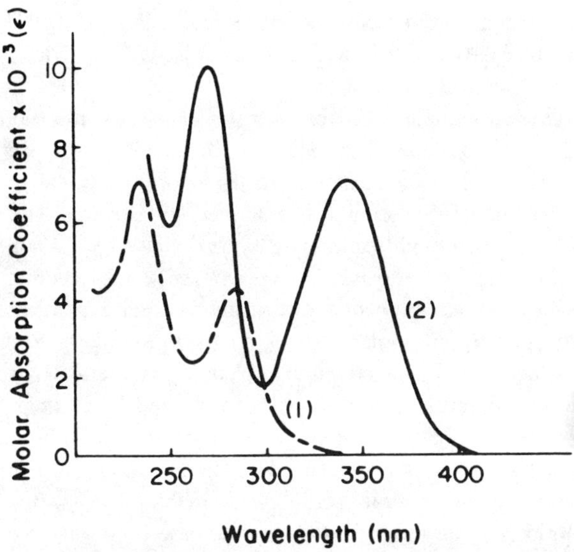

FIGURE 27. The UV absorption spectra of methyl-2-pyridyl disulfide (PySSMe) and 2-thiopyridone (PySH) in 0.050 *M* sodium phosphate, pH 7.5. Curve 1 is that for PySSMe and Curve 2 for PySH. (From Kimura, T. et al., *Anal. Biochem.*, 122, 274, 1982. With permission.)

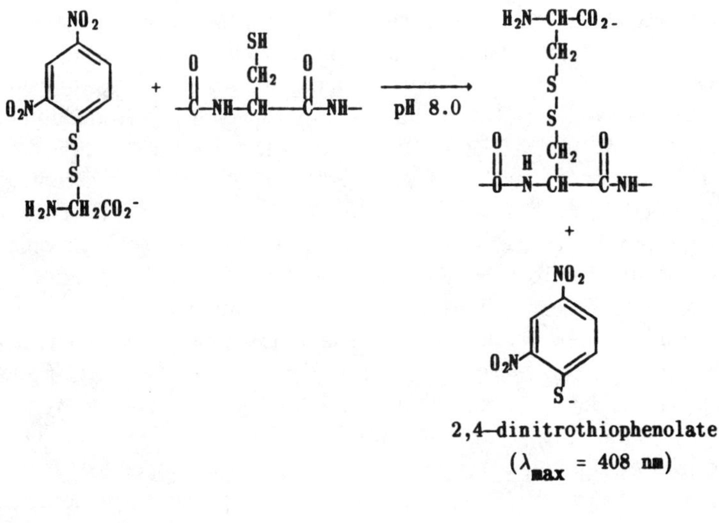

FIGURE 28. The reaction of 2,4-dinitrophenyl-^{14}C-cysteinyl disulfide with cysteine in proteins.

of papain was more rapid than that observed with DTNB. The resulting derivative can be easily reversed with thiols, but is stable to cyanogen bromide degradation and peptide purification.

p-Hydroxymercuribenzoate continues to be of use for the modification of sulfhydryl groups in proteins. The reagent is obtained as *p*-chloromercuribenzoate, but is instantaneously converted to the hydroxy derivative in aqueous solution. This reagent was originally described by Boyer.[83] The absorbance change at 255 nm upon modification is 6200 M^{-1} cm^{-1} at pH 4.6 and 7600 M^{-1} cm^{-1} at pH 7.0. Bai and Hayashi[84] have examined the reaction of organic mercurials with yeast carboxypeptidase (carboxypeptidase Y). Treatment of the modified enzyme with millimolar cysteine resulted in virtually complete recovery of catalytic activity. The inactivation of chalcone isomerase by *p*-chloromercuribenzoate and mercuric chloride has been studied by Bednar et al.[14] The modified protein could be readily reactivated by treatment with either thiols or KCN. The reactivation by KCN is based on the formation of a tight complex between cyanide and either organic or inorganic mercurials. The modification by mercuric chloride can be monitored by the increase in absorbance at 250 nm. Ojcius and Solomon[85] have examined the inhibition of erythrocyte urea and water transport by *p*-chloromercuribenzoate. Other studies with this reagent have included dissociation of erythrocyte membrane proteins[86] and NADH peroxidase.[34]

2-Chloromercuri-4-nitrophenol is a compound related to the organic mercurial described above. It has proved useful as a "reporter" group in the study of microenvironmental changes in the modified protein. An excellent example of this is provided from the studies of Marshall and Cohen[87] on the properties of ornithine transcarbamylase modified with 2-chloromercuri-4-nitrophenol. The enzyme from *Streptococcus faecalis* was modified in 0.1 *M* MOPS, 0.1 *M* KCl, pH 7.5 using changes in absorbance at 403 nm to follow the extent of modification. The bovine enzyme is carboxamido methylated on a nonessential sulfhydryl group before reaction with the organic mercurial. Modification of the bovine enzyme with 2-chloromercuri-4-nitrophenol is performed in 0.020 *M* MOPS, 0.1 *M* KCl, pH 7.11 at 25°C. The modification was followed by the change in absorbance at 405 nm. The effect of pH on the spectrum of the modified bovine enzyme is shown in Figure 29. Baines and Brocklehurst[88] have reported the synthesis and characterization of 2-(2′-pyridylmercapto)-mercuri-4-nitrophenol, a reagent which does have certain advantages.

A number of other modifications of sulfhydryl groups have proved useful. *O*-Methylisourea reacts with cysteinyl residues to form the *S*-methyl derivative (Figure 30).[89] Cyanate also can modify sulfhydryl groups as shown in Table 2.[90] The carbamoyl derivative of cysteine is stable at acid pH, but rapidly decomposes at alkaline pH. 4-Chloro-7-nitrobenzo-2-oxa-1,3-diazole (4-chloro-7-nitrobenzofurazan; Nbf-Cl) (Figure 31) is a reagent developed for the modification of amino groups.[91] It has also found application in the modification of sulfhydryl groups and is useful in that it introduces a fluorescent probe.[92–96]

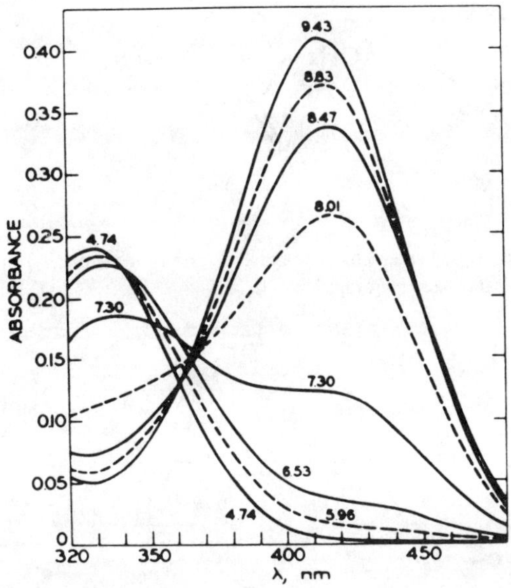

FIGURE 29. The effect of pH on the UV absorption spectrum of the 2-chloromercuri-4-nitrophenol derivative of the monocarboxamidomethyl bovine liver ornithine transcarbamylase. (From Marshall, M. and Cohen, P. P., *J. Biol. Chem.*, 255, 7296, 1980. With permission.)

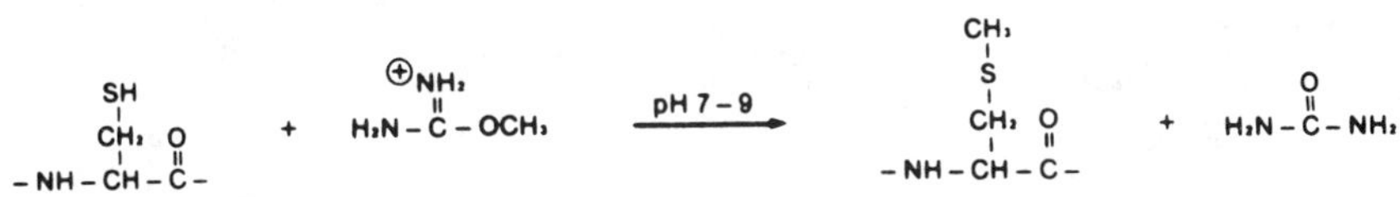

FIGURE 30. The reaction of *O*-methylisourea with cysteine.

Table 2
REACTION OF CYANATE WITH FUNCTIONAL GROUPS IN PROTEINS

Functional group	pKa	M^{-1} min^{-1}
α-Amino	7.8–8.2	1.4×10^{-1}
ε-Amino	10.5–10.8	2×10^{-3}
Sulfhydryl	8.3–8.5	4.0
Imidazole	7.0–7.2	1.8×10^{-1}

From Stark, G. R., *Meth. Enzymol.*, 11, 590, 1967. With permission.

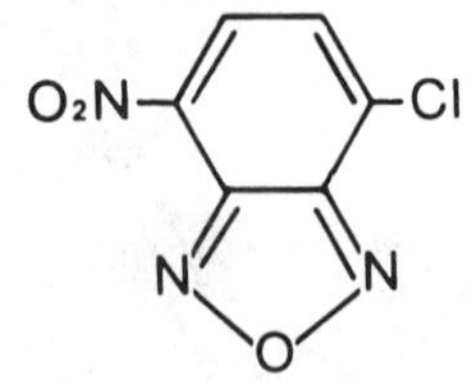

FIGURE 31. The structure of 4-chloro-7-nitrobenzo-2-oxa-1,3-diazole (4-chloro-7-nitrobenzofurazan; Nbf-Cl).

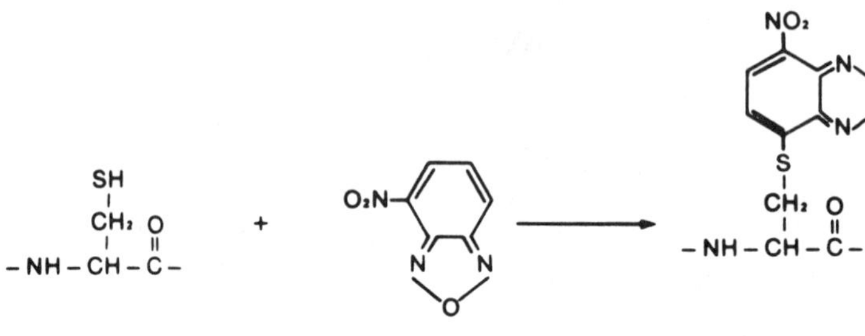

FIGURE 32. The reaction of 4-chloro-7-nitrobenzo-2-oxa-1,3-diazole with cysteine.

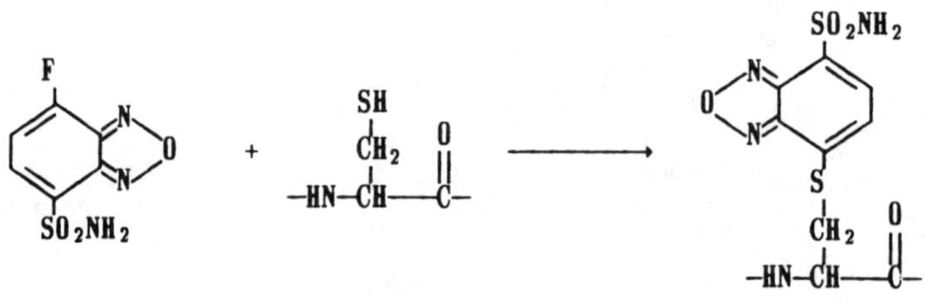

FIGURE 33. The structure of 4-(aminosulfonyl)-7-fluoro-2,1,3-benzoxadiazole (ABD-F), a fluorogenic reagent for the modification of cysteine residues in proteins.

Nitta and co-workers[95] have noted that there are other possible reaction products of Nbf-Cl, including the possibility of reaction products with sulfhydryl groups. The modification of the sulfhydryl group with concomitant reaction at the 4-position yields a derivative with molar absorption coefficient of 13,000 (Figure 32).[96] The reaction of Nbf-Cl with sulfhydryl groups in glutathione reductase and lipoamide dehydrogenase has also been reported.[97] Nitta and co-workers[95] have examined the chemistry of the reaction of Nbf-Cl with model sulfhydryl compounds in some detail.

Toyo'oka and Imai[98] have reported the synthesis of a related compound, 4-(aminosulfonyl)-7-fluoro-2,1,3-benzoxadiazole (Figure 33) which is a fluorogenic reagent for sulfhydryl groups. In a subsequent study, Kirley[99] has used this compound to label cysteinyl residues in proteins. Reaction readily occurred at pH 8.0 (100 m*M* borate-2 m*M* EDTA with 3% SDS).

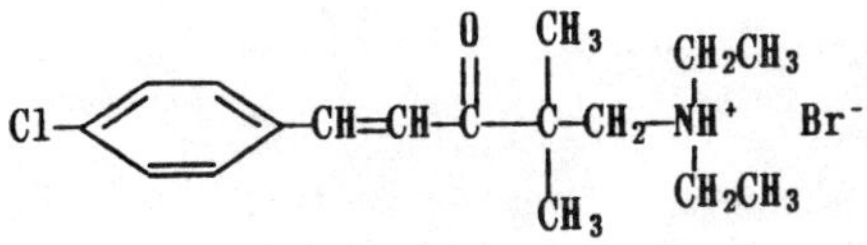

1-p-chlorophenyl-4,4-dimethyl-5-diethylamino-1-penten-3-one hydrobromide

FIGURE 34. The structure of a novel sulfydryl reagent.

The modification of cysteinyl residues in proteins with 2-bromoethane sulfonate has been reported.[100] This derivatization procedure was developed in response to a need for a strongly hydrophilic substituent in samples for the Edman degradation. The modification time is longer than for the corresponding carboxymethyl derivatives, taking 12 h for lysozyme, 24 h for insulin, and 48 h for glutathione. This derivative has considerable utility since the *S*-sulfoethylated lysozyme derivative is soluble between pH 5.0 and 10.0, while the *S*-carboxymethylated derivative is not. This procedure has potential for primary structure analysis.

Mutus et al.[101] have introduced 1-*p*-chorophenyl-4,4-dimethyl-5-diethylamino-1-penten-3-one hydrobromide (Figure 34) as a reagent for the modification of thiol groups. This reagent readily and reversibly reacts with low molecular weight thiols such as cysteine or glutathione with a large decrease in absorbance at 310 nM (= 21,000 M^{-1} cm^{-1}). The reaction with larger thiol-containing molecules such as proteins appears to be irreversible.

REFERENCES

1. **Liu, T.-Y.,** The role of sulfur in proteins, in *The Proteins,* Vol. 3, 3rd ed., Neurath, H. and Hill, R. L., Eds., Academic Press, New York, 1977, 240.
2. **Kenyon, G. L. and Bruice, T. W.,** Novel sulfydryl reagents, *Meth. Enzymol.,* 47, 407, 1977.
3. **Kanaya, S., Kimura, S., Katsuda, C., and Ikehara, M.,** Role of cysteine residues in ribonuclease H from *Escherichia coli.* Site-directed mutagenesis and chemical modification, *Biochem. J.,* 271, 59, 1990.
4. **Kato, H., Tanaka, T., Nishioka, T., Kimura, A., and Oda, J.,** Role of cysteine residues in glutathione synthetase from *Escherichia coli* B. Chemical modification and oligonucleotide site-directed mutagenesis, *J. Biol. Chem.,* 263, 11646, 1988.

4a. **Altenbach, C., Flitsch, S. L., Khorana, H. G., and Hubbell, W. L.,** Structural studies on transmembrane proteins. II. Spin labeling of bacteriorhodopsin mutants at unique cysteines, *Biochemistry,* 28, 7806, 1989.

4b. **Bech, L. M. and Breddam, K.,** Chemical modification of a cysteinyl residue introduced in the binding site of carboxypeptidase Y by site-directed mutagenesis, *Carlsberg Res. Commun.,* 53(Abstr.), 381, 1988.

4c. **Minard, P., Desmadril, M., Ballery, N., Perahia, D., Mouawad, L., Hall, L., and Yon, J. M.,** Study of the fast-reacting cysteines in phosphoglycerate kinase using chemical modification and site-directed mutagenesis, *Eur. J. Biochem.,* 185, 419, 1989.

4d. **Kanaya, S., Kimura, S., Katsuda, C., and Ikehara, M.,** Role of cysteine residues in ribonuclease H from *Escherichia coli*. Site-directed mutagenesis and chemical modification, *Biochem. J.*, 271, 59, 1990.

4e. **Ahmed, S. A., Kawasaki, H., Bauerle, R., Morita, H., and Miles, E. W.,** Site-directed mutagenesis of the alpha subunit of tryptophan synthase from Salmonella typhimurium, *Biochem. Biophys. Res. Commun.*, 151, 672, 1988.

4f. **Thrower, A. R., Byrd, J., Tarbet, E. B., Mehra, R. K., Hamer, D. H., and Winge, D. R.,** Effect of mutation of cysteinyl residues in yeast Cu-metallothionein, *J. Biol. Chem.*, 263, 7037, 1988.

4g. **Winther, J. R. and Breddam, K.,** The free sulfydryl group (CYS341) of carboxypeptidase Y: functional effects of mutational substitutions, *Carlsberg Res. Commun.*, 52, 263, 1987.

5. **Karlstrom, A. R., Shames, B. D., and Levine, R. L.,** Reactivity of cysteine residues in the protease from human immunodeficiency virus: identification of a surface-exposed region which affects enzyme function, *Arch. Biochem. Biophys.*, 304, 163, 1993.

6. **Okonjo, K. O. and Adejoro, I. A.,** Hemoglobins with multiple reactive sulphydryl groups: the reaction of dog hemoglobin with 5,5′-dithiobis(2-nitrobenzoate), *J. Prot. Chem.*, 12, 33, 1993.

7. **Nakayama, T., Tanabe, H., Deyashi, Y., Shinoda, M., Hara, A., and Sawada, H.,** Chemical modification of cysteinyl, lysyl and histidyl residues of mouse liver 17β-hydroxysteroid dehydrogenase, *Biochim. Biophys. Acta*, 1120, 144, 1992.

8. **Watala, C., Pietrurcha, T., Gwazdzinski, K., and Cierniewski, C. S.,** Do the spectra of maleimide spin-labelled whole blood platelets reflect the structure and conformation of membrane proteins, *J. Biochem. Biophys. Meth.*, 27, 157, 1993.

9. **Woodward, J., Tate, J., Herrmann, P. C., and Evans, B. R.,** Comparison of Ellman's reagent with N-(1-pyrenyl) maleimide for the determination of free sulfydryl groups in reduced cellobiohydrolase I from *Trichoderma reesei*, *J. Biochem. Biophys. Meth.*, 26, 121, 1993.

10. **Esmann, M., Sar, P. C., Hideg, K., and Marsh, D.,** Maleimide, iodoacetamide, indanedione, and chloromercuric spin label reagents with derivatized nitroxide rings as ESR reporter groups for protein conformation and dynamics, *Anal. Biochem.*, 213, 336, 1993.

11. **Rojas, M. C., Encinas, M. V., Kemp, R. G., Latshaw, S. P., and Cardemil, E.,** Identification of reactive vicinal cysteines in *Saccharomyces cerevisiae* (ATP) and cytosolic rat liver(GTP) phosphoenolpyruvate carboxykinases, *Biochim. Biophys. Acta*, 1164, 143, 1993.

12. **Poole, L. B. and Claiborne, A.,** The non-flavin redox center of the streptococcal NADH peroxidase. II. Evidence for a stabilized cysteine-sulfenic acid, *J. Biol. Chem.*, 264, 12330, 1989.

13. **Pihl, A. and Lange, R.,** The interaction of oxidized glutathione, cystamine monosulfoxide, and tetrathionate with the -SH groups of rabbit muscle D-glyceraldehyde 3-phosphate dehydrogenase, *J. Biol. Chem.*, 237, 1356, 1962.

14. **Bednar, R. A., Fried, W. B., Lock, Y. W., and Pramanik, B.,** Chemical modification of chalcone isomerase by mercurials and tetrathionate. Evidence for a single cysteine residue in the active site, *J. Biol. Chem.*, 264, 14272, 1989.

15. **Smith, D. J., Maggio, E. T., and Kenyon, G. L.,** Simple alkane thiol groups for temporary blocking of sulfydryl groups of enzymes, *Biochemistry*, 14, 766, 1975.

16. **Bruice, T. W. and Kenyon, G. L.,** Novel alkyl alkanethiolsulfonate sulfydryl reagents. Modification of derivatives of L-cysteine, *J. Protein Chem.*, 1, 47, 1982.

17. **Bech, L. M. and Breddam, K.,** Chemical modification of a cysteinyl residue introduced in the binding site of carboxypeptidase Y by site-directed mutagenesis, *Carlsberg Res. Commun.*, 53, 381, 1988.

18. **Gerwin, B. I.,** Properties of the single sulfhydryl group of streptococcal proteinase. A comparison of the rates of alkylation by chloroacetic acid and chloroacetamide, *J. Biol. Chem.*, 242, 451, 1967.

19. **Liu, T. Y.,** Demonstration of the presence of a histidine residue at the active site of streptococcal proteinase, *J. Biol. Chem.*, 242, 4029, 1967.
20. **Chaiken, I. M. and Smith, E. L.,** Reaction of chloroacetamide with the sulfhydryl group of papain, *J. Biol. Chem.*, 244, 5087, 1969.
21. **Chaiken, I. M. and Smith, E. L.,** Reaction of the sulfhydryl group of papain with chloroacetic acid, *J. Biol. Chem.*, 244, 5095, 1969.
22. **Jörnvall, H., Fowler, A. V., and Zabin, I.,** Probe of β-galactosidase structure with iodoacetate. Differential reactivity of thiol groups in wild-type and mutant forms of β-galactosidase, *Biochemistry*, 17, 5160, 1978.
23. **Kalimi, M. and Love, K.,** Role of chemical reagents in the activation of rat hepatic glucocorticoid-receptor complex, *J. Biol. Chem.*, 255, 4687, 1980.
24. **Kallis, G.-B. and Holmgren, A.,** Differential reactivity of the functional sulfhydryl groups of cysteine-32 and cysteine-35 present in the reduced form of thioredoxin from *Escherichia coli*, *J. Biol. Chem.*, 255, 10261, 1980.
25. **Mikami, B., Aibara, S., and Morita, Y.,** Chemical modification of sulfhydryl groups in soybean β-amylase, *J. Biochem.*, 88, 103, 1980.
26. **Hempel, J. D. and Pietruszko, R.,** Selective chemical modification of human liver aldehyde and dehydrogenases E_1 and E_2 by iodoacetamide, *J. Biol. Chem.*, 256, 10889, 1981.
27. **Crestfield, A. M., Moore, S., and Stein, W. H.,** The preparation and enzymatic hydrolysis of reduced and S-carboxymethylated proteins, *J. Biol. Chem.*, 238, 622, 1963.
28. **Friedman, M., Krull, L. H., and Cavins, J. F.,** The chromatographic determination of cystine and cysteine residues in proteins as *S*-β-(4-pyridyl-ethyl) cysteine, *J. Biol. Chem.*, 245, 3868, 1970.
29. **Mak, A. S. and Jones, B. L.,** Application of *S*-pyridylethylation of cysteine to the sequence analysis of proteins, *Anal. Biochem.*, 84, 432, 1978.

29a. **Pflouq, M., Stoffer, B., and Jensen, A. L.,** *In situ* alkylation of cysteine residues in a hydrophobic membrane protein immobilized on polyvinylidene difluoride membranes by electroblotting prior to microsequence and amino acid analysis, *Electrophoresis*, 13, 148, 1992.

30. **Dahl, K. S. and McKinley-McKee, J. S.,** The reactivity of affinity labels: a kinetic study of the reaction of alkyl halides with thiolate anions — a model reaction for protein alkylation, *Bioorg. Chem.*, 10, 329, 1981.
31. **Soper, T. S., Ueno, H., and Manning, J. M.,** Substrate-induced changes in sulfydryl reactivity of bacterial D-amino acid transaminase, *Arch. Biochem. Biophys.*, 240, 1, 1985.
32. **Kaslow, H. R., Schlotterbeck, J. D., Mar, V. L., and Burnette, W. N.,** Alkylation of cysteine 41, but not cysteine 200, decreases the ADP-ribosyltransferase activity of the S1 subunit of pertussis toxin, *J. Biol. Chem.*, 264, 6386, 1989.
33. **Makinen, A. L. and Nowak, T.,** A reactive cysteine in avian liver phosphoenolpyruvate carboxykinase, *J. Biol. Chem.*, 264, 12148, 1989.
34. **Poole, L. B. and Claiborne, A.,** The non-flavin redox center of the streptococcal NADH peroxidase. I. Thiol reactivity and redox behavior in the presence of urea, *J. Biol. Chem.*, 264, 12322, 1989.
35. **Wu, H., Yao, Q.-Z., and Tsou, C.-L.,** Creatine kinase is modified by 2-chloromercuri-4-nitrophenol at the active site thiols with complete inactivation, *Biochim. Biophys. Acta*, 997, 78, 1989.
36. **Wang, Z.-X., Preiss, B., and Tsou, C.-L.,** Kinetics of inactivation of creatine kinase during modification of its thiol groups, *Biochemistry*, 27, 5095, 1988.
37. **Seifried, S. E., Wang, Y., and Von Hippel, P. H.,** Fluorescent modification of the cysteine 202 residue of *Escherichia coli* transcription termination factor rho, *J. Biol. Chem.*, 263, 13511, 1988.
38. **Pardo, J. P. and Slayman, C. W.,** Cysteine 532 and cysteine 545 are the *N*-ethylmaleimide-reactive residues of the *Neurospora* plasma membrane H^+-ATPase, *J. Biol. Chem.*, 264, 9373, 1989.

39. **Miyanishi, T. and Borejdo, J.,** Differential behavior of two cysteine residues on the myosin head in muscle fibers, *Biochemistry,* 28, 1287, 1989.
40. **Bishop, J. E., Squier, T. C., Bigelow, D. J., and Inesi, G.,** (Iodoacetamido) fluorescein labels a pair of proximal cysteines on the Ca^{2+}-ATPase of sarcoplasmic reticulum, *Biochemistry,* 27, 5233, 1988.
41. **First, E. A. and Taylor, S. S.,** Selective modification of the catalytic subunit of cAMP-dependent protein kinase with sulfhydryl-specific fluorescent probes, *Biochemistry,* 28, 3598, 1989.
42. **Gregory, J. D.,** The stability of *N*-ethylmaleimide and its reaction with sulfhydryl groups, *J. Am. Chem. Soc.,* 77, 3922, 1955.
43. **Leslie, J.,** Spectral shifts in the reaction of *N*-ethylmaleimide with proteins, *Anal. Biochem.,* 10, 162, 1965.
44. **Gorin, G., Martic, P. A., and Doughty, G.,** Kinetics of the reaction of *N*-ethylmaleimide with cysteine and some congeners, *Arch. Biochem. Biophys.,* 115, 593, 1966.
45. **Smyth, D. G., Blumenfeld, O. O., and Konigsberg, W.,** Reaction of *N*-ethylmaleimide with peptides and amino acids, *Biochem. J.,* 91, 589, 1964.
46. **Gehring, H. and Christen, P.,** A diagonal procedure for isolating sulfhydryl peptides alkylated with *N*-ethylmaleimide, *Anal. Biochem.,* 107, 358, 1980.
47. **Brown, R. D. and Matthews, K. S.,** Chemical modification of lactose repressor proteins using *N*-substituted maleimides, *J. Biol. Chem.,* 254, 5128, 1979.
48. **Brown, R. D. and Matthews, K. S.,** Spectral studies on *Lac* repressor modified with *N*-substituted maleimide probes, *J. Biol. Chem.,* 254, 5135, 1979.
49. **Perussi, J. R., Tinto, M. H., Nascimento, O. R., and Tabak, M.,** Characterization of protein spin labeling by maleimide: evidence for nitroxide reduction, *Anal. Biochem.,* 173, 289, 1988.
50. **Marquez, J., Iriarte, A., and Martinez-Carrion, M.,** Covalent modification of a critical sulfhydryl group in the acetylcholine receptor: cysteine-222 of the α-subunit, *Biochemistry,* 28, 7433, 1989.
51. **Mills, J. S., Walsh, M. P., Nemcek, K., and Johnson, J. D.,** Biologically active fluorescent derivatives of spinach calmodulin that report calmodulin target protein binding, *Biochemistry,* 27, 991, 1988.
52. **Jezek, P. and Drahota, Z.,** Sulfhydryl groups of the uncoupling protein of brown adipose tissue mitochondria — distinction between sulfhydryl groups of the H^+-channel and the nucleotide binding site, *Eur. J. Biochem.,* 183, 89, 1989.
53. **Le-Quoc, K., Le-Quoc, D., and Gaudemer, Y.,** Evidence for the existence of two classes of sulfhydryl groups essential for membrane-bound succinate dehydrogenase activity, *Biochemistry,* 20, 1705, 1981.
54. **Rial, E., Aréchaga, I., Sainz-de-la-Maza, E., and Nicholls, D. G.,** Effect of hydrophobic sulphydryl reagents on the uncoupling protein and inner-membrane anion channel of brown-adipose-tissue mitochondria, *Eur. J. Biochem.,* 182, 187, 1989.
55. **Yee, A. S., Corley, D. E., and McNamee, M. G.,** Thiol-group modification of *Torpedo californica* acetylcholine receptor: subunit localization and effects on function, *Biochemistry,* 25, 2110, 1986.
56. **Pradier, L., Yee, A. S., and McNamee, M. G.,** Use of chemical modifications and site-directed mutagenesis to probe the functional role of thiol groups on the gamma subunit of *Torpedo californica* acetylcholine receptor, *Biochemistry,* 28, 6562, 1989.
57. **Abbott, R. E. and Schachter, D.,** Topography and functions of sulfhydryl groups of the human erythrocyte glucose transport mechanism, *Mol. Cell. Biochem.,* 82, 85, 1988.
58. **May, J. M.,** Reaction of an exofacial sulfhydryl group on the erythrocyte hexose carrier with an impermeant maleimide. Relevance to the mechanism of hexose transport, *J. Biol. Chem.,* 263, 13635, 1988.

59. **May, J. M.,** Interaction of a permeant maleimide derivative of cysteine with the erythrocyte glucose carrier. Differential labelling of an exofacial carrier thiol group and its role in the transport mechanism, *Biochem. J.,* 263, 875, 1989.
60. **Abbott, R. E. and Schachter, D.,** Impermeant maleimides. Oriented probes of erythrocyte membrane proteins, *J. Biol. Chem.,* 251, 7176, 1976.
61. **May, J. M.,** Selective labeling of the erythrocyte hexose carrier with a maleimide derivative of glucosamine: relationship of an exofacial sulfhydryl to carrier conformation and structure, *Biochemistry,* 28, 1718, 1989.
62. **Falke, J. J., Dernburg, A. F., Sternberg, D. A., Zalkin, N., Milligan, D. L., and Koshland, D. E., Jr.,** Structure of a bacterial sensory receptor. A site-directed sulfhydryl study, *J. Biol. Chem.,* 263, 14850, 1988.
63. **Lepke, S., Fasold, H., Pring, M., and Passow, H.,** A study of the relationship between inhibition of anion exchange and binding to the red blood cell membrane of 4,4′-dithiocyano stilbene-2,2′-disulfonic acid (DIDS) and its dihydro derivative (H_2DIDS), *J. Membr. Biol.,* 29, 147, 1976.
64. **Bettendorff, L., Wins, P., and Schoffeniels, E.,** Thiamine triphosphatase from *electrophorus* electric organ is anion-dependent and irreversibly inhibited by 4,4′-diisothiocyanostilbene-2,2′-disulfonic acid, *Biochem. Biophys. Res. Commun.,* 154, 942, 1988.
65. **Speth, M. and Schulze, H.-U.,** On the nature of the interaction between 4,4′-diisothiocyanostilbene 2,2′-disulfonic acid and microsomal glucose-6-phosphatase. Evidence for the involvement of sulfhydryl groups of the phosphohydrolase, *Eur. J. Biochem.,* 174, 111, 1988.
66. **Ellman, G. L.,** Tissue sulfhydryl groups, *Arch. Biochem. Biophys.,* 82, 70, 1959.
67. **Pecci, L., Cannella, C., Pensa, B., Costa, M., and Cavallini, D.,** Cyanylation of rhodanese by 2-nitro-5-thiocyanobenzoic acid, *Biochim. Biophys. Acta,* 623, 348, 1980.
68. **Vanaman, T. C. and Stark, G. C.,** A study of the sulfydryl groups of the catalytic subunit of *Escherichia coli* aspartate transcarbamylase. The use of enzyme-5-thio-2-nitrobenzoate mixed disulfides as intermediates in modifying enzyme sulfydryl groups, *J. Biol. Chem.,* 245, 3565, 1970.
69. **Habeeb, A. F. S. A.,** Reaction of protein sulfydryl groups with Ellman's reagent, *Meth. Enzymol.,* 25, 457, 1972.
70. **Talgoy, M. M., Bell, A. W., and Duckworth, H. W.,** The reactions of *Escherichia coli* citrate synthase with the sulfhydryl reagents 5,5′-dithiobis-(2-nitrobenzoic acid) and 4,4′-dithiodipyridine, *Can. J. Biochem.,* 57, 822, 1979.
71. **Soper, T. S., Jones, W. M., and Manning, J. M.,** Effects of substrates on the selective modification of the cysteinyl residues of D-amino acid transaminase, *J. Biol. Chem.,* 254, 10901, 1979.
72. **Lukas, R. J. and Bennett, E. L.,** Chemical modification and reactivity of sulfhydryls and disulfides of rat brain nicotinic-like acetylcholine receptors, *J. Biol. Chem.,* 255, 5573, 1980.
73. **Cockle, S. A., Epand, R. M., Stollery, J. G., and Moscarello, M. A.,** Nature of the cysteinyl residues in lipophilin from human myelin, *J. Biol. Chem.,* 255, 9182, 1980.
74. **Hallaway, B. E., Hedlund, B. E., and Benson, E. S.,** Studies of the effect of reagent and protein charges on reactivity of the β93 sulfhydryl group of human hemoglobin using selected mutations, *Arch. Biochem. Biophys.,* 203, 332, 1980.
75. **Luthra, M. P., Dunlap, R. B., and Odom, J. D.,** Characterization of a new sulfhydryl group reagent: 6,6′-diselenobis-(3-nitrobenzoic acid), a selenium analog of Ellman's reagent, *Anal. Biochem.,* 117, 94, 1981.
76. **Clayshulte, T. M., Taylor, D. F., and Henzl, M. T.,** Reactivity of cysteine 18 in oncomodulin, *J. Biol. Chem.,* 265, 1800, 1990.

77. **Kato, H., Tanaka, T., Nishioka, T., Kimura, A., and Oda, J.,** Role of cysteine residues in glutathione synthetase from *Escherichia coli* B. Chemical modification and oligonucleotide site-directed mutagenesis, *J. Biol. Chem.,* 263, 11646, 1988.
78. **Narasimhan, C., Lai, C.-S., Haas, A., and McCarthy, J.,** One free sulfhydryl group of plasma fibronectin becomes titratable upon binding of the protein to solid substrates, *Biochemistry,* 27, 4970, 1988.
79. **Poole, L. B. and Claiborne, A.,** Evidence for a single active-site cysteinyl residue in the streptococcal NADH peroxidase, *Biochem. Biophys. Res. Commun.,* 153, 261, 1988.
80. **Brandwein, H. J., Lewicki, J. A., and Murad, F.,** Reversible inactivation of guanylate cyclase by mixed disulfide formation, *J. Biol. Chem.,* 256, 2958, 1981.
81. **Kimura, T., Matsueda, R., Nakagawa, Y., and Kaiser, E. T.,** New reagents for the introduction of the thiomethyl group at sulfhydryl residues of proteins with concomitant spectrophotometric titration of the sulfhydryl: methyl 3-nitro-2-pyridyl disulfide and methyl 2-pyridyl disulfide, *Anal. Biochem.,* 122, 274, 1982.
82. **Drewes, G. and Faulstich, H.,** 2,4-Dinitrophenyl[^{14}C]cysteinyl disulfide allows selective radiolabeling of protein thiols under spectrophotometric control, *Anal. Biochem.,* 188, 109, 1990.
83. **Boyer, P. D.,** Spectrophotometric study of the reaction of protein sulfhydryl groups with organic mercurials, *J. Am. Chem. Soc.,* 76, 4331, 1954.
84. **Bai, Y. and Hayashi, R.,** Properties of the single sulfhydryl group of carboxypeptidase Y. Effects of alkyl and aromatic mercurials on activities toward various synthetic substrates, *J. Biol. Chem.,* 254, 8473, 1979.
85. **Ojcius, D. M. and Solomon, A. K.,** Sites of *p*-chloromercuribenzenesulfonate inhibition of red cell urea and water transport, *Biochim. Biophys. Acta,* 942, 73, 1988.
86. **Clark, S. J. and Ralston, G. B.,** The dissociation of peripheral proteins from erythrocyte membranes brought about by *p*-mercuribenzenesulfonate, *Biochim. Biophys. Acta,* 1021, 141, 1990.
87. **Marshall, M. and Cohen, P. P.,** The essential sulfhydryl group of ornithine transcarbamylases-pH dependence of the spectra of its 2-mercuri-4-nitrophenol derivative, *J. Biol. Chem.,* 255, 7296, 1980.
88. **Baines, B. S. and Brocklehurst, K.,** A thiol-labelling reagent and reactivity probe containing electrophilic mercury and a chromophoric leaving group, *Biochem. J.,* 179, 701, 1979.
89. **Banks, T. E. and Shafer, J. A.,** Inactivation of papain by S-methylation of its cysteinyl residue with *O*-methylisourea, *Biochemistry,* 11, 110, 1972.
90. **Stark, G.,** Modification of proteins with cyanate, *Meth. Enzymol.,* 11, 590, 1967.
91. **Ghosh, P. B. and Whitehouse, M. W.,** 7-Chloro-4-nitrobenzo-2-oxa-1,3-diazole: a new fluorigenic reagent for amino acids and other amines, *Biochem. J.,* 108, 155, 1968.
92. **Birkett, D. J., Price, N. D., Radda, G. K., and Salmon, A. G.,** The reactivity of SH groups with a fluorogenic reagent, *FEBS Lett.,* 6, 346, 1970.
93. **Birkett, D. J., Dwek, R. A., Radda, G. K., Richards, R. E., and Salmon, A. G.,** Probes for the conformational transitions of phosphorylase b. Effect of ligands studied by proton relaxation enhancement, fluorescence and chemical reactivities, *Eur. J. Biochem.,* 20, 494, 1971.
94. **Lad, P. M., Wolfman, N. M., and Hammes, G. G.,** Properties of rabbit muscle phosphofructokinase modified with 7-chloro-4-nitrobenzo-2-oxa-1,3-diazole, *Biochemistry,* 16, 4802, 1977.
95. **Nitta, K., Bratcher, S. C., and Kronman, M. J.,** Anomalous reaction of 4-chloro-7-nitrobenzofurazan with thiol compounds, *Biochem. J.,* 177, 385, 1979.
96. **Dwek, R. A., Radda, G. A., Richards, R. E., and Salmon, A. G.,** Probes for the conformational transitions of phosphorylase a. Effect of ligands studied by proton-relaxation enhancement, and chemical reactivities, *Eur. J. Biochem.,* 29, 509, 1972.
97. **Carlberg, I. and Mannervik, B.,** Interaction of 2,4,6-trinitrobenzenesulfonate and 4-chloro-7-nitrobenzo-2-oxa-1,3-diazole with the active sites of glutathione reductase and lipoamide dehydrogenase, *Acta Chem. Scand.,* B34, 144, 1980.

98. **Toyo'oka, T. and Imai, K.,** New fluorogenic reagent having halogenbenzofurazan structure for thiols: 4-(aminosulfonyl)-7-flouro-2,1,3-benzoxadiazole, *Anal. Chem.*, 56, 1984.

99. **Kirley, T. L.,** Reduction and fluorescent labeling of cyst(e)ine-containing proteins for subsequent structural analyses, *Anal. Biochem.*, 180, 231, 1989.

100. **Niketic, V., Thomsen, J., and Kristiansen, K.,** Modification of cysteine residues with 2-bromoethane-sulfonate. The application of *S*-sulfoethylated peptides in automatic Edman degradation, *Eur. J. Biochem.*, 46, 547, 1974.

101. **Mutus, B., Wagner, J. D., Talpas, C. J., Dimmock, J. R., Phillips, O. A., and Reid, R. S.,** 1-*p*-Chlorophenyl-4,4-dimethyl-5-diethylamino-1-penten-3-one hydrobromide, a sulfhydryl-specific compound which reacts irreversibly with protein thiols but reversibly with small molecular weight thiols, *Anal Biochem.*, 177, 237, 1989.

Chapter 7

THE MODIFICATION OF CYSTINE — CLEAVAGE OF DISULFIDE BONDS

There are several approaches to the cleavage of disulfide bonds in proteins. The majority of studies involve the cleavage of the disulfide bond of cystine to the free thiol group of cysteine by reduction. Reduction has been generally accomplished with a mild reducing agent such as β-mercaptoethanol, dithiothreitol, or cysteine. Gorin and co-workers[1] have examined the rate of reaction of lysozyme with various thiols. At pH 10.0 (0.025 *M* borate), the relative rates of reaction were β-mercaptoethanol (2-mercaptoethanol), 0.2; dithiothreitol, 1.0; 3-mercaptopropionate, 0.4; and 2-aminoethanol, 0.01. The results with aminoethanethiol were somewhat surprising since the reaction (disulfide exchange) involves the thiolate anion and 2-aminoethanethiol would be more extensively ionized than the other mercaptans. Dithiothreitol has been a useful reagent in the reduction of disulfide bonds in proteins[2] as introduced by Cleland. Dithiothreitol and the isomeric form, dithioerythritol, are each capable of the quantitative reduction of disulfide bonds in proteins. Furthermore, the oxidized form of dithiothreitol has an absorbance maximum at 283 nm ($\Delta\varepsilon = 273$), which can be used to determine the extent of disulfide bond cleavage.[3] The ultraviolet (UV) spectra of dithiothreitol and oxidized dithiothreitol are shown in Figure 1. Insolubilized dihydrolipoic acid has also been proposed for use in the quantitative reduction of disulfide bonds.[4]

Reduction of disulfide bonds can be controlled by various factors. Homandberg and Wai[5] demonstrated that the reduction of urokinase by dithiothreitol in the presence of arginine allows the selective reduction of a disulfide bond joining the catalytically active chain to a nonessential 13 amino acid peptide. A synthetic peptide may then be coupled to the free sulfydryl group.

In most proteins, the free sulfhydryl groups (cysteine) derived from the reduction of cystine will, at alkaline pH, fairly rapidly undergo reoxidation to form the original disulfide bonds. This process can be accelerated by the sulfhydryl-disulfide interchange enzyme[6,7] or sulfhydryl oxidase.[8] Thus, it is necessary to "block" the new sulfhydryl groups by alkylation, arylation, or reaction with dithionite (see Chapter 6). Although infrequently used, phosphorothioate had been demonstrated to effectively cleave disulfide bonds in proteins forming the *S*-phosphorothioate derivative.[9]

Light and co-workers have examined the susceptibility of disulfide bonds in trypsinogen to reduction.[10] At pH 9.0 (0.1 *M* sodium borate), a single disulfide bond (Cys_{179}–Cys_{203}) is cleaved in trypsinogen by 0.1 *M* $NaBH_4$. The resulting sulfhydryl groups are "blocked" by alkylation. The characterization of the modified protein has been performed by the same group.[11] The disulfide bond

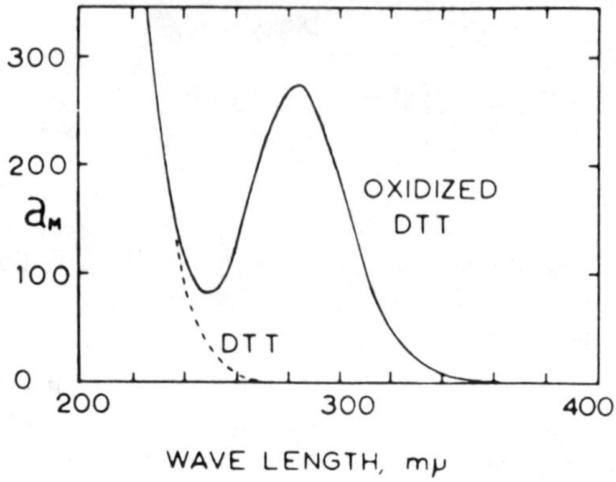

FIGURE 1. The absorption spectra of dithiothreitol (DTT) and oxidized dithiothreitol (oxidized DTT) in aqueous solution. (From Cleland, W. W., *Biochemistry*, 3, 480, 1964. With permission.)

which is modified under these conditions is critical in establishing the structure of the primary specificity site in trypsin.

From the above studies, there is little doubt that the various disulfide bonds in a protein show different reactivity toward reducing agents. These differences in reactivity can be explored with various reagents and can be utilized with the aid of partial reduction followed by alkylation with radiolabeled iodoacetate to determine the position of disulfide bonds in proteins.[12]

Gorin and Godwin[13] have reported that cystine can be quantitatively converted to cysteic acid by reaction with iodate in 0.1 to 1.0 *M* HCl. This reaction has been applied to insulin. The reaction product was not completely characterized, but given the relationship between iodate consumption and the cystine residues in insulin, the primary reaction is the oxidation of disulfide bonds. This reaction was complete in 15 to 30 min. After longer periods of reaction, the iodination of tyrosine residues occurred.

Disulfide bonds are somewhat unstable at alkaline pH (pH ≥13.0). This has been examined by Donovan in some detail.[14] With protein-bound cystine, there is change in the spectrum with an increase in absorbance at 300 nm. This problem has been more recently studied by Florence.[15] This investigation presented evidence to suggest that cleavage of disulfide bonds in proteins by base proceeds via β-elimination form dehydroalanine and a persulfide intermediate which can decompose to form several products.

The electrolytic reduction of proteins has been explored by Leach and coworkers.[16] These investigators recognized that although small peptides containing disulfide bonds could be reduced using cathodic reduction, there would likely be problems with proteins because of size and tertiary structure considerations. Therefore, a small thiol was used as a catalyst for the reduction.

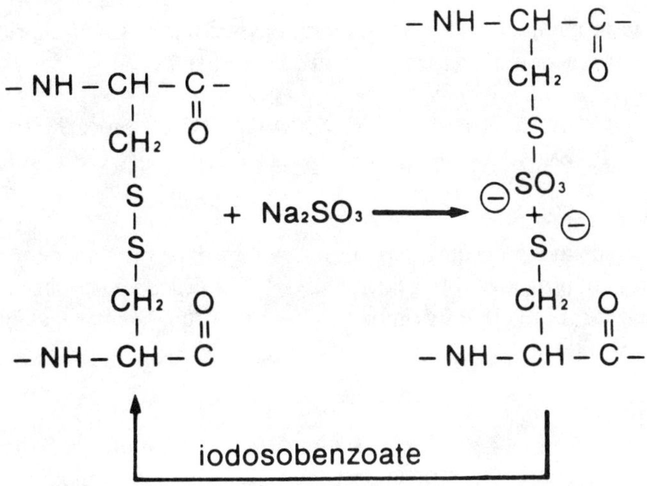

FIGURE 2. The cleavage of disulfide bonds by sodium sulfite to form the *S*-sulfo derivative.

Tri-*n*-butylphosphine will reduce disulfide bonds in proteins.[18] The reaction is generally performed under alkaline conditions (pH 8.0, 0.1 *M* Tris or 0.5 *M* bicarbonate). *n*-Propanol is added to dissolve the tri-*n*-butylphosphine. This procedure has recently been used to reduce disulfide bonds in various proteins prior to reaction with 4-(aminosulfonyl)-7-fluoro-2,1,3-benzodiazole.[18] Chin and Wold[19] used tributylphosphine and 4-(aminosulfonyl)-7-fluoro-2,1,3-benzoxadiazole (ABD-F) for the characterization of cysteine and cystine in proteins. Since the reagents do not react with each other, it is possible to distinguish free sulfydryl and cystine residue with two reactions; one with ABD-F and the second with ABD-F + tributylphosphine. Tris-(2-carboethoxy)-phosphine has been developed for the reduction of disulfide bonds in proteins.[20–22] This reagent has the advantage of being water soluble.

Finally, disulfide bonds can be cleaved by sulfite to form the *S*-sulfonate derivative as shown in Figure 2. The chemistry of this reaction has been reviewed by Cole.[23] The reaction proceeds optimally at alkaline pH (pH 9.0). It is necessary to include an oxidizing agent such as cupric ions or, as shown in Figure 2, *o*-iodosobenzoate to ensure effective conversion of all cystine residues to the corresponding *S*-sulfonate derivatives. The reaction is reversible to form cysteine upon treatment with a suitable mercaptan such as β-mercaptoethanol.

This reaction has been adapted to the controlled reduction of disulfide bonds in proteins in the absence of denaturing agents.[24] In the example presented, the disulfide bonds of bovine serum albumin were cleaved at pH 7.0 (0.1 *M* phosphate) at 40°C in the presence of 0.1 *M* sodium sulfite and simultaneously converted to the *S*-sulfo derivatives with oxygen and 0.4 m*M* cupric sulfate. The rate of reaction decreased markedly above pH 7.0.

APPENDIX I — MICRO PROCEDURE FOR THE REDUCTION AND CARBOXYMETHYLATION OF PROTEINS

General Comments

The following procedure has been developed in our laboratory for the preparation of proteins for tryptic hydrolysis. The procedure has been successfully used for 25 to 1000 µg amounts. The procedure has been adapted from previous work.[25,26]

Materials

Use urea-recrystallized-prepare saturated solutions of urea at 50°C and add 10 g Amberlite MB-1 cation/anion exchange resin. Filter rapidly through a scintered glass filter. Place warm filtrate in the cold and allow to stand overnight. Obtain crystalline material by filtration (Whatman #1) and dry by lyophilization. Store in brown bottle. Urea solutions must be freshly prepared for each procedure. Quality urea or guanidine hydrochloride/thiocyanate can be obtained from most first-tier suppliers and can be used without further purification. The problem of cyanate formation must be taken into consideration.[27] Mercaptoethanol can be obtained from the usual commercial sources (i.e., Eastman, Aldrich, etc.). With the trypsin-TPCK-treated solution, a 1.0 mg/ml solution in 0.001 *M* HCl is prepared for 2 to 4 h before use. For iodoacetic acid, the sodium salt is preferred, but this is frequently difficult to obtain. We usually use iodoacetic acid in NaOH. Dissolve 268 mg of iodoacetic acid (recrystallized from petroleum ether) in 1.0 ml 1.0 *M* NaOH. Note that 4-vinylpyridine[28] can also be used for blocking sulfydryl groups.

Procedure

1. The protein sample is dialyzed vs. 0.1 *M* ammonium bicarbonate and lyophilized (Speed-Vac, Savant Instruments, Farmingdale, New York) in the tube (12 × 75 polypropylene).
2. Place 200 µl 8.0 *M* urea-0.2% EDTA (1.80 g recrystallized urea) in a glass tube marked for 3.25 ml. Add 1.5 ml 1.4 *M* Tris, pH 8.6 and 0.15 ml 5% EDTA. The suspension is heated under the hot water tap until most of the urea is in the solution. The volume is taken to the 3.25 ml mark with deionized water added. The tube is flushed with nitrogen, and 5 µl mercaptoethanol is added. The tube is again flushed with nitrogen and placed at 37°C. A solvent blank tube is run with each series. This usually involves the initial processing of a volume of dialysis solvent equal to that of the protein samples.
3. The samples are transferred to microcentrifuge tubes, and the initial tubes are washed with 800 µl ETOH-HCl (ETOH:Conc.HCl; 98:2), with the washes being transferred to the microcentrifuge tubes. The samples

are placed overnight at –20°C. This step results in the precipitation of the reduced protein.

4. After centrifugation, the pellet fractions are dissolved in 200 µl Tris-urea containing 500 µg iodoacetic acid. Sodium salt and the reactions are placed at 37°C for 30 min. The Tris-urea-iodoacetate is prepared by adding 10 µl of the iodoacetic acid/NaOH (see above) per milliliter of the Tris-urea-EDTA (see above). After the incubation, the reduced carboxymethylated protein is precipitated as above with ETOH-HCl. Added is 800 µl of ETOH-HCl, and the reactions are placed at –20°C overnight.
5. The tubes from above are centrifuged and dried by lyophilization.
6. The reduced carboxymethylated proteins are suspended in 1000 µl 0.1 *M* ammonium bicarbonate, and 10 µl of trypsin is added. The tubes are placed at 37°C, and after 4 h, an additional 10 µl of trypsin are added. The digestion is allowed to continue for 16 h (overnight). The samples are then frozen subsequent to chromatographic analysis.

Notes

1. The temperature for the two precipitation steps is critical. We have found that ice bath temperatures are not adequate.
2. The urea solutions must be freshly prepared.
3. Avoid multiple transfers of samples. In the above procedure, only two tubes are used with the transfer step including a wash step as well.

REFERENCES

1. **Gorin, G., Fulford, R., and Deonier, R. C.,** Reaction of lysozyme with dithiothreitol and with other mercaptans, *Experientia,* 24, 26, 1968.
2. **Cleland, W. W.,** Dithiothreitol, a new protective reagent for SH groups, *Biochemistry,* 3, 480, 1964.
3. **Iyer, K. S. and Klee, W. A.,** Direct spectrophotometric measurement of the rate of reduction of disulfide bonds. The reactivity of the disulfide bonds of bovine α-lactalbumin, *J. Biol. Chem.,* 248, 707, 1973.
4. **Gorecki, M. and Patchornik, A.,** Polymer-bound dihydrolipoic acid: a new insoluble reducing agent for disulfides, *Biochim. Biophys. Acta,* 303, 36, 1973.
5. **Homandberg, G. A. and Wai, T.,** Reduction of disulfides in urokinase and insertion of a synthetic peptide, *Biochim. Biophys. Acta,* 1038, 209, 1990.
6. **Fuchs, S., DeLorenzo, F., and Anfinsen, C. B.,** Studies on the mechanism of the enzymic catalysis of disulfide interchange in proteins, *J. Biol. Chem.,* 242, 398, 1967.
7. **Creighton, T. E., Hillson, D. A., and Freedman, R. B.,** Catalysis by protein-disulphide isomerase of the unfolding and refolding of proteins with disulphide bonds, *J. Mol. Biol.,* 142, 43, 1980.
8. **Janolino, V. G., Sliwkowski, M. Y., Swaisgood, H. F., and Horton, H. R.,** Catalytic effect of sulfhydryl oxidase on the formation of three-dimensional structure in chymotrypsinogen A, *Arch. Biochem. Biophys.,* 191, 269, 1978.

9. **Neumann, H. and Smith, R. L.,** Cleavage of the disulfide bonds of cystine and oxidized glutathione by phosphorothioate, *Arch. Biochem. Biophys.,* 122, 354, 1967.
10. **Light, A., Hardwick, B. C., Hatfield, L. M., and Sondack, D. L.,** Modification of a single disulfide bond in trypsinogen and the activation of the carboxymethyl derivative, *J. Biol. Chem.,* 244, 6289, 1969.
11. **Knights, R. J. and Light, A.,** Disulfide bond-modified trypsinogen. Role of disulfide 179–203 on the specificity characteristics of bovine trypsin toward synthetic substrates, *J. Biol. Chem.,* 251, 222, 1976.
12. **Mise, T. and Bahl, O. P.,** Assignment of disulfide bonds in the α-subunit of human chorionic gonadotropin, *J. Biol. Chem.,* 255, 8516, 1980.
13. **Gorin, G. and Godwin, W. E.,** The reaction of iodate with cystine and with insulin, *Biochem. Biophys. Res. Commun.,* 25, 227, 1966.
14. **Donovan, J. W.,** Spectrophotometric observation of the alkaline hydrolysis of protein disulfide bonds, *Biochem. Biophys. Res. Commun.,* 29, 734, 1967.
15. **Florence, T. M.,** Degradation of protein disulphide bonds in dilute alkali, *Biochem. J.,* 189, 507, 1980.
16. **Leach, S. J., Meschers, A., and Swanepoel, O. A.,** The electrolytic reduction of proteins, *Biochemistry,* 4, 23, 1965.
17. **Ruegg, U. Th. and Rudingder, J.,** Reductive cleavage of cystine disulfides with tributylphosphine, *Meth. Enzymol.,* 47, 111, 1977.
18. **Kirley, T. L.,** Reduction and fluorescent labeling of cyst(e)ine-containing proteins for subsequent structural analysis, *Anal. Biochem.,* 180, 231, 1989.
19. **Chin, C. C. Q. and Wold, F.,** The use of tributylphosphine and 4-(aminosulfonyl)-7-fluoro-2,1,3-benzoxadiazole in the study of protein sulfydryls and disulfides, *Anal. Biochem.,* 214, 128, 1993.
20. **Burns, J. A., Butler, J. C., Moran, J., and Whitesides, G. M.,** Selective reduction of disulfides by tris-(2-carboethoxyethyl)-phosphine, *J. Org. Chem.,* 56, 2648, 1991.
21. **Gray, W. R.,** Disulfide structures of highly bridged peptides: a new strategy for analysis, *Protein Sci.,* 2, 1732, 1993.
22. **Gray, W. R.,** Echistatin disulfide bridges: selective reduction and linkage assignment, *Protein Sci.,* 2, 1749, 1993.
23. **Cole, R. D.,** Sulfitolysis, *Meth. Enzymol.,* 11, 206, 1967.
24. **Kella, N. K. D. and Kinsella, J. E.,** A method for the controlled cleavage of disulfide bonds in proteins in the absence of denaturants, *J. Biochem. Biophys. Meth.,* 11, 251, 1985.
25. **Canfield, R. E. and Anfinsen, C. B.,** Chromatography of pepsin and chymotrypsin digests of egg white lysozyme on phosphocellulose, *J. Biol. Chem.,* 238, 2684, 1963.
26. **Crestfield, A. M., Moore, S., and Stein, W. H.,** The preparation and enzymatic hydrolysis of reduced and S-carboxymethylated proteins, *J. Biol. Chem.,* 238, 622, 1963.
27. **Stark, G. R., Stein, W. H., and Moore, S.,** Reaction of the cyanate present in aqueous urea with amino acids and proteins, *J. Biol. Chem.,* 235, 214, 1960.
28. **Ploug, M., Stoffer, B., and Jensen, A. L.,** *In situ* alkylation of cysteine residues in a hydrophobic membrane protein immobilized on polyvinylidene diflouride membranes by electroblotting prior to microsequence and amino acid analysis, *Electrophoresis,* 13, 148, 1992.

Chapter 8

THE MODIFICATION OF METHIONINE

The modification of methionine in a protein is generally accomplished with considerable difficulty. This is possibly a reflection of the fact that, as a relatively hydrophobic residue, methionine is frequently buried in a protein. As a reflection of the relatively drastic solvent conditions which are generally required for the modification of methionine, group-specific modification is difficult to obtain. Since the dissociation of a proton from the sulfur nucleophile is unnecessary, relatively specific derivatization by alkylating agents can be accomplished at low pH.[1] Table 1 describes reagents used to modify methionine.

Oxidation of methionine to methionine sulfoxide (Figure 1) can occur under a variety of conditions. Reagents for the "selective" oxidation of methionine which have attracted recent attention include chloramine T[2,3] (0.1 *M* phosphate, pH 7.0 or 0.1 *M* Tris, pH 8.4), sodium periodate[3] (0.1 *M* sodium acetate, pH 5.0), and hydrogen peroxide.[4] The reaction of methionine with chloramine T can be followed spectrophotometrically.[2] The reaction of chloramine T with methionine (Figure 2) results in a significant change in the spectrum of chloramine T as shown in Figure 3. The use of this spectral change in the determination of methionine is shown in Figure 4. Cysteine interfered with this determination, but other amino acids (i.e., tyrosine, tryptophan, histidine, serine) did not have any effect on the accuracy of analysis for methionine. It is noted that the oxidation of methionine is a possible side-reaction of the treatment of proteins with *N*-bromosuccinimide.[5] Oda and Tokushige[1] have studied the oxidation of tryptophanyl residues in tryptophanase by chloramine T. When the native enzyme is treated with chloroamine T (20 m*M* potassium phosphate, pH 8.5, 0°C), sulfhydryl groups and methionine residues are oxidized with loss of catalytic activity. With prior modification of the sulfhydryl groups with 5,5′-dithiobis-(2-nitrobenzoic acid), 4-5/16 methionyl residues are modified with further loss of catalytic activity. In an earlier study, Sakurai and Nagahama[6] compared the relative sensitivity of amino acids in ε-toxin to oxidation by *N*-bromosuccinimide (pH 5.0, 0.05 *M* acetate), *N*-chlorosuccinimide (pH 8.5, 0.05 *M* Tris), and chloramine T (pH 8.5, 0.05 *M* Tris). Methionine was totally lost with both chloramine T and *N*-chlorosuccinimide, but 21% remained in the *N*-bromosuccinimide-treated sample. The opposite was found for tryptophan with total loss with either *N*-bromosuccinimide or *N*-chlorosuccinimide, but no loss with chloramine T.

It is possible to convert methionine sulfoxide to methionine under relatively mild conditions,[7] thus providing for the reversibility of the oxidative reactions described above (Figure 5). This can be accomplished through both nonenzymatic and enzymatic methods. The nonenzymatic approaches, in general, have

Table 1
REAGENTS USED FOR THE MODIFICATION OF METHIONINE IN PROTEINS

Reagent	Other AA modified	Ref.
Chloramine T	Cysteine, cystine, histidine, tyrosine, tryptophan	1–3
Hydrogen peroxide	Cysteine, cystine, histidine, tryptophan	4
Iodoacetate	Cysteine, histidine, lysine	5
N-Chlorosuccinimide		6

References for Table 1

1. **Oda, T. and Tokushige, M.,** Chemical modification of tryptophanase by chloramine T: a possible involvement of the methionine residue in enzyme activity, *J. Biochem.,* 104, 178, 1988.
2. **Cutruzzola, F., Ascenzi, P., Barra, D., Bolognesi, M., Menegatti, E., Sarti, P., Schenebli, H.-P., Tomova, S., and Amiconi, G.,** Selective oxidation of Met-192 in bovine alpha-chymotrypsin. Effect on catalytic and inhibitor binding properties, *Biochim. Biophys. Acta,* 1161, 201, 1993.
3. **Hussain, A. A., Awad, R., Crooks, P. A., and Dittert, L. W.,** Chloramine T in radiolabeling techniques. I. Kinetics and mechanism of the reaction between chloramine-T and amino acids, *Anal. Biochem.,* 214, 495, 1993.
4. **Drozdz, R., Naskalski, J. W., and Sznajd, J.,** Oxidation of amino acids and peptides in reaction with myeloperoxidase, chloride and hydrogen peroxide, *Biochim. Biophys. Acta,* 957, 47, 1988.
5. **Kleanthous, C., Campbell, D. G., and Coggins, J. R.,** Active site labeling of the shikimate pathway enzyme, dehydroquinase. Evidence for a common substrate binding site within dehydroquinase and dehydroquinate synthase, *J. Biol. Chem.,* 265, 10929, 1990.
6. **Padrines, M., Rabaud, M., and Bieth, J. G.,** Oxidized alpha-1-proteinase inhibitor: a fast-acting inhibitor of human pancreatic elastase, *Biochim. Biophys. Acta,* 1118, 174, 1992.

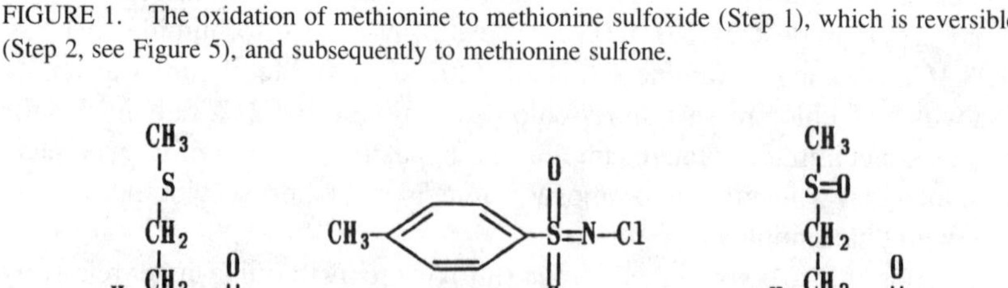

FIGURE 1. The oxidation of methionine to methionine sulfoxide (Step 1), which is reversible (Step 2, see Figure 5), and subsequently to methionine sulfone.

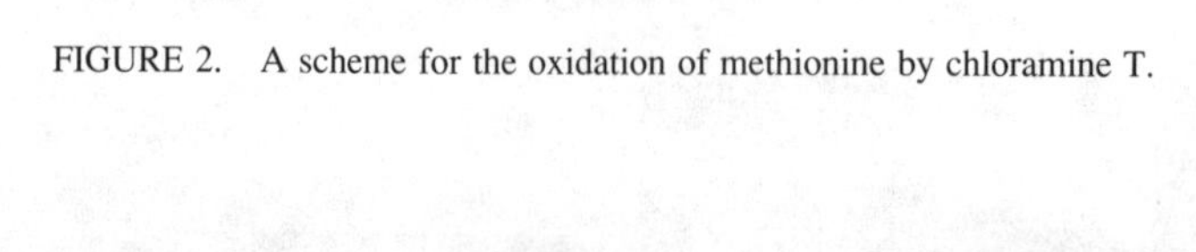

FIGURE 2. A scheme for the oxidation of methionine by chloramine T.

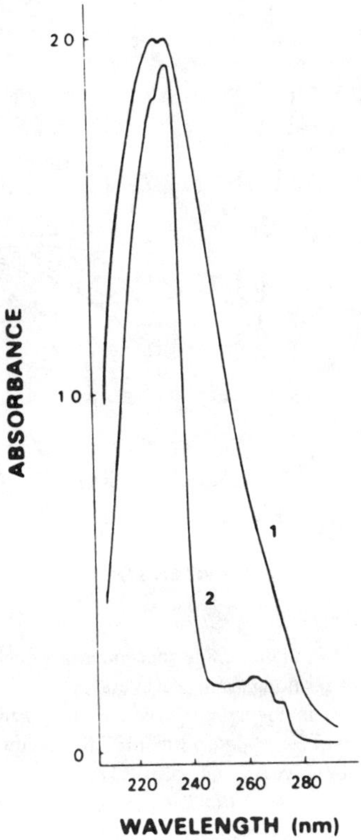

FIGURE 3. The UV absorption spectrum of chloramine T (93.1 μg) in 3.0 ml 0.1 *M* sodium phosphate, pH 7.0 before (Curve 1) and after (Curve 2) the addition of 99.4 μg methionine. The major change in the spectrum of chloramine T resulting from the interaction with methionine is a decrease in the width of the peak with λ_{max} 234 nm resulting in a decrease in absorbance which is maximal between 244 and 248 nm. (From Trout, G. E., *Anal. Biochem.,* 93, 419, 1979. With permission.)

proved to be of greater value. A systematic study has shown that of four reducing agents tested, mercaptoacetic acid, β-mercaptoethanol, dithiothreitol, and *N*-methylmercaptoacetamide, the latter reagent, *N*-methylmercaptoacetamide, was the most effective. The reactions demonstrated little pH dependence, but did not proceed well at concentrations of acetic acid above 50% (v/v). Complete regeneration of methionine could be accomplished with 0.7 to 2.8 *M* reagent at 37°C for 21 h. An enzymatic system for the reduction of methionine sulfoxide has been reported.[8]

Methionine can be modified with various alkylation agents such as the α-halo acetic acids and their derivatives (Figure 6). The reaction of iodacetate with methionine has been examined in some detail by Gundlach et al.[9] The reaction of iodacetate with methionine does not appear to be pH dependent and proceeds much slower than the reaction with cysteine under the mildly

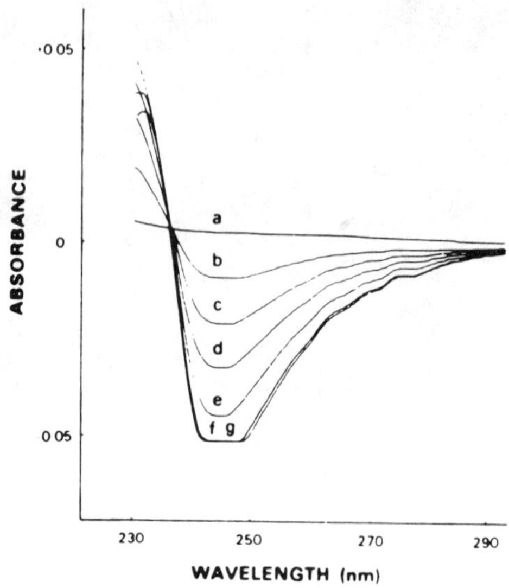

FIGURE 4. The effect of increasing concentration of chloramine T on the difference spectrum resulting from the interaction of chloramine T with methionine as shown in Figure 2. Curve a is the spectrum of 24.9 μg methionine in 3.0 ml of 0.1 *M* phosphate, pH 7.0. Spectra b through g were obtained from the successive additions of 10-μl portions of 10.0 m*M* chloramine T. (From Trout, G. E., *Anal. Biochem.*, 93, 419, 1979. With permission.)

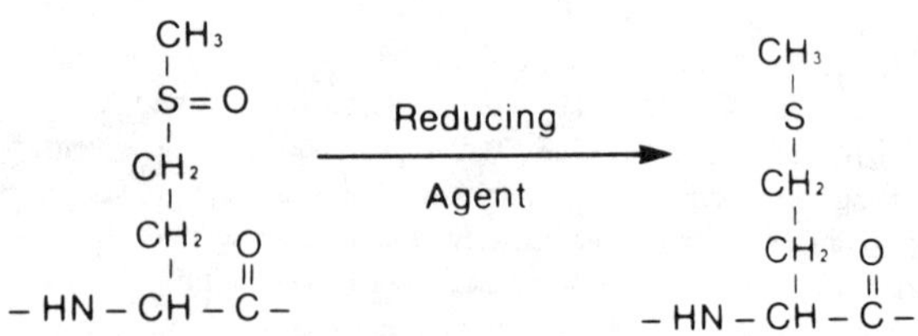

FIGURE 5. The conversion of methionine sulfoxide to methionine in the presence of a reducing agent.

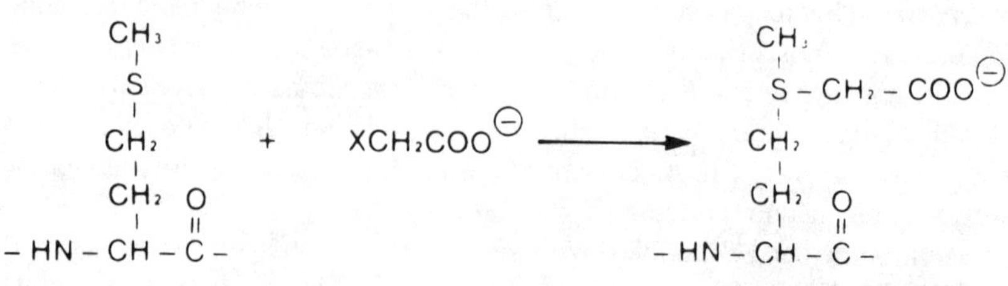

FIGURE 6. The alkylation of methionine with an α-halo acid, such as iodoacetic acid, to form the sulfonium salt derivative.

alkaline conditions used for reduction and carboxymethylation. The resulting sulfonium salt yields homoserine and homoserine lactone when heated at 100°C at pH 6.5. On acid hydrolysis (6 *N* HCl, 110°C, 22 h), a mixture of methionine and *S*-carboxymethyl homocysteine together with a small amount of homoserine lactone was obtained. In general, methionine residues only react with the α-halo acids after the disruption of the secondary and tertiary structure of a protein.[10] Selectivity in the modification of methionine in proteins by α-halo acids can be achieved by performing the reaction at acid pH (pH 3.0 or less). The modification of methionine by ethyleneimine has been reported in a reaction producing a sulfonium salt derivative.[11] The modification of methionine in azurin with bromoacetate has been reported.[12] In this protein, four of six methionine residues were modified at pH 4.0, while all methionine residues were reactive at pH 3.2. These modification reactions were performed in 0.1 *M* sodium formate at ambient temperatures for 24 h with 0.16 *M* bromoacetate. The modification of methionine in porcine kidney acyl CoA-dehydrogenase occurs with iodoacetate (0.030 *M*) in 0.1 *M* phosphate, pH 6.6 at ambient temperature.[13] The identification of methionine as the residue modified by iodoacetate in this protein was supported by the comparison of the chromatogram of the acid hydrolyzate of the modified protein (reacted with ^{14}C-iodacetate) with that of the acid hydrolyzate of synthetic *S*-([l-^{14}C]carboxymethyl)-methionine.[9] This is necessary since the *S*-carboxymethyl derivative yielded several different compounds on acid hydrolysis.[9,14]

Naider and Bohak[15] have reported that the sulfonium salt derivatives of methionine (e.g., *S*-carboxymethyl methionine, the reaction product of methionine and iodoacetic acid) can be converted to methionine by reaction with a suitable nucleophile. For example, reaction of *S*-carboxamidomethyl methionine (in the peptide Gly-Met-Gly) with a sixfold molar excess of mercaptoethanol at pH 8.9 at a temperature of 30°C resulted in the complete regeneration of methionine after 24 h of reaction. The *S*-phenacyl derivative of methionine (in the peptide Gly-Met-Gly) was converted to methionine in 1 h under the same reaction conditions. These investigators also showed that chymotrypsin previously treated with phenacyl bromide under conditions which inactivate the enzyme concomitant with the alkylation of methionine-192[16] could be reactivated by treatment with β-mercaptoethanol at pH 7.5 (sodium phosphate). It is of interest that the *S*-phenacyl methionine in chymotrypsin is converted to methionine at a substantially faster rate than the tripeptide derivative. The authors speculate that the increased reactivity of the chymotrypsin derivative is a reflection of interaction of the phenacyl moiety with the substrate-binding site.

Alkylation of methionyl residues in pituitary thyrotropin and lutropin with iodoacetic acid has been reported.[17] Differential reactivity of various methionyl residues was reported on reaction with iodoacetate in 0.2 *M* formate, pH 3.0 for 18 h at 37°C.

The reversible alkylation of methionine by iodoacetate in dehydroquinase has been reported by Kleanthous et al.[18] In this reaction, iodoacetate behaves kinetically as an affinity label with a K_i of 30 μM and a k_{inact} of 0.014 min^{-1}, pH 7.0 (50 m*M* potassium phosphate). There is no reaction with iodoacetamide. Two methionyl residues are modified during the reaction of dehydroquinase with iodoacetate. In a companion study, Kleanthous and Coggins[19] demonstrated that 2-mercaptoethanol treatment under alkaline conditions (0.5% ammonium bicarbonate, 37°C) could reverse modification at one of the two residues. If the modified protein is denatured, there is no reversal of modification at either residue. The results are interpreted in terms of the proximity of a positive charge (i.e., lysine) in close proximity to one of the two methionyl residues which (1) provides the basis for the affinity labeling and (2) provides the basis for the 2-mercaptoethanol-mediated reversal of modification.

REFERENCES

1. **Oda, T. and Tokushige, M.,** Chemical modification of tryptophanase by chloramine T: a possible involvement of the methionine residue in enzyme activity, *J. Biochem.*, 104, 178, 1988.
2. **Trout, G. E.,** The estimation of microgram amounts of methionine by reaction with chloroamine-T, *Anal. Biochem.*, 93, 419, 1979.
3. **de la Llosa, P., El Abed., A., and Roy, M.,** Oxidation of methionine residues in lutropin, *Can. J. Biochem.*, 58, 745, 1980.
4. **Caldwell, P., Luk, D. C., Weissbach, H., and Brot, N.,** Oxidation of the methionine residues of *Escherichia coli* ribosomal protein L12 decreases the protein's biological activity, *Proc. Natl. Acad. Sci. U.S.A.*, 75, 5349, 1978.
5. **Spande, T. F. and Witkop, B.,** Determination of the tryptophan content of proteins with N-bromosuccinimide, *Meth. Enzymol.*, 11, 498, 1967.
6. **Sakurai, J. and Nagahama, M.,** Role of one tryptophan residue in the lethal activity of *Clostridium perfringens* epsilon toxin, *Biochem. Biophys. Res. Commun.*, 128, 760, 1985.
7. **Houghten, R. A. and Li, C. H.,** Reduction of sulfoxides in peptides and proteins, *Anal. Biochem.*, 98, 36, 1979.
8. **Brot, N., Weissbach, L., Werth, J., and Weissbach, H.,** Enzymatic reduction of protein-bound methionine sulfoxide, *Proc. Natl. Acad. Sci. U.S.A.*, 78, 2155, 1981.
9. **Gundlach, H. G., Moore, S., and Stein, W. H.,** The reaction of iodoacetate with methionine, *J. Biol. Chem.*, 234, 1761, 1959.
10. **Gurd, F. R. N.,** Carboxymethylation, *Meth. Enzymol.*, 11, 532, 1967.
11. **Schroeder, W. A., Shelton, J. R., and Robberson, B.,** Modification of methionyl residues during aminoethylation, *Biochim. Biophys. Acta,* 147, 590, 1967.
12. **Marks, R. H. L. and Miller, R. D.,** Chemical modification of methionine residues in azurin, *Biochem. Biophys. Res. Commun.*, 88, 661, 1979.
13. **Mizzer, J. P. and Thorpe, C.,** An essential methionine in pig kidney general acyl-CoA dehydrogenase, *Biochemistry,* 19, 5500, 1980.
14. **Goren, H. J., Glick, D. M., and Barnard, E. A.,** Analysis of carboxymethylated residues in proteins by an isotopic method and its application to the bromoacetate-ribonuclease reaction, *Arch. Biochem. Biophys.*, 126, 607, 1968.

15. **Naider, F. and Bohak, Z.,** Regeneration of methionyl residues from their sulfonium salts in peptides and proteins, *Biochemistry,* 11, 3208, 1972.
16. **Schramm, H. J. and Lawson, W. B.,** Über das activ Zentrum von Chymotrypsin. II. Modifizierung eines Methioninrestes in Chymotrypsin durch einfache Benzolderivate, *Hoppe-Seyler's Z. Physiol. Chem.,* 332, 97, 1963.
17. **Goverman, J. M. and Pierce, J. G.,** Differential effects of alkylation of methionine residues on the activities of pituitary thyrotropin and lutropin, *J. Biol. Chem.,* 256, 9431, 1981.
18. **Kleanthous, C., Campbell, D. G., and Coggins, J. R.,** Active site labeling of the shikimate pathway enzyme, dehydroquinase. Evidence for a common substrate binding site within dehydroquinase and dehydroquinate synthase, *J. Biol. Chem.,* 265, 10929, 1990.
19. **Kleanthous, C. and Coggins, J. R.,** Reversible alkylation of an active site methionine residue in dehydroquinase, *J. Biol. Chem.,* 265, 10935, 1990.

Chapter 9

THE MODIFICATION OF HISTIDINE RESIDUES

Since many enzymes contain a histidine residue, which is critical for the catalytic process, the site-specific modification of this residue has been the subject of many studies. Although several approaches are described below, with the exception of peptide chloromethylketones which are used for the modification of active site histidyl residues in serine proteinases,[1–4] only diethylpyrocarbonate (ethoxyformic anhydride) is currently used for the modification of histidine in proteins.[5–9]

The technique of photooxidation has not proved to be of extensive value because of problems with the specificity of modification. Histidine, methionine, and tryptophan are quite sensitive to photooxidation, while tyrosine, serine, and threonine are somewhat less sensitive.[10] Photooxidation was used to identify the protein(s) at the peptidyltransferase site of a bacterial ribosomal subunit.[11,12] Rose bengal dye was used in these experiments. Histidine is the only amino acid modified under the reaction conditions. The time course for the loss of the various biological activities of reconstituted ribosomes on photooxidation is shown in Figure 1. Figure 2 shows the rate of histidine loss during photooxidation. The rates of histidine loss were then compared to the rate of biological activity loss during photooxidation as shown in Figure 3. With the exception of EF-G·GTP binding activity, the loss of biological activity (see Figure 1 for symbols) is most closely related to the "fast" histidine loss. In subsequent experiments, methylene blue dye (Eastman, dye content 91%) was used.[13] Peptidyltransferase activity was lost at a more rapid rate in the presence of methylene blue than rose bengal, but data are not presented regarding any differences in residues modified or whether amino acid residues other than histidine are modified in the presence of methylene blue. Other investigators have also explored the effects of photooxidation on peptidyltransferase activity in *Escherichia coli* ribosomes.[14] These experiments were performed in 0.030 *M* Tris, 0.020 *M* $MgCl_2$, 0.220 KCl, pH 7.5 (9 mg ribosomes in 0.300 ml) with either eosin or rose bengal as the photooxidation agent. Irradiation was performed at 0 to 4°C using a 500-W slide projector (26 cm from condenser lens to sample) for 20 min. Photooxidation has also been used to study the role of histidine residues in polypeptide chain elongation factor Tu from *E. coli*.[15] The reaction is performed in 0.05 *M* Tris, 0.010 *M* Hg $(OAc)_2$, 0.005 *M* β-mercaptoethanol, 10% glycerol, pH 7.9. Irradiation is performed at 0 to 4°C with gentle stirring using a 375-W tungsten lamp at a distance of 15 cm. A glass plate was placed in the light beam to eliminate ultraviolet (UV) irradiation. The rose bengal dye is removed after 5 to 30 min from the reaction by chromatography on DEAE-Sephadex A-25 or A-50 equilibrated with 0.050 *M* Tris, pH 7.9, 0.010 *M* Mg

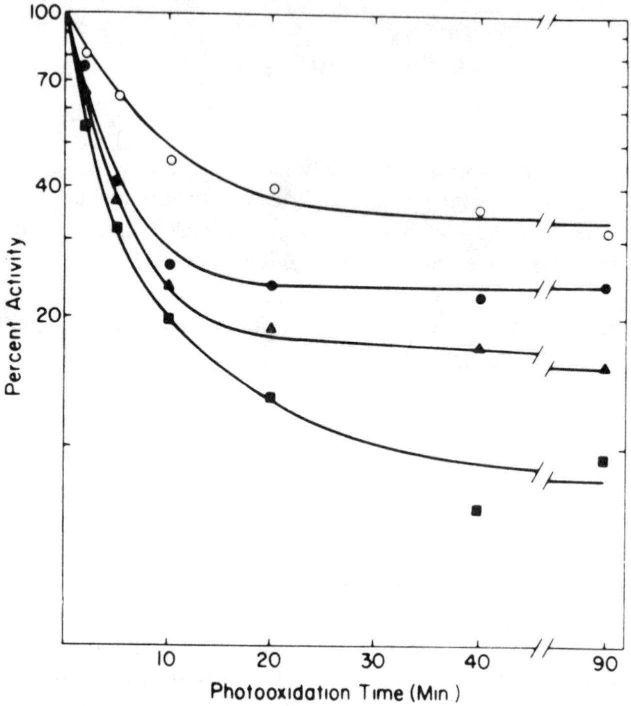

FIGURE 1. The inactivation of biological activity of ribosome reconstituted with photooxidized ribosomal protein BL3. Ribosomal protein BL3 was photooxidized as the B-13–23 S RNA complex in the presence of rose bengal. Samples of the purified reconstituted ribosomes were assayed for EF-G·GTP binding (○), polyphenylalanine synthesis (●), peptidyltransferase activity (▲), and the ability to bind phenylalanyl-tRNA (■). (From Auron, P. E., Erdelsky, K. J., and Fahnestock, S. R., *J. Biol. Chem.*, 253, 6893, 1978. With permission.)

$(OAc)_2$, 0.005 *M* β-mercaptoethanol, 10% glycerol. Amino acid analysis after acid hydrolysis (6 *N* HCl, 22 h, 110°C, or 4 *M* methanesulfonic acid, 0.2% 2-aminoethylindol, 115°C, 24 h for the determination of tryptophan) demonstrated that only histidine is modified (approximately 5 out of 10 residues are modified; only one residue is modified in the presence of guanosine diphosphate). Photooxidation with methylene blue (25 m*M* Tris, pH 7.9, 0.05% methylene blue; 8°C) abolished placental anticoagulant protein activity with loss only of histidine residues (based on amino acid analysis).[16] Diol dehydrase was inactivated with first-order kinetics by photooxidation in the presence of either rose bengal or methylene blue.[17] In these experiments, the substitution of a helium atmosphere markedly decreased the rate of enzyme inactivation. There was a difference in the pH dependence of the photooxidation reaction performed in the presence of rose bengal (optimum pH 6.2) as compared to the reaction in the presence of methylene blue

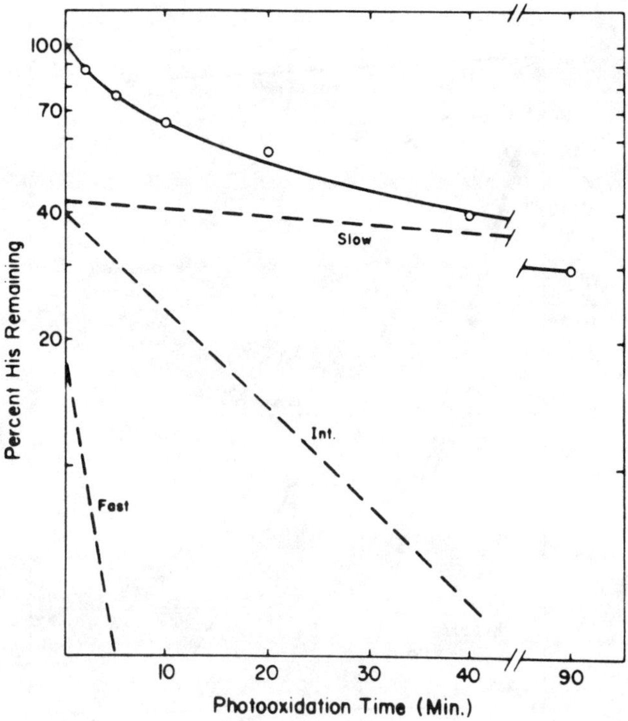

FIGURE 2. The loss of histidine residues occurring with photooxidation of BL3 in the presence of rose bengal dye. The loss of histidine was assessed by amino acid analysis after acid hydrolysis. Loss of amino acids other than histidine was not observed under these experimental conditions. The solid line (○) shows the rate of loss of histidine residues. These data can be fit to represent three different classes of histidine residues as indicated by the broken lines. (From Auron, P. E., Erdelsky, K. J. and Fahnestock. S. R., *J. Biol. Chem.*, 253, 6893, 1978. With permission.)

(optimum >pH 8.0). The pH dependence for the rose bengal reaction was suggested to reflect the charge status of this compound.

Histidine residues can be modified by α-halo carboxylic acids and amides (Figure 4) (i.e., iodoacetate and iodocetamide). In general, the histidine residue must have either enhanced nucleophilic character[18] or have been located in a unique microenvironment such as in ribonuclease.[19–22]

A compound related to α-halo carboxylic acids is *p*-bromophenacyl bromide, which has been demonstrated in several instances to modify histidyl residues in proteins. *p*-Bromophenacyl bromide modifies a single histidine residue in taipoxin with a 350-fold decrease in neurotoxicity.[23] The modification was performed in 0.1 *M* sodium cacodylate, pH 6.0, 0.1 *M* NaCl with an eightfold molar excess of *p*-bromophenacyl bromide* at 30°C for 22 h. The

* It has been the authors' experience that this reagent is somewhat unstable and preparations *must* be recrystallized.

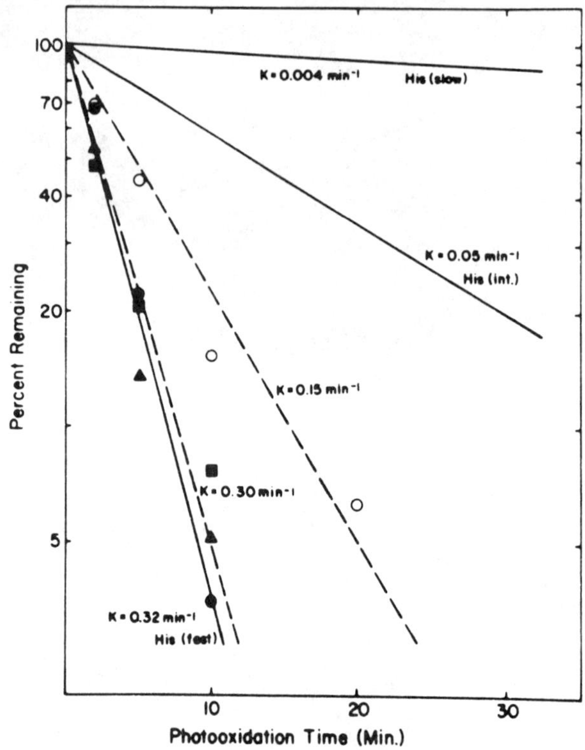

FIGURE 3. A comparison of the rates of the loss of activity and histidine as a function of time of photooxidation. The solid lines represent the first-order rates for the loss of histidine replotted as a percentage of each class of reactive histidyl residues (see Figure 2). The broken line indicates the loss of activity (for symbols see Figure 1 caption). (From Auron, P. E., Erdelsky, K. J., and Fahnestock, S. R., *J. Biol. Chem.*, 253, 6893, 1978. With permission.)

reaction mixture was concentrated by lyophilization and subjected to gel filtration (G-25 Sephadex in 0.1 *M* ammonium acetate) to remove excess reagent and buffer salts. The protein fraction was taken to dryness as a salt-free preparation and subjected to a second reaction with *p*-bromophenacyl bromide. The extent of modification was assessed by both amino acid analysis (loss of histidine) and spectral analysis ($\Delta\varepsilon$ 271 = 17,000 M^{-1} cm^{-1}).[24] Two of seven histidine residues are modified (1 mol/mol in α-subunit; 1 mol/mol in β-subunit) under these reaction conditions.

The basic phospholipase A_2 from *Naja nigrocollis* venom has been modified with *p*-bromophenacyl bromide.[25] The modification was performed in 0.025 *M* Tris, pH 8.0 with a tenfold molar excess of reagent at 30°C. After 40 min of reaction, the mixture was taken to pH 4.0 with glacial acetic acid and taken through a G-25 Sephadex column. Amino acid analysis after acid hydrolysis showed the loss of 1 mol of histidine per mole of enzyme with no other

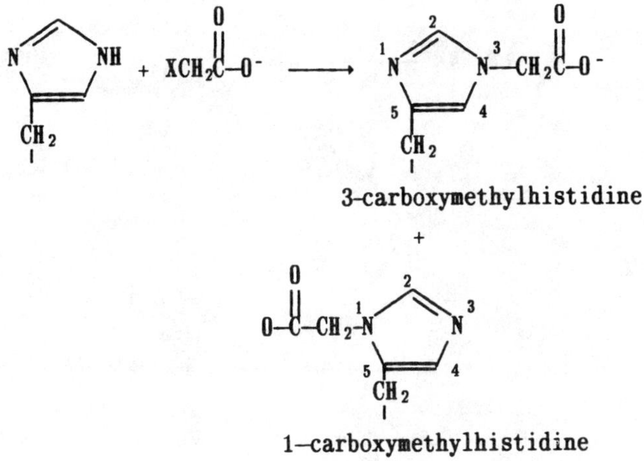

FIGURE 4. A scheme for the carboxymethylation of histidine with iodoacetate resulting in the formation of 1-carboxymethylhistidine and 3-carboxymethylhistidine.

significant changes in composition. Subsequent analysis identified His[47] as the residue modified.

The reaction of *p*-bromophenacyl bromide with pancreatic phospholipase A_2 has also been studied.[26] The reaction was performed in 0.1 *M* sodium cacodylate, pH 6.0.[27] Only histidine residues are modified under these conditions, and it has been established that His[48] is the residue modified. Under these reaction conditions, the second-order rate constant for the reaction of *p*-bromophenacyl bromide with porcine pancreatic phospholipase A_2 is 125 M^{-1} min^{-1} as compared to 79 M^{-1} min^{-1} for phenacyl bromide and 75 M^{-1} min^{-1} for 1-bromooctan-2-one. No reaction was observed with iodoacetamide under these reaction conditions. This same study[26] also explored the methylation of His[48] at the N′-position on the imidazole ring with either methyl *p*-toluenesulfonate or methyl *p*-nitrobenzenesulfonate. Reaction with the latter reagent at pH 6.0 (0.050 *M* cacodylate) (40°C) is more rapid than with the former reagent.

Methyl *p*-nitrobenzenesulfonate has also been used to methylate histidine residue(s) in ribosomal peptidyl transferase.[28] In these experiments, the ribosome preparation was modified by a 300-fold molar excess of methyl *p*-nitrobenzenesulfonate (from a stock solution dissolved in acetonitrile). The reaction took place in 0.01 *M* Tris, pH 7.4, 0.008 *M* $MgCl_2$, 0.05 *M* NH_4Cl, 1 μ*M* puromycin at 24°C for 45 min. The author suggests only histidine residues are modified, but definitive evidence on this point is absent.

Another example of the modification of histidine by reagents which, in general, react more avidly with other residues is the reaction of D-amino acid oxidase with dansyl chloride.[29] In this study, D-amino acid oxidase was allowed to react with a fivefold molar excess of dansyl chloride (from a stock solution

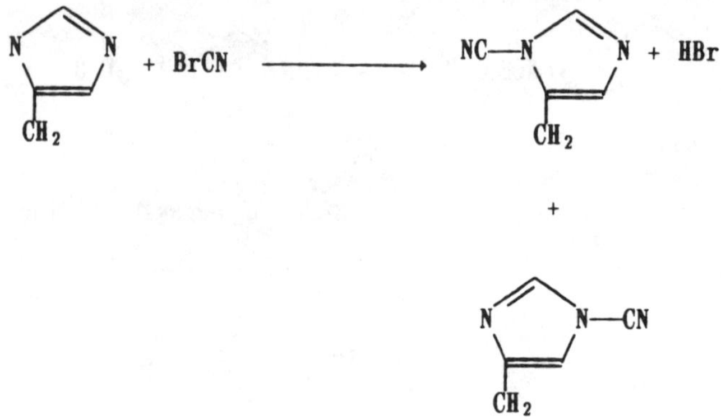

FIGURE 5. The reversible modification of histidine with cyanogen bromide.

in acetone; final concentration of acetone in the reaction mixture did not exceed 5% final volume) in 0.05 *M* phosphate, pH 6.6. The reaction was terminated by the addition of benzoate, insoluble material was removed by centrifugation, and the mixture was passed through a G-25 Sephadex column equilibrated with 0.06 *M* phosphate, 0.010 *M* benzoate, pH 6.6. Reaction with dansyl chloride under these conditions resulted in virtually complete inactivation of the enzyme with the incorporation of 1.7 mol of reagent per mole of enzyme. Substantially complete reactivation occurred with 0.5 *M* hydroxylamine (NH_2OH) at pH 6.6. This reactivation excluded reaction with primary amino functional groups such as lysine, and amino acid analysis suggested the reaction had not occurred with an oxygen nucleophile such as tyrosine. Treatment of the enzyme with diethylpyrocarbonate also resulted in the loss of catalytic activity and reduced the amount of dansyl groups incorporated in a subsequent reaction, suggesting that dansyl chloride reacts with the same functional group that reacted with diethylpyrocarbonate.

The cyanation of histidine residues (Figure 5) in myoglobin using an equimolar ratio of cyanogen bromide and protein at pH 7.0 has been reported.[30,31] This derivative is somewhat unstable, but it has proved useful in spectral studies (NMR, IR, UV-VIS) of this protein.

Competitive labeling (trace labeling; see Chapter 10 for full conceptual description of this technique) of the amino-terminal histidine residue in secretin with 1-fluoro-2,4-dinitrobenzene has been used to study the reactivity of this residue vs. other nucleophiles[32] in this protein. The amino-terminal functional group has a pKa of 8.83 and fivefold greater reactivity than the model compound (histidyl-glycine), while the imidazolium ring has a pKa value of 8.24 and a 26-fold greater reactivity than the model compound. These results were interpreted as reflecting a conformational state where the histidine is interacting with a carboxylate function.

Diethylpyrocarbonate (ethoxyformic anhydride) is the most useful reagent for the specific modification of histidine. In the pH range from 5.5 to 7.5,

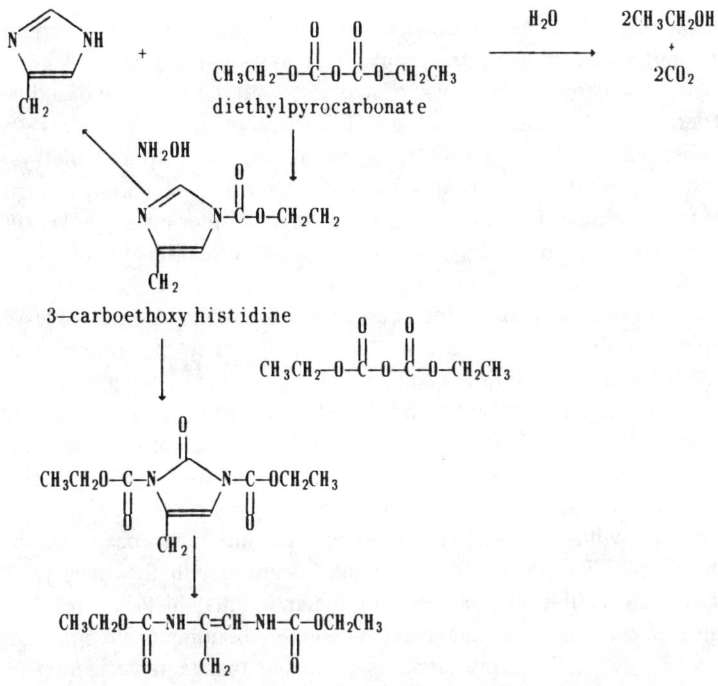

FIGURE 6. A scheme for the reaction of diethylpyrocarbonate with histidine.

diethylpyrocarbonate is reasonably specific for reaction with histidyl residues. There are several studies of the reaction under more acidic conditions.[33,34] In one of the studies, polymerization of ribonuclease was observed both in deionized water (presumably at acidic pH) and in 0.1 *M* Tris, pH 7.2. Maleylation of ribonuclease obviated polymerization, suggesting that the amino groups were involved in this cross-linking reaction. Work with diethylpyrocarbonate through 1975 has been reviewed by Miles.[35]

Reaction of diethylpyrocarbonate with histidine residues at a *moderate excess* of diethylpyrocarbonate results in substitution at one of the nitrogen positions on the imidazole ring (Figure 6). This reaction is associated with an increase in absorbance at 240 nm ($\Delta\varepsilon = 3200\ M^{-1}\ cm^{-1}$). The modification is readily reversed at alkaline pH and, in particular, in the presence of nucleophiles such as Tris. Generally, treatment with neutral hydroxylamine (0.1 to 1.0 *M*, pH 7.0) is used to regenerate histidine. As with the deacylation of *O*-acetyl tyrosine by neutral hydroxylamine, the higher the concentration of hydroxylamine, the more rapid the process of decarboxyethylation. Disubstitution on the imidazole ring and carboxyethylation at both the N_1 and N_3 positions results in a derivative with altered spectral properties compared to the monosubstituted derivative. This derivative does not regenerate histidine and treatment with neutral hydroxylamine or base results in scission of the imidazole ring. With disubstitution, a loss of histidine is detected by amino acid analysis after acid hydrolysis. Sequence analysis using Edman degradation chemistry also shows the absence of histidine with the presence of a disubstituted derivative.[36] In these studies, a PTH derivative eluting near PTH-glycine was observed and the structure verified by mass spectrometry. The monosubstituted derivative is unstable under conditions of acid hydrolysis and yields free histidine. Reaction can also occur with other nucleophiles such as cysteine, tyrosine, and primary amino groups. Modification at sulfhydryl residues, not well documented with protein-bound cysteine, can be determined by a decrease in free sulfhydryl groups. Reaction of tyrosine is easily assessed by a decrease in absorbance at 275 to 280 nm, similar to that observed on *O*-acetylation with *N*-acetylimidazole (see Chapter 13). This modification is reversed by neutral hydroxylamine. Reaction at primary amino groups (α-amino groups; ε-amino groups of lysine) results in a derivative which is stable to hydroxylamine. An elegant study[37] has examined the reaction of diethylpyrocarbonate with histidyl residues in cytochrome b_5. Using (NMR) spectroscopy with this well-characterized protein, it has been possible to identify factors influencing histidine modification with this reagent; three major factors include (1) the pKA of the individual histidine residue, (2) solvent exposure of the residue, and (3) hydrogen bonding of the imidazolium ring. Furthermore, these investigators point out that tautamerization of the imidazolium ring leads to heterogeneity of modification which, in turn, explains differences in the spectral properties of modified proteins.[35] Site-specific mutagenesis studies of subtilisin[38] have demonstrated the influence of neighboring charged groups on histidine ionization (and, hence, reactivity).

As described in some detail by Miles,[35] the reagent is very sensitive to base-catalyzed hydrolysis. At ambient temperature, the $T_{1/2}$ for the hydrolysis of diethylpyrocarbonate at pH 7.0 (phosphate) is less than 10 min and is markedly shorter with increasing pH. Increasing the pH not only decreases reagent stability (and thus the concentration of one component of a second-order reaction over the time period studied), but it also increases the possibility of reaction at primary amine functional groups. In our laboratory, we have found it convenient to use dilute (0.025 to 0.100 *M*) phosphate buffer, pH 6.0 for our studies. We prepare stock solutions of diethylpyrocarbonate in *anhydrous* ethanol. These solutions are used within a few hours, and the actual concentration of reagent is obtained by the stoichiometry of reaction with imidazole in the pH 6.0 buffer using the increase in absorbance at 230 nm to monitor the reaction ($\Delta\varepsilon = 3 \times 10^3$ M^{-1} cm^{-1})[33,35] both before and after a given series of experiments. Morjana and Scarborough[39] have presented data on the rate of diethylpyrocarbonate hydrolysis and discussed the necessity of correcting for actual reagent concentration. It should be noted that there is not complete agreement regarding the magnitude of the spectral change as a result of carboxyethylation. The value $\Delta\varepsilon = 3200$ M^{-1} cm^{-1} at 242 nm is given above. Other investigators have used the value of $\Delta\varepsilon = 3600$ M^{-1} cm^{-1} at 240 nm.[40,41] A value of $\Delta\varepsilon = 3500$ M^{-1} cm^{-1} at 242 nm has also been reported.[42] The greater

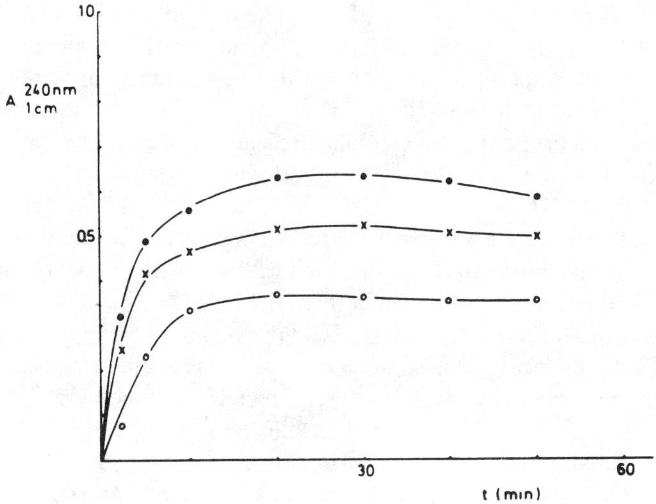

FIGURE 7. The time course for the increase in absorbance at 240 nm occurring upon reaction between histidine and diethylpyrocarbonate (DEP) as a function of diethylpyrocarbonate concentration. The reaction mixtures were 0.1 m*M* with respect to histidine in 0.1 *M* acetate, pH 6.0 at 2.3 m*M* DEP (○), 11.5 m*M* DEP (x) and 23 m*M* DEP (●). (From Roosemont, J. L., *Anal. Biochem.*, 88, 314, 1978. With permission.)

the excess of diethylpyrocarbonate used the less reliable the value for $\Delta\varepsilon$ obtained with *known* stoichiometric modification of model histidine (imidazole derivative) compounds.[42] This is shown in Figure 7 where increasing the ratio of diethylpyrocarbonate to histidine results in species with increased absorbance. It is suggested that the commonly used $\Delta\varepsilon = 3200\ M^{-1}\ cm^{-1}$ can only be used at low concentrations of diethylpyrocarbonate.

A single histidine residue essential for catalysis by D-xylose isomerase has been identified by reaction with diethylpyrocarbonate.[43] Instability of *N*-carboethoxyhistidine has made unequivocal identification of the histidyl residues modified by diethylpyrocarbonate difficult. In this study, the protein was first denatured in 6.0 *M* guanidine hydrochloride (pH 7.0) and then digested with subtilisin in 2.0 *M* guanidine hydrochloride, pH 7.0 at 30°C for 2 h. A single peptide containing the modified histidine residue was purified by HPLC using dual wavelength detection. In this technique, effluent is monitored by absorbance at 238 (the maximum in the difference spectrum between the modified and native protein) and 214 nm (peptide bond absorbance). The ratio of A_{238} to A_{214} was used to identify peptides containing the modified histidine residues.

Selected examples of the use of diethylpyrocarbonate to study the function of histidyl residues in proteins are presented in Table 1.

Horiike and co-workers[44] have examined the reaction of diethylpyrocarbonate with pyridoxamine (pyridoxine)-5′-phosphate oxidase. The modification reaction was performed at pH 7.0 (0.1 *M* potassium phosphate containing 5% (v/v) EtOH) at 25°C generally in the presence of flavin mononucleotide (FMN). Figure 8 shows the time course for the modification of the oxidase under these experimental conditions. Also shown in Figure 8 is the dependence of the observed first-order rate constant on diethylpyrocarbonate concentration (panel B). The inset plot shows the reaction to be second-order with a rate constant of 12.5 M^{-1} s^{-1}.

Figure 9 shows the pH dependence for the inactivation process. The loss of catalytic activity appeared to be correlated with the modification of one of the four histidyl residues in this protein (Figure 10).

Saluja and McFadden[45] have explored the reaction of diethylpyrocarbonate with spinach ribulose bisphosphate carboxylase. One interesting observation is that the plot of half-inactivation time vs. the reciprocal of diethylpyrocarbonate concentration suggested that saturation kinetics existed consistent with the "affinity" binding of reagent prior to protein modification. The study of Bloxham[46] on the reactivity of the active-site histidine in lactate dehydrogenase is particularly fascinating. The rate of reaction of the histidine residue in the native enzyme was compared to the thiomethyl derivative (prepared by reaction with methyl methanethiosulfonate) as shown in Figure 11. There is a substantial decrease in the nucleophilic character of the active-site histidine (histidine-195). Cromartie[47] has examined the modification of alcohol oxidase with diethylpyrocarbonate in 0.050 *M* sodium phosphate, pH 7.5, 0°C. The UV difference spectrum of the enzyme before (a) and after (b) with the addition of diethylpyrocarbonate is shown in Figure 12. No evidence for tyrosine modification is seen under these reaction conditions. Treatment with neutral hydroxylamine (0.010 to 0.100 *M*) did not result in the recovery of catalytic activity, although most of the radiolabeled reagent was removed ([1-^{14}C]-diethylpyrocarbonate). It was observed that 0.010 *M* hydroxylamine caused an 80% loss of alcohol oxidase activity at 0°C. As mentioned above, reaction of diethylpyrocarbonate at pH values above 7.0 does increase the possibility of amino group modification. This is demonstrated by the observation of Bond and co-workers[48] on the reaction of diethylpyrocarbonate with a bacterial collagenase. Reaction of the enzyme with diethylpyrocarbonate results in the loss of catalytic activity. Reaction with hydroxylamine did not markedly restore enzymatic activity. Hydroxylamine by itself did not have a deleterious effect on catalytic activity, as treatment with this reagent did restore activity lost on the modification of tyrosyl residues with *N*-acetylimidazole. These investigators consider modification of the ε-amino group(s) of lysine with diethylpyrocarbonate to be a more likely cause for the irreversible loss of enzymic activity. Daron and Aull[49] have studied the reaction of diethylpyrocarbonate with dihydrofolate reductase *(Lactobacillus casei)*. Catalytic activity is lost on reaction with diethylpyrocarbonate, but it is partially recovered on reaction with 1.0 *M* hydroxylamine (pH 7.5) but not with 0.1 *M* hydroxylamine (pH 7.5.). Clearly, the reaction of proteins with hydroxylamine after modification with diethylpyrocarbonate must be carefully studied in order to obtain meaningful results.

Table 1
THE REACTION OF DIETHYLPYROCARBONATE WITH HISTIDYL RESIDUES IN PROTEINS

Protein	Solvent/temp.	Reagent excess[a]	Extent of modification[b]	Second-order rate constant[c] ($1\ M^{-1}\ min^{-1}$)	Other amino acid modified	Ref.
Thermolysin	0.05 *M* $CaCl_2$, 0.1 *M* NaCl, 0.025 *M* HEPES, pH 7.2 at 25°C	—	2[d]	—	Tyr (2), Lys (8)	1
Thermolysin	0.05 *M* $CaCl_2$, 0.025 *M* HEPES, pH 5.7 at 25°C	—	1[d]	—	Tyr (1), Lys (2)	1
Bacterial luciferase	0.1 *M* phosphate, pH 6.1 at 0°C	600	3	146[e]		2
Crotoxin	Half-saturated NaOAc	1,000		—	—	3
Prostatic acid phosphatase	0.025 *M* sodium barbital, 0.15 *M* NaCl, pH 6.9 at 25°C	5,000	12/27	7[e]	—[f]	
Pyridoxamine-5′-phosphate oxidase	0.1 *M* phosphate, pH 7.0 at 25°C	500	4/7	750[g]	Cys (1/6)[h]	5
Yeast enolase	0.050 *M* ADA, 0.001 *M* $MgCl_2$, 0.01 m*M* EDTA, pH 6.1 at 0°C	1,390	6[i]	55.0[j] 10.7[t]	—[k]	6
Fructose bisphosphatase	0.050 *M* acetate, pH 6.5	4,000 8,000 40,000	5.3/13 8.8/13 13/13[l]	—	—[l]	7
Ribulose bisphosphate carboxylase	0.050 or Tris, 0.001 *M* EDTA, 0.020 *M* $MgCl_2$, pH 7.0 at 30°C	—	2/2	—[m]	—[n]	8
Escherichia coli elongation factor Tu	—[o]	—	—	—	—	10
L-α-Hydroxy acid oxidase	0.020 *M* MES, pH 7.0 at 25°C	—	2[p]	690[q]	—[r]	11
Thiamin binding protein from *Saccaromyces cerevisiae*	0.050 *M* sodium phosphate, pH 7.0 at 25°C	—	—	120.5[s]	—	12
Clostridium histolyticum collagenase	0.050 *M* HEPES, 0.010 *M* $CaCl_2$, pH 17.5 at 22°C	—	—	—	—	13

Table 1 (continued)
THE REACTION OF DIETHYLPYROCARBONATE WITH HISTIDYL RESIDUES IN PROTEINS

Protein	Solvent/temp.	Reagent excess[a]	Extent of modification[b]	Second-order rate constant[c] (1 M^{-1} min^{-1})	Other amino acid modified	Ref.
Lactate dehydrogenase	0.1 *M* sodium pyrophosphate, pH 7.2 at 10°C	—	—	10,380 522[t]	—	14
Mitochondrial nicotinamide nucleotide transhydrogenase	0.025 *M* sucrose, 0.020 *M* MES, pH 5.9 at 23°C	—	—	—	—	15
Alcohol oxidase	0.050 *M* sodium[v] phosphate, pH 7.5 at 0°C	—	4	25.2	—[w]	16
D-β-hydroxybutyrate dehydrogenase	0.020 MES, 5 *M* rotenone, pH 6.0, 20°C	—	—	—	—	17
Benzodiazepine receptor	0.01 *M* sodium phosphate, 0.2 *M* NaCl, pH 6.0, 0°C	—	—	—	—	19
Scrapie agent	0.020 *M* Tris, pH 7.4 containing 1 m*M* EDTA and 0.2% Sarkosyl at 23°C	—	—	—	—[x]	20
Succinate dehydrogenase	0.24 *M* sucrose, 0.100 *M* potassium phosphate, pH 6.0 at 0°C	—	—	—	—[y]	21
Ribulose bis-phosphate carboxylase/oxygenase	0.050 *M* Tris, 0.001 *M* EDTA, 0.020 *M* $MgCl_2$, pH 7.0 at 21°C	—	4.5[z]	2,340	—	22
Transferrin	0.01 *M* potassium phosphate, 0.05 KCl, pH 6.1	—	54–72%[aa]	—	—	23
Dihydrofolate reductase	0.05 *M* Tris, pH 7.5 at 10°C	—	6/7	29	—[bb]	24
RNA polymerase	100 m*M* phosphate, pH 6.0 at ambient	50–500	—	—	—	25
A_1 adenosine receptor	20 m*M* potassium phosphate, pH 7.0 at ambient	—	—	—	—	26
Lysyl oxidase	200 m*M* potassium phosphate, pH 7.0 at ambient	—	—	2.5 min^{-1} mM^{-1}	—	27

Malic enzyme	50 m*M* acetate, pH 6.0 with 1.0 m*M* EDTA at 25°C	—	2.8/5.0[cc]	—	—	28
		—	1.8/5.0[cc]	—	—	28
Neurospora membrane ATPase	50 m*M* HEPES, pH 6.9 with 30% (V/V) glycerol	—	—	385–420 min^{-1} M^{-1}[dd]	—	29
Asparaginase	50 m*M* MES, pH 6.0 at 25°C	—	—	—	—	30
Ampicillin acylase	10 m*M* sodium phosphate, pH 7.0	—	—	—	—	31
Pancreatic lipase	100 m*M* phosphate, pH 6.0 at ambient	—	—	—	—	32
Lipase fragment	100 m*M* phosphate, pH 6.0 at ambient	—	—	—	—	33
Phenol hydroxylase	50 m*M* potassium phosphate, pH 6.0 at 4°C[ff]	—	—	—[ee]	—	34
Lysolecithin-lysolecithin acyltransferase	100 m*M* phosphate, pH 6.5	—	—	1.17 min^{-1} mM^{-1}[gg] 0.56 min^{-1} mM^{-1}	—	35

[a] Moles per mole of protein.

[b] Residues His modified per total His in protein.

[c] For reaction with histidine.

[d] Inactivation was demonstrated to result from the modification of a single histidine residue.

[e] Assuming loss of activity is a direct indication of a single histidine modification.

[f] There was only partial recovery of activity upon treatment with hydroxylamine (0.2 *M*, pH 7.0, 25°C). Two residues of histidine were lost as assessed by amino acid analysis after acid hydrolysis without loss of other amino acids, suggesting that disubstitution has occurred on the imidazole ring of certain histidine residues.

[g] For inactivation of catalytic activity. A value of 51.6 M^{-1} s^{-1} (3096 M^{-1} min^{-1}) was calculated for the pH-independent second-order rate constant.

[h] No direct determination of primary amino modification is reported. Activity is recovered by neutral hydroxylamine (0.09 *M*). Direct determination of tryptophan and tyrosine revealed no loss of these residues.

[i] Obtained from reaction at either 0 or 25°C.

[j] Data obtained at pH 6.6 [0.050 *M* *N*-(2-acetamido)iminodiacetic acid, ADA], 0°C. Two-phase reaction was observed.

[k] Spectral analysis did not indicate tyrosine modification. Possible primary amine modification was not determined. The loss in catalytic activity was reversed by 0.25 *M* hydroxylamine, pH 7.0.

[l] Addition of more diethylpyrocarbonate results in further increases in absorbance at 242 nm, suggesting disubstitution on the imidazole ring of histidyl residues. Spectral analysis did not suggest modification of tyrosine under the reaction conditions. Possible modification of primary groups was not assessed.

Table 1 (continued)
THE REACTION OF DIETHYLPYROCARBONATE WITH HISTIDYL RESIDUES IN PROTEINS

m The reaction of diethylpyrocarbonate with ribulose bisphosphate carboxylase shows saturation kinetics (k = 7.3 m*M*), suggesting "specific" binding of diethylpyrocarbonate to the enzyme prior to the reaction resulting in inactivation. Data are not presented to show a similar phenomena with the actual reaction of histidyl residues in the enzyme.

n Activity was recovered by treatment with hydroxylamine (0.4 *M* NH_2OH, pH 7.0 for 48 h at 4°C increased activity from 55 to 89%; similar treatment at 25°C resulted in similar activity recovery in 1 h). The authors note that reaction of diethylpyrocarbonate with cysteine (*N*-acetylcysteine) also results in an increase in absorbance at 240 nm that is reversed by hydroxylamine. This reaction apparently occurs only in carboxylate buffers (e.g., acetate or succinate) and has been noted by other investigators.[9] The reaction product of diethylpyrocarbonate with cysteine is considerably less stable than *N*-carbethoxyimidazole derivatives.

o The crystalline enzyme preparation (7 nmol) [washed with 41% $(NH_4)SO_4$] was dissolved in 0.600 ml, 0.010 *M* Tris, pH 7.0 containing 5 m*M* $MgCl_2$, 0.100 *M* KCl, and 10 μ*M* guanosine diphosphate. The pH of this solution was then adjusted to 6.0 with 1.0 *M* sodium cacodylate — 0.050 *M* $MgCl_2$.

p Per FMN (hence per dimer, therefore this would be four residues per tetramer).

q Determined from rate of loss of catalytic activity.

r No reaction at cysteine, tryptophan, or tyrosine is observed under these reaction conditions.

s Determined from rate of loss of thiamine binding activity.

t Thiomethylated at cysteine-165 (reaction with methyl methanethiosulfonate). Enzyme remains catalytically active, but with reduced affinity for pyruvate and lactate.

u The second-order rate constant for the native enzyme is 10,920 M^{-1} min^{-1}. These values were obtained from the measurement of the rate of loss of enzyme activity. Virtually identical values were obtained from direct measurement of histidine modification by spectroscopy.

v Diethylpyrocarbonate introduced as acetonitrile solution.

w Incorporation of radiolabeled diethylpyrocarbonate was closely related to extent of histidine modification as assessed by spectroscopy ($\Delta\varepsilon$ = 3900 M^{-1} cm^{-1} for monosubstituted derivative). Hydroxylamine treatment did not result in recovery of enzyme activity, although radiolabel was lost. The enzyme was inactivated by 10 m*M* hydroxylamine at neutral pH at 0°C. This is not an infrequent observation from our consideration of the literature in this area. Although very few investigators (we have not found any report) have examined the possibility of peptide bond cleavage with hydroxylamine at neutral pH, the possibility cannot be disregarded considering the cleavage of Asn-Gly bonds under more alkaline conditions.[17]

x Inactivation reversed by 0.100 to 0.5 *M* hydroxylamine. The pH of this reaction is not specified.

y Submitochondrial particles were used in this study. The inactivation produced by diethylpyrocarbonate is partially reversed by neutral (pH 7.0) hydroxylamine. The extent of activity recovery was dependent on hydroxylamine with maximum activity recovery at 0.020 *M* hydroxylamine decreasing significantly at 0.115 *M* hydroxylamine.

[z] Stoichiometry determined by spectral analysis ($\Delta\varepsilon = 3200\ M^{-1}\ cm^{-1}$ at 240 nm) (3.4 residues modified) is in reasonable agreement with amount of radiolabeled diethylpyrocarbonate incorporated (4.2).

[aa] Varied with species source of transferrin: human, 14:7; rabbit, 14:18; human lactotransferrin, 7:10; bovine lactotransferrin, 7:9; chicken ovotransferrin, 9:14.

[bb] There is no reaction with tyrosine under these conditions. Reaction at primary amine functions was not excluded. Only partial reactivation is obtained upon treatment with hydroxylamine (approximately 50% recovery with 1.0 *M* hydroxylamine; no reaction at 0.1 *M* hydroxylamine).

[cc] Differences were noted between enzymes isolated from aged animals (old) and young animals (young). The difference in the extent of modification (5.0 residues are present in the native enzyme) is ascribed to oxidation occurring during the aging process.

[dd] Partial reactivation with hydroxylamine, but no detectable modification at tyrosyl residues.

[ee] Fast (A) and slow (B) reaction and reaction at nonhistidine residues as being involved in the loss of activity. K_A, 0.46 min^{-1}; K_B,0.011 min^{-1}; and K_C,0.031 min^{-1}.

[ff] Alternatively, 20 m*M* MES (pH 5.0 to 5.5) was used with 50 m*M* sodium phosphate.

[gg] Inactivation of hydrolytic reaction, 1.17 $min^{-1}\ mM^{-1}$; inactivation of transacylation, 0.56 $min^{-1}\ mM^{-1}$. It was concluded that two different histidine residues are necessary for hydrolysis and only one histidine residue for transacylation.

References for Table 1

1. **Burstein, Y., Walsh, K. A., and Neurath, H.,** Evidence of an essential histidine residue in thermolysin, *Biochemistry,* 13, 205, 1974.
2. **Cousineau, J. and Meighen, E.,** Chemical modification of bacterial luciferase with ethoxyformic anhydride: evidence for an essential histidyl residue, *Biochemistry,* 15, 4992, 1976.
3. **Jeng, T.-W. and Fraenkel-Conrat, H.,** Chemical modification of histidine and lysine residues of crotoxin, *FEBS Lett.,* 87, 291, 1978.
4. **McTigue, J. J. and van Etten, R. L.,** An essential active-site histidine residue in human prostatic acid phosphatase. Ethoxyformylation by diethylpyrocarbonate and phosphorylation by a substrate, *Biochim. Biophys. Acta,* 523, 407, 1978.
5. **Horiike, K., Tsuge, H., and McCormick, D. B.,** Evidence for an essential histidyl residue at the active site of pyridoxamine (pyridoxine)-5′-phosphate oxidase from rabbit liver, *J. Biol. Chem.,* 254, 6638, 1979.
6. **George, A. L., Jr. and Borders, C. L., Jr.,** Chemical modification of histidyl and lysyl residues in yeast enolase, *Biochim. Biophys. Acta,* 569, 63, 1979.
7. **Demaine, M. M. and Benkovic, S. J.,** Selective modification of rabbit liver fructose bisphosphatase, *Arch. Biochem. Biophys.,* 205, 308, 1980.
8. **Saluja, A. K. and McFadden, B. A.,** Modification of histidine at the active site of spinach ribulose bisphosphate carboxylase, *Biochem. Biophys. Res. Commun.,* 94, 1091, 1980.
9. **Garrison, C. K. and Himes, R. H.,** The reaction between diethylpyrocarbonate and sulfhydryl groups in carboxylate buffers, *Biochem. Biophys. Res. Commun.,* 67, 1251, 1975.
10. **Jonák, J. and Rychik, I.,** Chemical evidence for the involvement of histidyl residues in the functioning of *Escherichia coli* elongation factor Tu, *FEBS Lett.,* 117, 167, 1980.
11. **Meyer, S. E. and Cromartie, T. H.,** Role of essential histidine residues in L-α-hydroxy acid oxidase from rat kidney, *Biochemistry,* 19, 1874, 1980.
12. **Nishimura, H., Sempuku, K., and Iwashima, A.,** Possible functional roles of carboxyl and histidine residues in a soluble thiamine-binding protein of *Saccharomyces cerevisiae, Biochim. Biophys. Acta,* 668, 333, 1981.
13. **Bond, M. D., Steinbrink, D. R., and Van Wart, H. E.,** Identification of essential amino acid residues in *Clostridium histolyticum* collagenase using chemical modification reactions, *Biochem. Biophys. Res. Commun.,* 102, 243, 1981.
14. **Bloxham, D. P.,** The chemical reactivity of the histidine-195 residue in lactate dehydrogenase thiomethylated at the cysteine-165 residue, *Biochem. J.,* 193, 93, 1981.
15. **Phelps, D. C. and Hatefi, Y.,** Inhibition of the mitochondrial nicotinamide nucleotide transhydrogenase by dicyclohexylcarbodiimide and diethylpyrocarbonate, *J. Biol. Chem.,* 256, 8217, 1981.
16. **Cromartie, T. H.,** Sulfhydryl and histidinyl residues in the flavoenzyme alcohol oxidase from *Candida boidinii, Biochemistry,* 20, 5416, 1981.
17. Deleted in proof.
18. **Phelps, D. C. and Hatefi, Y.,** Inhibition of D-β-hydroxybutyrate dehydrogenase by butanedione, phenylglyoxal and diethyl pyrocarbonate, *Biochemistry,* 20, 459, 1981.
19. **Burch, T. P. and Ticku, M. K.,** Histidine modification with diethylpyrocarbonate shows heterogeneity of benzodiazepine receptors, *Proc. Natl. Acad. Sci. U.S.A.,* 78, 3945, 1981.
20. **McKinley, M. P., Masiarz, F. R., and Prusiner, S. B.,** Reversible chemical modification of the scrapie agent, *Science,* 214, 1259, 1981.
21. **Vik, S. B. and Hatefi, Y.,** Possible occurrence and role of an essential histidyl residue in succinate dehydrogenase, *Proc. Natl. Acad. Sci. U.S.A.,* 78, 6749, 1981.
22. **Saluja, A. K. and McFadden, B. A.,** Modification of the active site histidine in ribulosebisphosphate carboxylase/oxygenase, *Biochemistry,* 21, 89, 1982.

References for Table 1 (continued)

23. **Mazurier, J., Leger, D., Tordera, V., Montreuil, J., and Spik, G.,** Comparative study of the iron-binding properties of transferrins. Differences in the involvement of histidine residues as revealed by carbethoxylation, *Eur. J. Biochem.,* 119, 537, 1981.
24. **Daron, H. H. and Aull, J. L.,** Inactivation of dihydrofolate reductase from *Lactobacillus casei* by diethylpyrocarbonate, *Biochemistry,* 21, 737, 1982.
25. **Abdulwajid, A. W. and Wu, F. Y.-H.,** Chemical modification of *Escherichia coli* RNA polymerase by diethyl pyrocarbonate: evidence of histidine requirement for enzyme activity and intrinsic zinc binding, *Biochemistry,* 25, 8167, 1986.
26. **Klotz, K.-N., Lohse, M. J., and Schwabe, U.,** Chemical modification of A_1 adenosine receptors in rat brain membranes. Evidence for histidine in different domains of the ligand binding site, *J. Biol. Chem.,* 263, 17522, 1988.
27. **Gacheru, S. N., Trackman, P. C., and Kagan, H. M.,** Evidence for a functional role for histidine in lysyl oxidase catalysis, *J. Biol. Chem.,* 263, 16704, 1988.
28. **Gordillo, E., Ayala, A., F.-Lobato, M., Bautista, J., and Machado, A.,** Possible involvement of histidine residues in the loss of enzymatic activity of rat liver malic enzyme during aging, *J. Biol. Chem.,* 263, 8053, 1988.
29. **Morjana, N. A. and Scarborough, G. A.,** Evidence for an essential histidine residue in the *Neurospora crassa* plasma membrane H^+-ATPase, *Biochim. Biophys. Acta,* 985, 19, 1989.
30. **Bagert, U. and Röhm, K.-H.,** On the role of histidine and tyrosine residues in *E. coli* asparaginase. Chemical modification and ^{1}H-nuclear magnetic resonance studies, *Biochim. Biophys. Acta,* 999, 36, 1989.
31. **Kim, D. J. and Byun, S. M.,** Evidence for involvement of 2 histidine residues in the reaction of ampicillin acylase, *Biochem. Biophys. Res. Commun.,* 166, 904, 1990.
32. **De Caro, J. D., Guidoni, A. A., Bonicel, J. J., and Rovery, M.,** The histidines reacting with ethoxyformic anhydride in porcine pancreatic lipase: their relationships with enzyme activity, *Biochimie,* 71, 1211, 1989.
33. **De Caro, J. D., Rouimi, P., and Rovery, M.,** Hydrolysis of *p*-nitrophenyl acetate by the peptide chain fragment (335–449) of porcine pancreatic lipase, *Eur. J. Biochem.,* 158, 601, 1986.
34. **Sejlitz, T. and Neujahnr, H. Y.,** Chemical modification of phenol hydroxylase by ethoxyformic anhydride, *Eur. J. Biochem.,* 170, 351, 1987.
35. **Kinnunen, P. M., DeMichele, A., and Lange, L. G.,** Chemical modification of acyl-CoA:cholesterol *O*-acyltransferase. 1. Identification of acyl-CoA:cholesterol *O*-acyltransferase subtypes by differential diethyl pyrocarbonate sensitivity, *Biochemistry,* 27, 7344, 1988.

In addition to the direct modifications of lysine, cysteine, and tyrosine by carboethoxylation, which are occasionally observed as side reactions in the use of diethylpyrocarbonate for the modification of histidine residues in proteins, an additional side reaction involving lysyl residues has been observed. Sams and Matthews[50] reported isopeptide bond formation between the ε-amino group of lysine and an adjacent carboxylic acid. This reaction has been previously observed with ribonuclease.[34] A reaction mechanism has not been described, but it likely involves the transient diethylpyrocarbonate modification of the carboxyl group.[51]

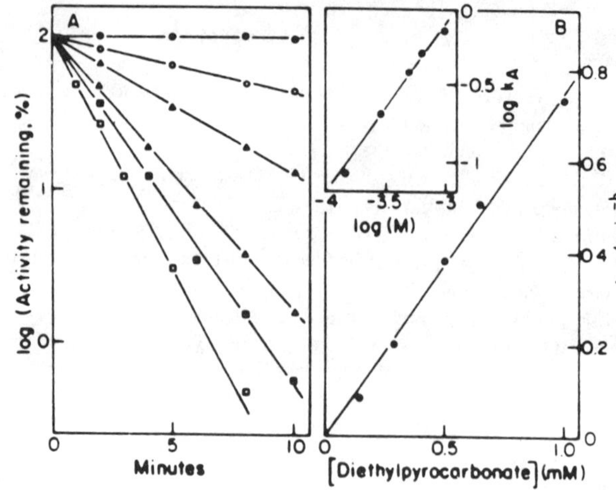

FIGURE 8. The inactivation of pyridoxamine-5′-phosphate oxidase by diethylpyrocarbonate. Panel A describes an experiment where the enzyme (2.32 μ*M*) was incubated in 0.1 *M* potassium phosphate, pH 7.0 containing 3.75 μ*M* riboflavin 5′-phosphate in the absence (●) or presence of diethylpyrocarbonate at the following concentrations: 0.144 m*M* (○), 0.287 m*M* (▲), 0.501 m*M* (△), 0.644 m*M* (■), or 1.00 m*M* (□). Panel B shows the dependence of the pseudo-first-order rate constants for inactivation on the concentration of diethylpyrocarbonate. Values for k_A were determined from the slopes of the semilogarithmic plots of panel A. The inset in panel B shows a plot of k_A vs. log concentration of diethylpyrocarbonate. (From Horiike, K., Tsuge, H., and McCormick, D. B., *J. Biol. Chem.*, 254, 6638, 1979. With permission.)

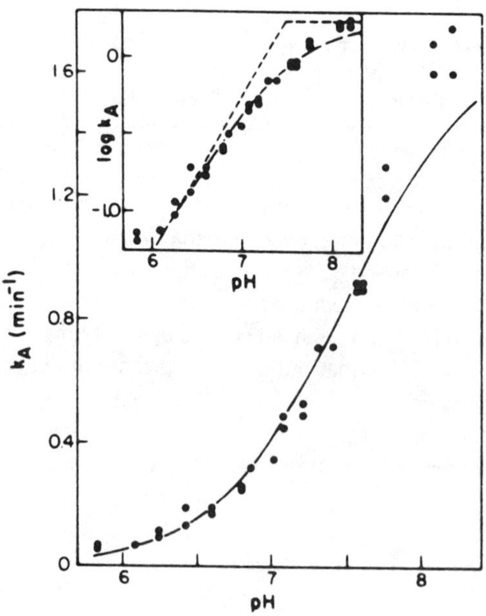

FIGURE 9. The dependence of the rate constant (k_A) for the inactivation of pyridoxamine-5′-phosphate oxidase by diethylpyrocarbonate. The pseudo-first-order rate constants (k_A) were calculated from the slopes of the semilogarithmic plots of activity vs. time. The solid curve is drawn assuming there was no reaction between the protonated residue (pKa = 7.5). The second-order rate constant for the reaction of diethylpyrocarbonate with the unprotonated enzyme was 51.6 M^{-1} s^{-1}. The inset shows a plot of k_A vs. pH. (From Horiike, K., Tsuge, H., and McCormick, D. B., *J. Biol. Chem.*, 254, 6638, 1979. With permission.)

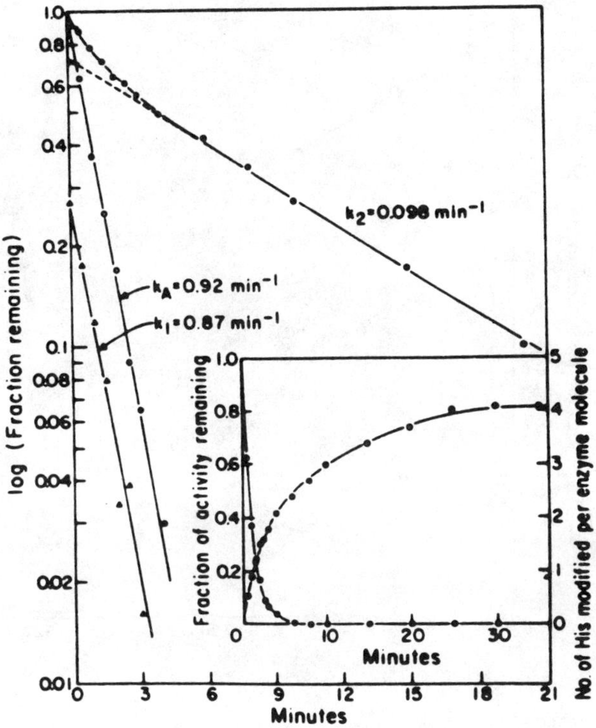

FIGURE 10. The relationship between the extent of histidine modification and the loss of catalytic activity of pyridoximine-5′-phosphate oxidase upon reaction with diethylpyrocarbonate. The holoenzyme (2.4 μ*M*) was incubated with 1.2 m*M* diethylpyrocarbonate in 0.1 *M* potassium phosphate, pH 7.0. At the indicated times, the number of *N*-carboethoxyhistidyl residues was determined by spectrophotometric analysis and the fraction of remaining histidine residues (●) calculated by taking the total number of such modifiable residues (4.1) as unity and enzyme activity (○) determined. The fast phase of the modification (▲) was obtained by subtracting the contribution of the slow phase (dashed line) and replotting the differences. The solid curve is calculated on the basis of the following equation

$$x = (n - m)/n = 0.27\,e^{-0.87t} + 0.73\,e^{-0.098t}$$

where x is the total fraction of residues remaining after reaction, n is the total number of modifiable residues, m is the number of residues actually modified, and t is time. The inset shows a plot of activity remaining (○) and the number of histidyl residues modified per enzyme molecule (●) as a function of reaction time. (From Horiike, K., Tsuge, H., and McCormick, D. B., *J. Biol. Chem.*, 254, 6638, 1979. With permission.)

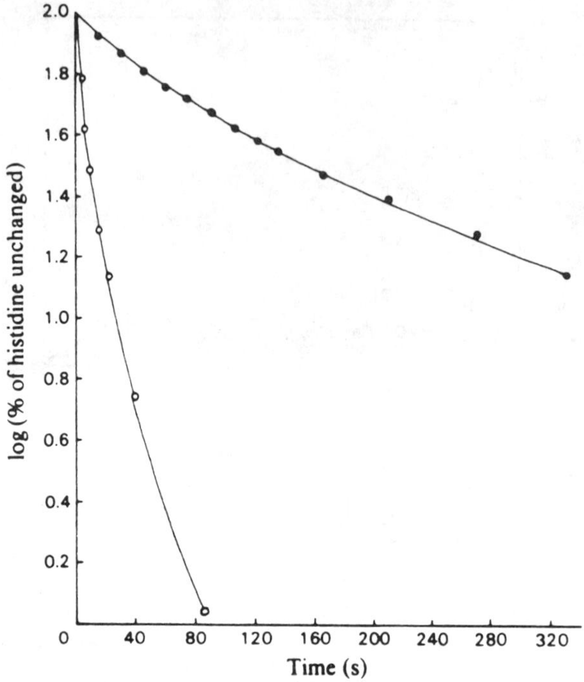

FIGURE 11. A comparison of the reaction rate of native (○) and thiomethylated (●) lactate dehydrogenase with diethylpyrocarbonate. The thiomethylated enzyme was prepared by reaction with methyl methanethiosulfonate. The proteins were in 0.1 *M* sodium phosphate, pH 7.2 at 10°C and allowed to react with 1 m*M* diethylpyrocarbonate. The extent of histidine modification was assessed by the increase in absorbance at 240 nm. ($\Delta\varepsilon = 3600\ M^{-1}\ cm^{-1}$.) (From Bloxham, D. P., *Biochem. J.*, 193, 93, 1981. With permission.)

REFERENCES

1. **Kettner, C. and Shaw, E.,** Inactivation of trypsin-like enzymes with peptides of arginine chloromethyl ketone, *Meth. Enzymol.*, 80, 826, 1981.
2. **Bock, P. E.,** Active-site-selective labeling of blood coagulation proteinases with fluorescence probes by the use of thioester peptide chloromethyl ketones. II. Properties of thrombin derivatives as reporters of prothrombin fragment 2 binding and specificity of the labeling approach for other proteinases, *J. Biol. Chem.*, 267, 14974, 1992.
3. **Bock, P. E.,** Active site selective labeling of serine proteases with spectroscopic probes using thioester peptide chloromethyl ketones: demonstration of thrombin labeling using N^{α}[T-[(acetylthio)acetyl]-D-Phe-Pro-Arg-CH_2Cl, *Biochemistry,* 27, 6633, 1988.
4. **Williams, E. B., Krishnaswamy, S., and Mann, K. G.,** Zymogen/enzyme discrimination using peptide chloromethyl ketones, *J. Biol. Chem.*, 264, 7536, 1989.
5. **Turk, T. and Macek, P.,** The role of lysine, histidine and carboxyl residues in biological activity of equinatoxin II, a pore forming polypeptide from the sea anemone *Actinia equina* L., *Biochim. Biophys. Acta,* 1119, 5, 1992.

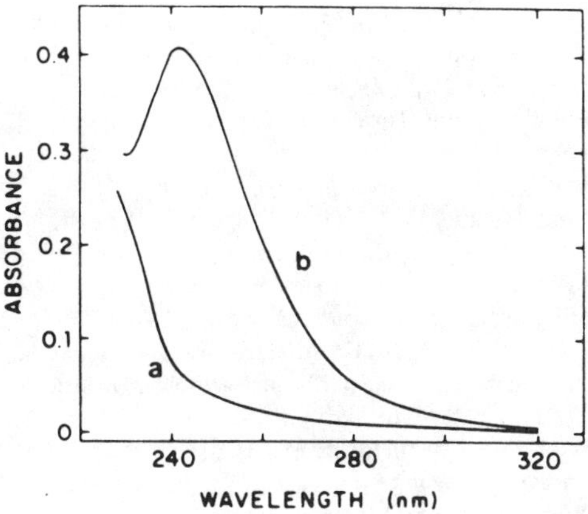

FIGURE 12. UV difference spectrum for the reaction of *Candida boidinii* flavoenzyme alcohol oxidase with diethylpyrocarbonate in 0.05 *M* sodium phosphate, pH 7.5. Curve a was obtained before the addition of diethylpyrocarbonate, while curve b was obtained after 30 min of reaction of protein in the sample cuvette with diethylpyrocarbonate. (From Cromartie, T. H., *Biochemistry,* 20, 5416, 1981. With permission.)

6. **Mattsson, P., Pohjalainen, T., and Korpela, T.,** Chemical modification of cyclomaltodextrin glucanotransferase from *Bacillus circulans* var. *alkalophilus, Biochim. Biophys. Acta,* 1122, 33, 1992.
7. **Park, I.-S. and Hausinger, R. P.,** Diethylpyrocarbonate reactivity of *Klebsiella aerogenes* urease: effect of pH and active site ligands on the rate of inactivation, *J. Prot. Chem.,* 12, 51, 1993.
8. **Bhattacharyya, D. K., Bandyopadhyay, U., and Bannerjee, R. K.,** Chemical and kinetic evidence for an essential histidine residue in the electron transfer from aromatic donor to horseradish peroxidase compound I, *J. Biol. Chem.,* 268, 22292, 1993.
9. **Basuki, W., Furuichi, K., Iizuka, M., Ito, K., and Minamiura, N.,** Differences between isozymes of glucose isomerase of *Streptomyces phaeochromogenes* by chemical modification, *Biosci. Biotech. Biochem.,* 57, 1341, 1993.
10. **Bond, J. S., Francis, S. H., and Park, J. H.,** An essential histidine in the catalytic activities of 3-phosphoglyceraldehyde dehydrogenase, *J. Biol. Chem.,* 245, 1041, 1970.
11. **Fahnestock, S. R.,** Evidence of the involvement of a 50S ribosomal protein in several active sites, *Biochemistry,* 14, 5321, 1975.
12. **Auron, P. E., Erdelsky, K. J., and Fahnestock, S. R.,** Chemical modification studies of a protein at the peptidyltransferase site of the *Bacillus stearothermophilus* ribosome. The 50S ribosomal subunit is a highly integrated functional unit, *J. Biol. Chem.,* 253, 6893, 1978.
13. **Dohme, F. and Fahnestock, S. R.,** Identification of proteins involved in the peptidyl transferase activity of ribosomes by chemical modification, *J. Mol. Biol.,* 129, 63, 1979.
14. **Cerna, J. and Rychlik, I.,** Photoinactivation of peptidyl transferase binding sites, *FEBS Lett.,* 102, 277, 1979.
15. **Nakamura, S. and Kaziro, Y.,** Selective photooxidation of histidine residues in polypeptide chain elongation factor Tu from *E. coli, J. Biochem. (Tokyo),* 90, 1117, 1981.

16. **Funakoshi, T., Abe, M., Sakata, M., Shoji, S., and Kubota, Y.,** The functional site of placental anticoagulant protein: essential histidine residue of placental anticoagulant protein, *Biochem. Biophys. Res. Commun.*, 168, 125, 1990.
17. **Kuno, S., Fukui, S., and Toraya, T.,** Essential histidine residues in coenzyme B_{12}-dependent diol dehydrase: dye-sensitized photooxidation and ethoxycarbonylation, *Arch. Biochem. Biophys.*, 277, 211, 1990.
18. **Inagami, T. and Hatano, H.,** Effect of alkylguanidines on the inactivation of trypsin by alkylation and phosphorylation, *J. Biol. Chem.*, 244, 1176, 1969.
19. **Stark, G. R., Stein, W. H., and Moore, S.,** Relationships between the conformation of ribonuclease and its reactivity toward iodoacetate, *J. Biol. Chem.*, 236, 436, 1961.
20. **Heinrikson, R. L., Stein, W. H., Crestfield, A. M., and Moore, S.,** The reactivities of the histidine residues at the active site of ribonuclease toward halo acids of different structures, *J. Biol. Chem.*, 240, 2921, 1965.
21. **Fruchter, R. G. and Crestfield, A. M.,** The specific alkylation by iodoacetamide of histidine 12 in the active site of ribonuclease, *J. Biol. Chem.*, 242, 5807, 1967.
22. **Lin, M. C., Stein, W. H., and Moore, S.,** Further studies on the alkylation of the histidine residues of pancreatic ribonuclease, *J. Biol. Chem.*, 243, 6167, 1968.
23. **Fohlman, J., Eaker, D., Dowdall, M. J., Lüllmann-Rauch, R., Sjödin, T., and Leander, S.,** Chemical modification of taipoxin and the consequences for phospholipase activity, pathophysiology, and inhibition of high-affinity choline uptake, *Eur. J. Biochem.*, 94, 531, 1979.
24. **Halpert, J., Eaker, D., and Karlsson, E.,** The role of phospholipase activity in the action of a presynaptic neurotoxin of *Notechis scutatus scutatus* (Australian Tiger Snake), *FEBS Lett.*, 61, 72, 1976.
25. **Yang, C. C. and King, K.,** Chemical modification of the histidine residue in basic phospholipase A_2 from the venom of *Naja nigricollis*, *Biochim. Biophys. Acta*, 614, 373, 1980.
26. **Verheij, H. M., Volwerk, J. J., Jansen, E. H. J. M., Puyk, W. C., Dijkstra, B. W., Drenth, J., and de Haas, G. H.,** Methylation of histidine-48 in pancreatic phospholipase A_2. Role of histidine and calcium ion in the catalytic mechanism, *Biochemistry*, 19, 743, 1980.
27. **Volwerk, J. J., Pieterson, W. A., and de Haas, G. H.,** Histidine at the active site of phospholipase A_2, *Biochemistry*, 13, 1446, 1974.
28. **Glick, B. R.,** The chemical modification of *Escherichia coli* ribosomes with methyl *p*-nitrobenzenesulfonate. Evidence for the involvement of a histidine residue in the functioning of the ribosomal peptidyl transferase, *Can. J. Biochem.*, 58, 1345, 1980.
29. **Nishino, T., Massey, V., and Williams, C. H., Jr.,** Chemical modifications of D-amino acid oxidase. Evidence for active site histidine, tyrosine, and arginine residues, *J. Biol. Chem.*, 255, 3610, 1980.
30. **Morishima, I., Shiro, Y., Adachi, S., Yano, Y., and Orii, Y.,** Effect of the distal histidine modification (cyanation) of myoglobin on the ligand binding kinetics and the heme environmental structures, *Biochemistry*, 28, 7582, 1989.
31. **Shiro, Y. and Morishima, I.,** Modification of the heme distal side chain in myoglobin by cyanogen bromide. Heme environmental structures and ligand binding properties of the modified myoglobin, *Biochemistry*, 23, 4879, 1984.
32. **Hefford, M. A. and Kaplan, H.,** Chemical properties of the histidine residue of secretin: evidence for a specific intramolecular interaction, *Biochim. Biophys. Acta*, 998, 267, 1989.
33. **Melchior, W. B., Jr. and Fahrney, D.,** Ethoxyformylation of proteins. Reaction of ethoxyformic anhydride with α-chymotrypsin, pepsin and pancreatic ribonuclease at pH 4, *Biochemistry*, 9, 251, 1970.
34. **Wolf, B., Lesnaw, J. A., and Reichmann, M. E.,** A mechanism of the irreversible inactivation of bovine pancreatic ribonuclease by diethylpyrocarbonate. A general reaction of diethylpyrocarbonate with proteins, *Eur. J. Biochem.*, 13, 519, 1970.

35. **Miles, E. W.,** Modification of histidyl residues in proteins by diethylpyrocarbonate, *Meth. Enzymol.*, 47, 431, 1977.
36. **Welsch, D. J. and Nelsestuen, G. L.,** Irreversible degradation of histidine-96 of prothrombin fragment 1 during protein acetylation: another unusually reactive site in the kringle, *Biochemistry*, 27, 7513, 1988.
37. **Altman, J., Lipka, J. J., Kuntz, I., and Waskell, L.,** Identification by proton nuclear magnetic resonance of the histidines in cytochrome b_5 modified by diethyl pyrocarbonate, *Biochemistry*, 28, 7516, 1989.
38. **Bycroft, M. and Fersht, A. R.,** Assignment of histidine resonances in the ^{1}H NMR (500 MHz) spectrum of subtilisin BPN′ using site-directed mutagenesis, *Biochemistry*, 27, 7390, 1988.
39. **Morjana, N. A. and Scarborough, G. A.,** Evidence for an essential histidine residue in the *Neurospora crassa* plasma membrane H^+-ATPase, *Biochim. Biophys. Acta*, 985, 19, 1989.
40. **Holbrook, J. J. and Ingram, V. A.,** Ionic properties of an essential histidine residue in pig heart lactate dehydrogenase, *Biochem. J.*, 131, 729, 1973.
41. **Cousineau, J. and Meighen, E.,** Chemical modification of bacterial luciferase with ethoxyformic anhydride: evidence for an essential histidyl residue, *Biochemistry*, 15, 4992, 1976.
42. **Roosemont, J. L.,** Reaction of histidine residues in proteins with diethylpyrocarbonate: differential molar absorptivities and reactivities, *Anal. Biochem.*, 88, 314, 1978.
43. **Vangrysperre, W., Ampe, C., Kersters-Hilderson, H., and Tempst, P.,** Single active-site histidine in D-xylose isomerase from *Streptomyces violaceoruber*. Identification by chemical derivatization and peptide mapping, *Biochem. J.*, 263, 195, 1989.
44. **Horiike, K., Tsuge, H., and McCormick, D. B.,** Evidence for an essential histidyl residue at the active site of pyridoxamine (pyridoxine)-5′-phosphate oxidase from rabbit liver, *J. Biol. Chem.*, 254, 6638, 1979.
45. **Saluja, A. K. and McFadden, B. A.,** Modification of histidine at the active site of spinach ribulose bisphosphate carboxylase, *Biochem. Biophys. Res. Commun.*, 94, 1091, 1980.
46. **Bloxham, D. P.,** The chemical reactivity of the histidine-195 residue in lactate dehydrogenase thiomethylated at the cysteine-165 residue, *Biochem. J.*, 193, 93, 1981.
47. **Cromartie, T. H.,** Sulfhydryl and histidinyl residues in the flavoenzyme alcohol oxidase from *Candida boidinii*, *Biochemistry*, 20, 5416, 1981.
48. **Bond, M. D., Steinbrink, D. R., and Van Wart, H. E.,** Identification of essential amino acid residues in *Clostridium histolyticum* collagenase using chemical modification reactions, *Biochem. Biophys. Res. Commun.*, 102, 243, 1981.
49. **Daron, H. H. and Aull, J. L.,** Inactivation of dihydrofolate reductase from *Lactobacillus casei* by diethyl pyrocarbonate, *Biochemistry*, 21, 737, 1982.
50. **Sams, C. F. and Matthews, K. S.,** Diethyl pyrocarbonate reaction with the lactose repressor protein affects both inducer and DNA binding, *Biochemistry*, 27, 2277, 1988.
51. **Wold, F.,** Bifunctional reagents, *Meth. Enzymol.*, 25, 423, 1972.

Chapter 10

THE MODIFICATION OF LYSINE

Although there are a number of approaches to the group-specific modification of lysine in proteins, it is, however, somewhat more difficult to obtain site-specific modification. Lysine residues must be unprotonated to function as satisfactory nucleophiles, so alkaline pH conditions are required. In general, modification reactions are performed at pH values of 8.0 or greater, although it has been possible to obtain site-specific modification at a lower pH. A listing of reagents commonly used to modified lysine residues in proteins is presented in Table 1.

Lysine residues can be modified by reaction with α-ketoalkyl halides such as iodoacetic acid.[1] Acylation can occur at pH >7.0, but the rate of reaction is much slower than reaction with cysteinyl residues. Both the mono- and disubstituted derivatives have been reported. The monosubstituted derivative migrates close to methionine on amino acid analysis, while the disubstituted derivative migrates near aspartic acid. It should be noted that reaction with α-ketoalkyl halides is not considered particularly useful for the modification of primary amino groups. This reaction can be a possible side reaction occurring during the reduction and carboxymethylation of proteins. The reactivity of a given lysyl residue is affected by the nature of surrounding amino residues.

Both fluoronitrobenzene and fluorodinitrobenzene have been of considerable value in protein chemistry since Sanger and Tuppy's work on the structure of insulin.[2] Carty and Hirs[3] developed the use of 4-sulfonyl-2-nitrofluorobenzene for the modification of amino groups in pancreatic ribonuclease. This reagent also is more stable than, for example, fluorodinitrobenzene under alkaline conditions, permitting more accurate measurement at pH >9.0. The lysine residue at position 41 is the site of major substitution which is a reflection of the lower pKa for the ε-amino group of this residue. Use of this compound did not present the solubility and reactivity problems posed by the fluoronitrobenzene compounds. It was possible to qualitatively determine the classes of amino groups in ribonuclease; these were the α-amino group, nine "normal" amino groups, and lysine 41. The reactivity of lysine 41 was influenced by neighboring functional groups. This effect was lost at pH >11 or on thermal denaturation of the protein. The reaction of 1-dimethylaminonaphthalene-5-sulfonyl chloride (dansyl chloride) has been useful both in the structural analysis and amino group modification with proteins. In one study,[4] dansyl chloride (in acetone) is added to a solution of trypsin in 0.1 *M* phosphate, pH 8.0. The reaction is terminated after 24 h at 25°C by acidification to pH 3.0 with 1.0 *M* HCl. Insoluble material is removed by centrifugation, and the supernatant fraction is placed in dialysis. These investigators reported modification of the amino-terminal isoleucine and one lysine residue. The extent of modification was determined by absorbance at 336 nm ($\varepsilon_m = 3.4 \times 10^4\ M^{-1}\ cm^{-1}$).

Table 1
SOME REAGENTS USED TO OBTAIN SITE-SPECIFIC MODIFICATION OF LYSYL RESIDUES IN PROTEINS

Reagent	Other AA modified	Ref.
Acetic anhydride	α-Amino groups, tyrosine	1, 2
Methyl acetyl phosphate	α-Amino groups	3–7
Pyridoxal-5′-phosphate		8–10
Reductive alkylation		11–13
2,4,6-Trinitrobenzenesulfonic acid	α-Amino groups	14–18

References for Table 1

1. **Shen, S. and Strobel, H. W.,** Role of lysine and arginine residues of cytochrome P450 in the interaction between cytochrome P4502B1 and NADPH-cytochrome P450 reductase, *Arch. Biochem. Biophys.,* 304, 257, 1993.
2. **Suckau, D., Mak, M., and Przybylski, M.,** Protein surface topology-probing by selective chemical modification and mass spectrometric peptide mapping, *Proc. Natl. Acad. Sci. U.S.A.,* 89, 5630, 1992.
3. **Kluger, R. and Tsui, W.-C.,** Methyl acetyl phosphate. A small anionic acetylating agent, *J. Org. Chem.,* 45, 2723, 1980.
4. **Ueno, H., Pospischil, M. A., Manning, J. M., and Kluger, R.,** Site-specific modification of hemoglobin by methyl acetyl phosphate, *Arch. Biochem. Biophys.,* 244, 795, 1986.
5. **Ueno, H., Pospischil, M. A., and Manning, J. M.,** Methyl acetyl phosphate as a covalent probe for anion-binding sites in human and bovine hemoglobins, *J. Biol. Chem.,* 264, 12344, 1989.
6. **Ueno, H. and Manning, J. M.,** The functional, oxygen-linked chloride binding sites of hemoglobin are contiguous within a channel in the central cavity, *J. Prot. Chem.,* 11, 177, 1992.
7. **Ueno, H., Popowicz, A. M., and Manning, J. M.,** Random chemical modification of the oxygen-linked chloride-binding sites of hemoglobin: those in the central dyan axis may influence the transition between deoxy- and oxy-hemoglobin, *J. Prot. Chem.,* 12, 561, 1993.
8. **Lo Bello, M., Petruzzelli, R., Reale, L., Ricci, G., Barra, D., and Federici, G.,** Chemical modification of human placental glutathione transferase by pyridoxal-5′-phosphate, *Biochim. Biophys. Acta,* 1121, 167, 1992.
9. **Paine, L. J., Perry, N., Popplewell, A. G., Gore, M. G., and Atkinson, T.,** The identification of a lysine residue reactive to pyridoxal-5-phosphate in the glycerol dehydrogenase from the thermophile *Bacillus stearothermophilus, Biochim. Biophys. Acta,* 1202, 235, 1993.
10. **Basu, S., Basu, A., and Modak, M. J.,** Pyridoxal 5′-phosphate mediated inactivation of *Escherichia coli* DNA polymerase I: identification of lysine-635 as an essential residue for the processive mode of DNA synthesis, *Biochemistry,* 27, 6710, 1988.
11. **Brown, E. M., Pfeffer, P. E., Kumosinski, T. F., and Greenberg, R.,** Accessibility and mobility of lysine residues in β-lactoglobulin, *Biochemistry,* 27, 5601, 1988.
12. **Fujita, Y. and Noda, Y.,** Effect of alkylation with different sized substituents on thermal stability of lysozyme, *Int. J. Pept. Protein Res.,* 40, 103, 1992.
13. **Rypniewski, W. R., Holden, H. M., and Rayment, I.,** Structural consequences of reductive methylation of lysine residues in hen egg white lysozyme: an X-ray analysis at 1.8 Å resolution, *Biochemistry,* 32, 9851, 1993.

References for Table 1 (continued)

14. **Coffee, C. J., Bradshaw, R. A., Goldin, B. R., and Frieden, C.,** Identification of the sites of modification of bovine liver glutamate dehydrogenase reacted with trinitrobenzenesulfonate, *Biochemistry,* 10, 3516, 1971.
15. **Whitson, P. A., Burgumn, A. A., and Matthews, K. S.,** Trinitrobenzenesulfonate modification of the lysine residues in lactose repressor protein, *Biochemistry,* 23, 6046, 1984.
16. **Sams, C. F. and Matthews, K. S.,** Diethyl pyrocarbonate reaction with the lactose repressor protein affects both inducer and DNA binding, *Biochemistry,* 27, 2277, 1988.
17. **Haniu, M., Yuan, H., Chen, S., Iyanagi, T., Lee, T. D., and Shively, J. E.,** Structure-function relationship of NAD(P)H:quinone reductase: characterization of NH_2-terminal blocking group and essential tyrosine and lysine residues, *Biochemistry,* 27, 6877, 1988.
18. **Xia, C., Meyer, D. J., Chen, H., Reinemer, P., Huber, P., and Ketterer, B.,** Chemical modification of GSH transferase P1-1 confirms the presence of Arg-13, Lys-44 and one carboxylate group in the GSH-binding domain of the active site, *Biochem. J.,* 293, 357, 1993.

Reaction of dansyl chloride with phosphoenolpyruvate carboxylase has been used to introduce a fluorescent probe into this protein.[5] A somewhat specific modification of one of the eight lysine residues was achieved. The extent of modification was determined by spectral analysis at 355 nm using an extinction coefficient of 3400 M^{-1} cm^{-1}.

The reaction of 2-carboxy-4,6-dinitrochlorobenzene with proteins has been explored.[6,7] This reagent reacts with amino, sulfhydryl, and amino groups. This reagent has recently been used for the modification of specific lysine residues in cytochrome c.[8,9] The modification reaction (approximately sixfold molar excess of reagent) was performed in 0.2 *M* sodium bicarbonate, pH 9.0 at ambient temperature for 24 h. The extent of modification was determined as described by Brautigan et al.[6] The absorbance maximum of derivatives formed with various alkylamines was 436 nm with an extinction coefficient of 6.9×10^3 M^{-1} cm^{-1}. Chromatographic fractionation of the modified protein (sulfoethyl-cellulose) yielded six fractions with major lysine group modification. Hiratsuka and Uchida[10] examined the reaction of *N*-methyl-2-anilino-6-naphthalenesulfonyl chloride with lysyl residues in cardiac myosin. There was a difference in the nature of the reaction in the presence and absence of a divalent cation. *N*-Methyl-2-anilino-6-naphthalenesulfonyl chloride has been suggested for use as a fluorescent probe for hydrophobic regions of protein molecules.[11–13] The extent of incorporation of the *N*-methyl-2-anilino-6-naphthalenesulfonyl moiety into protein can be determined by spectral analysis at 327 nm ($\Delta\varepsilon = 2.0 \times 10^4$ M^{-1} cm^{-1}).[11,12] Modification of protein amino groups with isothiocyanate derivatives of various dyes has proved to be an effective means of introducing structural probes into proteins at specific sites.[13] Eosin isothiocyanate has been used to modify the lysyl residues in phosphoenol-pyruvate carboxylase.[5] The reagent was dissolved in dimethylsulfoxide/50 m*M* HEPES, pH 8.0 (50:50) immediately prior to use and added to the protein (in 50 m*M* HEPES, pH 8.0). The modified derivatives were used to determine the spatial proximity of the modified lysine residues using resonance energy transfer. Fluorescein isothiocyanate has been used to modify cytochrome P-450 (reaction performed in 30 m*M* Tris, pH 8.0 containing 0.1% Tween 80; 2 h at 0°C in the dark),[14] actin (2 m*M* borate, pH 8.5; 3 h at ambient temperature then at 4°C for 16 h),[15] and ricin (pH 8.1, 6°C for 4 h).[16] The extent of modification with fluorescein isothiocyanate can be determined by spectroscopy using an extinction coefficient of 80,000 M^{-1} cm^{-1} at 495 nm (1% SDS with 0.1 *M* NaOH)[14] or 74,500 M^{-1} cm^{-1} (0.1 *M* Tris, pH 8.0).[15] Antibodies labeled with fluorescein have been used as targeted phototoxic agents.[17] In this approach, the fluorescein moiety is iodinated, resulting in a photodynamic sensitizer. The reader is directed to an elegant study on the effect of microenvironment on the fluorescence of arylaminophthalenesulfonates.[18]

Welches and Baldwin[19] have examined the reaction of bacterial luciferase with 2,4-dinitrofluorobenzene. Modification was associated with inactivation at the rate of 157 M^{-1} min^{-1}, pH 7.0 (0.05 *M* phosphate). Both lysyl and cysteinyl residues can be modified under the experimental conditions (0.05 *M* phosphate, pH 7.0 at 25°C) used in these studies. In order to assess the significance of reaction at primary amino groups, the cysteinyl residues were "blocked" with methyl methanethiosulfonate. Reaction of luciferase with methyl methanethiosulfonate resulted in greater than 95% loss of catalytic activity (twofold molar excess of methyl methanethiosulfonate in 0.02 *M* phosphate, pH 7.0 at 25°C). The loss of activity can be completely reversed with β-mercaptoethanol (97 m*M*). The small amount of residual activity present after treatment with methyl methanethiosulfonate is further reduced on treatment with 2,4-dinitrofluorobenzene, and the recovery of activity subsequent to β-mercaptoethanol is greatly reduced. Quantitative analysis was not performed, but qualitative analysis suggested that the modification occurred at the α-amino group of methionine or the α- and/or β-subunits. A combination of site-specific mutagenesis and site-specific chemical modification with 2,4-dinitroflurobenzene was used to study lysine residues in angiogenin.[20]

Acylation of amino groups in proteins by reaction with carboxylic acid anhydrides has been extensively used.[21] Acetylation is generally carried out with acetic anhydride at alkaline pH in either a pH-stat or in saturated sodium acetate. Performing the modification reaction under these latter conditions (saturated sodium acetate) results in increased specificity, since *O*-acetyl tyrosine is unstable in sodium acetate. Acetylation has been used to study calcitonin[22] and a bacterial cytochrome.[23] Acetic and maleic anhydride have been used to study elastase.[24] In these studies, the reaction was carried out in a pH-stat to maintain alkaline pH. Reaction occurred at both lysyl and tyrosyl residues. It is relatively easy to differentiate between the two sites of modification, since *O*-acyl tyrosyl residues are unstable at pH ≥9.0. Studies with maleic anhydride showed that the amino terminal valine was not available for modification at pH 8.0 to 9.0, but could be modified at pH 11.0. Modification of this residue could be achieved in the presence of urea at a lower pH.

Competitive labeling (trace labeling) is a technique for determining the ionization state or constant and intrinsic reactivity of individual amino groups in a protein.[25] The method is based on the hypothesis that the individual amino groups will compete for a trace amount of radiolabeled reagent (the reagent is selected on the basis of nonselective reactivity with amino groups; with most studies, acetic anhydride has been the reagent of choice). The extent of radiolabel incorporation into the protein at a given site will then be a function of the pKa, microenvironment, and inherent nucleophilicity of that particular amino group.[25] After the reaction with the radiolabeled reagent is complete, the protein is denatured, and complete modification at each amino group is achieved by the addition of an excess of unlabeled reagent. A reproducible digestion method (i.e., tryptic or chymotryptic hydrolysis) is used to obtain peptides from the completely modified protein. The peptides are separated by a chromatographic technique, and the extent of radiolabel at each site is determined. The extent of radiolabel incorporation at a given site is a function of the reactivity of that individual amino group under the reaction conditions used at the radiolabel step. An alternative approach[26,27] involves a "trace" labeling step with tritiated acetic anhydride followed by complete modification with unlabeled acetic anhydride under denaturing conditions. This modified protein is then mixed with a preparation of the same protein which has been uniformly labeled with the ^{14}C-labeled acetic anhydride. Digestion and separation of peptide is performed by conventional techniques (see above), and the extent of radiolabeling is determined. The ratio of ^{3}H to ^{14}C in peptides containing amino groups is an indication of functional group reactivity. This method is somewhat more sensitive than the original method. Reductive methylation has also been used.[28]

Although this is a laborious technique, the data obtained are excellent and provide considerable insight into the solution structure of proteins. There has been a consistent use of this technique for the study of troponin-T,[29] troponin-C,[30] troponin-I,[31] calmodulin,[32–34] and tropomyosin.[35] In particular, studies[32–35] which have used this technique to assess conformational change in solution have been particularly rewarding.

Trifluoroacetylated derivatives have been of interest in the study of protein structure. In these studies, ethylthiotrifluoroacetate was used to modify cytochrome c in 0.14 *M* sodium phosphate, pH 8.0.[36,37] The pH was maintained at 8.0 using a pH-stat. Singly substituted derivatives of cytochrome c can be separated by chromatography on anion-exchange resin (Bio Rex 70) and carboxymethylcellulose. It is critical to avoid lyophilization during the preparation of the various derivatives. These derivatives have been subjected to further investigation,[38,39] including the use of ^{19}F-containing derivatives for nuclear magnetic resonance (NMR) probes.[40]

Succinic anhydride has also proved useful in the modification of lysine.[41] Modification of lysine residues with succinic anhydride results in charge reversal. Reaction with succinic anhydride frequently results in the dissociation of multimeric proteins and has also been used to "solubilize" insoluble proteins. Meighen and co-workers[42] have produced a "variant" form of bacterial luciferase through reaction with succinic anhydride. The succinylated protein retained the dimeric subunit structure of the native enzyme. By complementation experiments involving the mixing/hybridization of the modified and native enzyme, it was determined that succinylation of bacterial luciferase resulted in the inactivation of the α-subunit without markedly affecting the function of the β-subunit. Shetty and Rao[43] studied the reaction of succinic anhydride with arachin. In this study, reaction of the protein was performed in 0.1 *M* sodium phosphate, pH 7.8 with the pH maintained over the course of the reaction by the addition of 2.0 *M* NaOH. The extent of modification was determined by reaction of the unmodified primary amino groups on the protein with trinitrobenzenesulfonic acid (see below). With a 200:1 molar excess of succinic anhydride, 82% of the available amino groups were succinylated with concomitant dissociation of the subunits of this protein. The reaction of chymotrypsinogen with succinic anhydride has been studied.[44] In these experiments, the reaction was performed under ambient conditions in 0.05 *M* sodium phosphate, pH 7.5. During the course of the reaction, the pH was maintained at 7.5 by the addition of 1.0 *M* NaOH. Chymotrypsinogen (1 g) was dissolved in the sodium phosphate buffer, and 50 mg of succinic anhydride was added over a 30-min period. Under these conditions, 8 of the 14 lysine residues were modified. A related reaction involves the trimesylation of amino groups in proteins.[45] This reaction involves the modification of the protein with di(trimethysilyethyl)trimesic acid. Removal of the blocking groups results in an extremely polar derivative. The procedure is suggested to have value in the solubilization of membrane proteins.

Citraconic anhydride has proved useful since the modification of lysine residues with this reagent is a reversible reaction (Figure 1). Reaction conditions for the modification of lysine residues in proteins are similar to those described above for other carboxylic acid anhydrides. Atassi and Habeeb[46] have discussed the use of this reagent in some detail. As an example, the reaction of egg white lysozyme with citraconic anhydride has been studied.[47] With multiple additions of reagent, all primary amino groups were modified at pH 8.2 (the pH of the reaction mixture was maintained with a pH-stat). The product of the reaction was heterogeneous as judged by polyacrylamide gel electrophoresis. All citraconyl groups could be removed by treatment with 1.0 *M* hydroxylamine, pH 10.0. This treatment also resulted in an electrophoretically homogeneous species. Complete removal of the citraconyl groups could also be achieved by incubation at pH 4.2 for 3 h at 40°C.

Reaction with citraconic anhydride has been used to dissociate nucleoprotein complexes.[48] Modification of the lysine residues with citraconic anhydride (pH 8.0 to 9.0 maintained with pH-stat) resulted in a marked change in the charge relationship between the ε-amino groups of lysine and the phosphate backbone of the nucleic acid, allowing subsequent separation of protein from

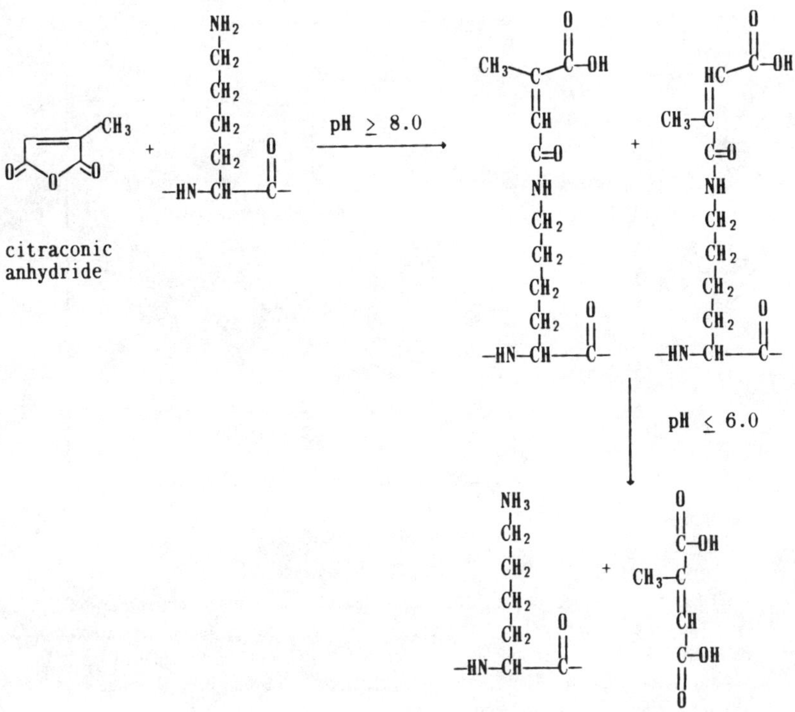

FIGURE 1. A scheme for the reversible reaction of lysine residues with citraconic anhydride.

nucleic acid. The citraconyl groups were subsequently removed from this protein by incubation at pH 3.0 to 4.0 at 30°C for 3 h (Figure 2).

Mahley and co-workers have prepared the acetoacetyl derivatives of lipoproteins by reaction with diketene in 0.3 borate, pH 8.5.[49,50] The modification of tyrosyl and seryl residues also can occur under these conditions, but the *O*-acetoacetyl groups can be removed by dialysis against a mild alkaline buffer such as bicarbonate. The modification at lysyl residues can be reversed by 0.5 *M* hydroxylamine, pH 7.0 at 37°C. A 0.06 *M* solution of diketene was prepared by taking 50 µl diketene into 10 ml of 0.1 *M* sodium borate, pH 8.5. The extent of modification was determined by subsequent titration with fluorodinitrobenzene. The effect of the modification of lysine residues on the *in vivo* clearance of lipoproteins in rats has been investigated.[50]

Urabe and co-workers[51] prepared various mixed carboxylic acid anhydrides of tetradecanoic acid and oxa derivatives which varied in their "hydrophobicity". This represented an attempt to change the surface properties of the enzyme molecule, in this case, thermolysin. The carboxylic acid anhydrides were formed *in situ* from the corresponding acid and ethylchloroformate in dioxane with triethylamine. The modification reaction was performed in 0.013 *M* barbital, 0.013 *M* $CaCl_2$, pH 8.5 containing 39% (v/v) dioxane and was terminated

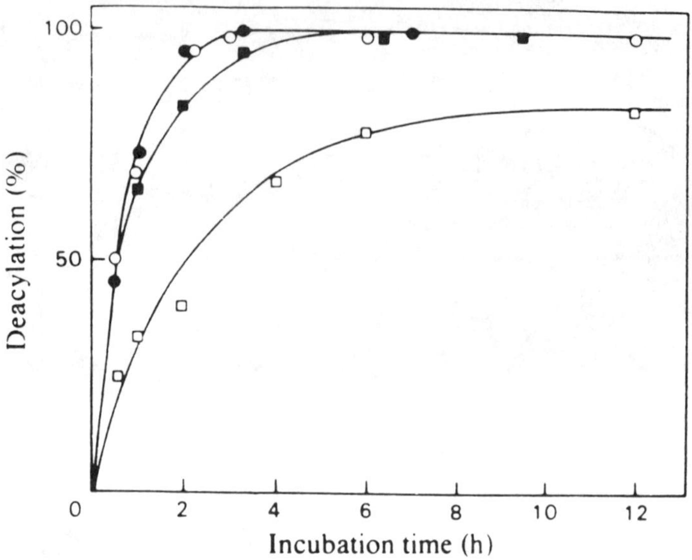

FIGURE 2. The effect of pH on the rate of deacylation of the ε-amino groups of lysine in the citraconylated protein prepared from the nucleoprotein complex prepared from yeast. The reactions were performed at 30°C at pH 3.0 (●), pH 4.0 (○), pH 5.0 (■), and pH 6.0 (□). (From Shetty, J. K. and Kinsella, J. E., *Biochem. J.*, 191, 269, 1980. With permission.)

with neutral hydroxylamine which also served to remove *O*-acyl derivatives. The extent of reaction was determined by titration with trinitrobenzenesulfonic acid. Derivatives obtained with tetradecanoic acid and 4-oxatetradecanoic acid were insoluble. Derivatives obtained with 4,7,10-trioxatetradecanoic acid and 4,7,10,13-tetraoxotetradecanoic acid both had approximately seven amino groups modified per mole of enzyme, showed little if any loss in either proteinase or esterase activity, and possessed enhanced thermal stability. Howlett and Wardrop[52] were able to dissociate the components of human erythrocyte membrane by the use of 3,4,5,6-tetrahydrophthalic anhydride. The reaction was performed in 0.02 *M* Tricine, pH 8.5. The 3,4,5,6-tetrahydrophthalic anhydride was introduced into the reaction mixture as a dioxane solution (a maximum of 0.10 ml/5 ml reaction mixture). The pH was maintained at pH 8.0 to 9.0 with 1.0 *M* NaOH. The reaction was considered complete when no further change in pH was observed. The extent of modification was determined by titration with trinitrobenzenesulfonic acid. The reaction could be reversed by incubation for 24 to 48 h at ambient temperature following the addition of an equal volume of 0.1 *M* potassium phosphate, pH 5.4 (the final pH of the reaction mixture was 6.0).

The reaction of primary amino groups in proteins with cyanate (Figure 3) has been a useful procedure for several decades. Stark and co-workers[53] pursued the observation that ribonuclease was inactivated by urea in a time-

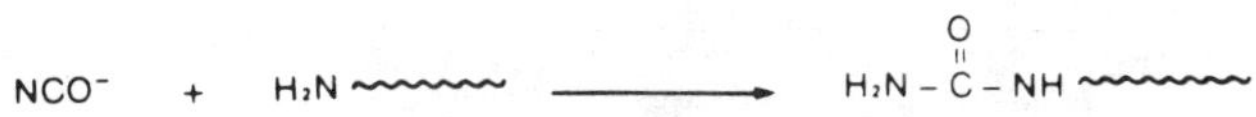

FIGURE 3. The reaction of primary amino groups with cyanate.

FIGURE 4. The formation of cyanate from urea.

dependent reaction. It was established that this inactivation was a reflection of the content of cyanate in the urea preparation (Figure 4). This observation was subsequently developed into a method for the quantitative determination of amino-terminal residues in peptides and proteins.[54] The reaction of cyanate with amino acid residues has been reported by Stark.[55] The ε-amino group of lysine is the least reactive ($k = 2.0 \times 10^{-3}\ M^{-1}\ min^{-1}$) as compared to the α-amino group of glycylglycine ($k = 1.4 \times 10^{-1}\ M^{-1}\ min^{-1}$). The carbamyl derivative of histidine is quite unstable as is the corresponding derivative of cysteine. Concern should be given to reaction at residues other than amines. For example, the reaction of chymotrypsin with cyanate results in loss of catalytic activity associated with the carbamylation of the active-site serine residue.[56]

Manning and co-workers[57–60] established that the modification of sickle cell hemoglobin with cyanate increased the oxygen affinity of this protein. It has been established that the amino-terminal value of hemoglobin is more reactive to cyanate in deoxygenated blood than in partially deoxygenated blood. At pH 7.4, the amino-terminal valyl residues of oxyhemoglobin S are carbamylated 50 to 100 times faster than lysyl residues.[50] The same laboratory has examined the carbamylation of α- and β-chains in some detail.[59] With the deoxy protein, the ratio of radiolabel from ^{14}C-cyanate on the α-chain as compared to the β-chain is 1.7:1.0, while it is 1:1 with the oxy protein. The carbamylation of the amino-terminal valine residues of hemoglobin is approximately 2.5-fold greater in partially deoxygenated media as compared to fully oxygenated media. Thus, it would appear that the reactivity of the amino-terminal valine is a sensitive index of conformational change.[60] It is also of interest that removal of Arg^{141} (α) with carboxypeptidase B abolishes the enhancement of carbamylation observed with the removal of oxygen from hemoglobin.

Weisgraber and co-workers[49] used carbamylation to explore the role of lysyl residues in the binding of plasma lipoprotein to fibroblasts. The reaction was performed in 0.3 *M* sodium borate, pH 8.0. The extent of modification was determined in two ways. In the first, the modified protein was subjected to acid hydrolysis. The amount of homocitrulline, the product of the reaction of the ε-amino group of lysine with cyanate, was considered equivalent to the number of lysine residues modified. However, homocitrulline is partially degraded on acid hydrolysis to produce lysine (17 to 30%). In order to obviate this diffi-

Derivative	Substituent	$CaCl_2$ used in preparation	No. of amino groups modified	Activity
		mM		%
Guanidino	$H_2N—C(=NH_2^{\oplus})—$	0	8.8 ε, 0.1 α	40
		5	8.9 ε, 0.2 α	75
Picolinimidyl	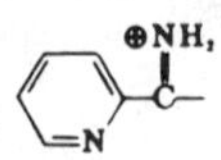	0	9.5 (ε + α)	100
		1	9 ε, 0.7 α	100
α-Picolinimidyl, ε-guanidino		5	9.0 ε, 0.9 α	65
Carbamyl	$H_2N—C(=O)—$	0	7 ε, 1 α	55[a]
		0 or 10	7 ε, 1 α	90
		0 or 10	9 ε, 1 α	0
Trinitrophenyl	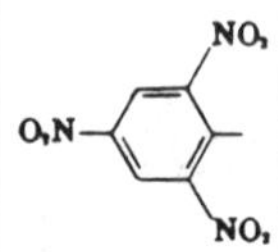	0	1	95[a]
		0	7–8	0[a]
		5	4–5	100
		5	7	0

[a] Assayed in the absence of Ca^{++}

FIGURE 5. The modification of bovine pancreatic DNase I by various reagents specific for the modification of lysine residues. The extent of lysine modification was determined by homocitrulline formation for reaction with *O*-methylisourea (guanidation), radiolabeled sodium cyanate for carbamylation and spectroscopy for picolinimydilation or trinitrophenylation. Enzymatic activity is expressed as a percent of that of a control preparation of DNase. (From Plapp, B. V., Moore, S., and Stein, W. H., *J. Biol. Chem.*, 246, 939, 1971. With permission.)

culty, these investigators removed a portion of the modified protein and reacted it under denaturing conditions with 2,4-dinitrifluorobenzene, yielding an acid-stable derivative. The number of lysine residues modified was therefore the sum of free lysine and homocitrulline obtained on amino acid analysis following acid hydrolysis.

In an elegant study by Plapp and co-workers,[61] the modification of lysyl residues in bovine pancreatic deoxyribonuclease A by several different reagents, including cyanate, was examined as shown in Figure 5. The modification with cyanate is performed at 37°C in 1.0 *M* triethanolamine hydrochloride, pH 8.0. The extent of modification was determined by analysis for homocitrulline following acid hydrolysis. A time course of hydrolysis was utilized to provide for the accurate determination of homocitrulline, since this amino acid slowly decomposes to form lysine during acid hydrolysis (see above).

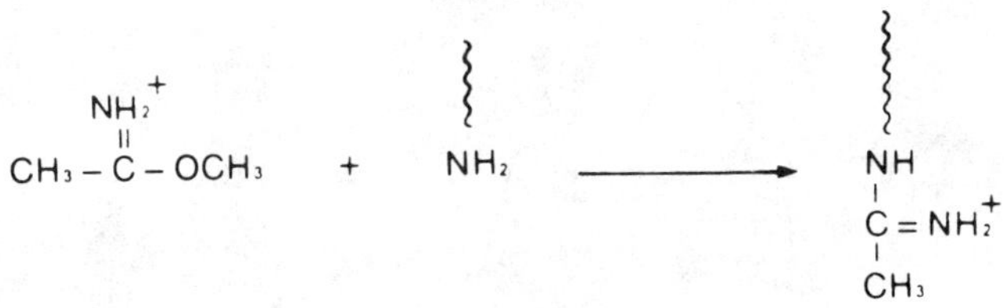

FIGURE 6. The reaction of lysine with methyl acetimidate.

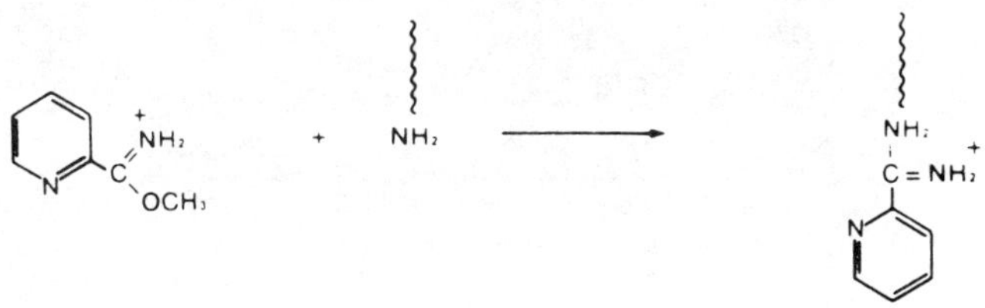

FIGURE 7. The reaction of amino groups with methyl picolinimidate.

The reaction of imidoesters with the primary amino groups of proteins has been the subject of considerable investigation in the past 10 to 20 years. The most extensive use of this class of reagents has been the covalent cross-linking of proteins. These reagents have the particular advantage that the charge of the lysine residue is maintained during the modification as shown for the reaction of lysine with methyl acetimidate in Figure 6. The reaction was performed with a 1000-fold molar excess of reagent in 0.02 *M* sodium borate, 0.15 *M* NaCl, pH 8.5. Amino acid analysis indicated that approximately 80% of the lysyl residues were modified under these conditions. The modification of a glutamine synthetase from *Bacillus stearothermophilus* with ethyl acetimidate has been studied by Sekiguchi and co-workers.[62] The modification was performed at pH 9.5 with 0.2 *M* phosphate for 1 h at 35°C and terminated by dialysis at pH 7.2. The extent of modification was determined by titration of the modified protein with trinitrobenzenesulfonic acid. As these investigators suggest, consideration must be given to the possibility of cross-linking occurring with this reagent under the conditions used.[63] Monneron and d'Alayer[64] examined the reaction of either methyl acetamidate or dimethyl suberimidate with particulate adenylate cyclase. The reaction was performed in 0.05 *M* triethanolamine, 10% (w/v) sucrose, 0.005 *M* $MgCl_2$, pH 8.1. Plapp and co-workers[61] examined the reaction of methyl picolinimidate with pancreatic deoxyribonuclease. Methyl picolinimidate is an imidoester which reacts with the primary amino groups in proteins (Figure 7). The reaction was performed in 0.5 *M* triethanolamine hydrochloride, pH 8.0 containing 1 m*M* $CaCl_2$ with 0.1 *M* methyl picolinimidate for 22 h at 25°C, then with 0.2 *M* methyl picolinimidate for an additional 8 h. The extent of modification of a protein by methyl picolinimidate can be determined by spectral analysis (see Figure 8). Under these conditions, essen-

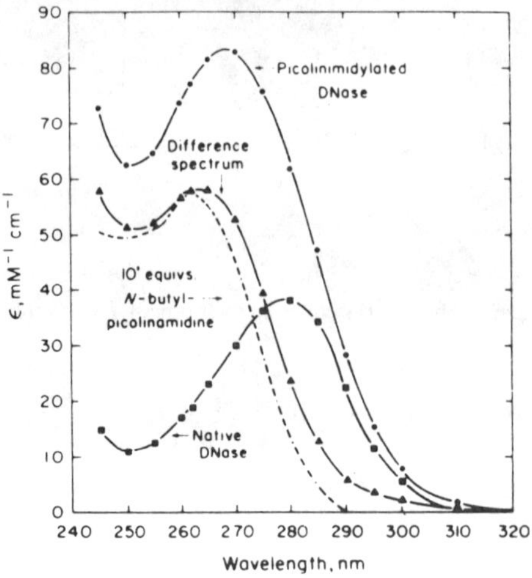

FIGURE 8. The UV spectra of native DNase and the picolinimidylated derivative. The spectra of picolinimidylated DNase (●) and native DNase (■) are presented together with the difference spectrum (▲). From the absorbance at 262 nm and the extinction coefficient for *N*-butylpicolinamidine (5700 M^{-1} cm^{-1}), it was calculated that the modified enzyme contained 10 picolinimidyl groups (the theoretical difference spectrum for this extent of modification is shown by the dashed line). The proteins were dissolved in 0.05 *M* sodium acetate, 1 m*M* $CaCl_2$ and clarified by centrifugation prior to analysis. A molecular weight of 31,000 was assumed in the calculations. (From Plapp, B. V., Moore, S., and Stein, W. H., *J. Biol. Chem.*, 246, 939, 1971. With permission.)

tially all of the primary amino groups in deoxyribonuclease (nine lysine and one amino-terminal amino group) were modified, but there was no change in biological activity. Plapp has also studied the reaction of methyl picolinimidate with horse liver alcohol dehydrogenase.[65] This study was somewhat unique in that modification of the enzyme resulted in enhanced catalytic activity reflecting more rapid dissociation of the enzyme-coenzyme complex. It should be noted that the derivatized lysine reverts to lysine (60% yield) under the normal conditions of acid hydrolysis.

A number of investigators have used pyridoxal-5′-phosphate to modify lysyl residues in proteins. Pyridoxal-5′-phosphate is the cofactor form of vitamin B_6 and plays an important role in biological catalysis.[66] Pyridoxal phosphate is useful for the modification of lysine because of the selectivity of reaction, spectral properties of the modified residue, reversibility of reaction, and the establishment of stereochemistry by use of radiolabeled sodium borohydride (sodium borotritiide) to reduce the Schiff base initially formed on the reaction of pyridoxal phosphate with a primary amine. Pyridoxal phosphate will react with all primary amines (both ε-amino groups of lysine and the

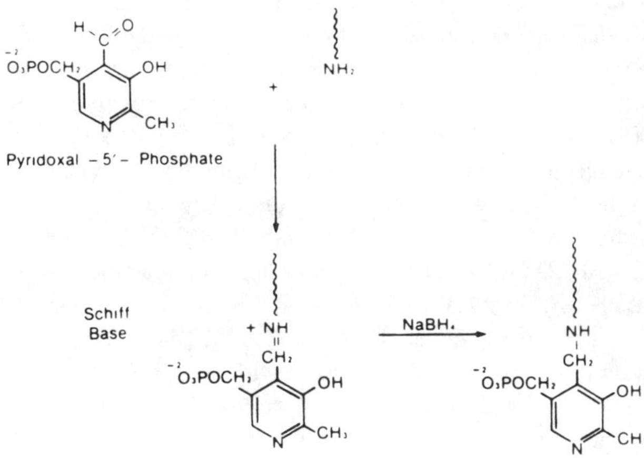

FIGURE 9. The reaction of pyridoxal-5′-phosphate with amino groups in proteins.

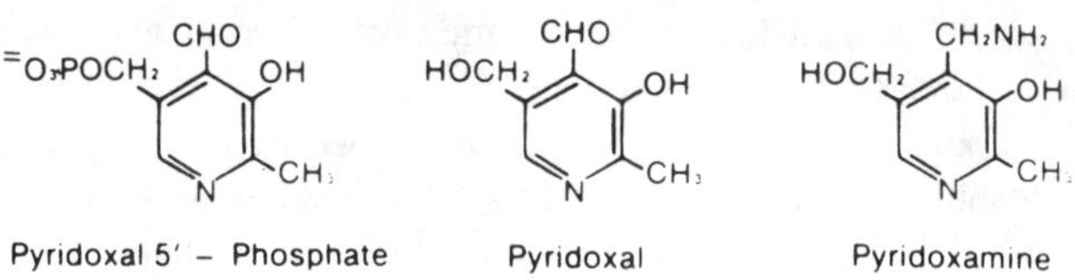

FIGURE 10. The structure of pyridoxal-5′-phosphate, pyridoxal, and pyridoxamine.

amino-terminal α-amino function) in a protein (Figure 9). In general, pyridoxal-5′-phosphate is far more reactive than pyridoxal because of intramolecular hemiacetal formation and the neighboring group effect of the phosphate moiety (Figure 10). Shapiro and co-workers investigated the reaction of pyridoxal phosphate with rabbit muscle aldolase.[67] The initial reaction produced a species with an absorbance maximum at 430 to 435 nm reflecting the protonated Schiff base form of the pyridoxal phosphate-protein complex. After reduction with sodium borohydride, the absorbance maximum was at 325 nm which is characteristic of the reduced Schiff base. This is a quite useful study in that the difference in reactivity between pyridoxal and pyridoxal-5′-phosphate is demonstrated as is the reversible nature of the initial complex. Schnackerz and Noltmann[68] compared the reaction of pyridoxal-5′-phosphate and other aldehydes in reaction with rabbit muscle phosphoglucose at pH 8.0. Pyridoxal-5′-phosphate (0.19 m*M*) resulted in 82% inactivation, while the following results were obtained with other aldehydes: pyridoxal (8.4 m*M*),

16% inactivation; acetaldehyde (75 m*M*), 75% inactivation; and acetone (75 m*M*), 31% inactivation. This last reaction is of interest as many investigators are unaware that acetone can react with amino groups in proteins. The reaction of acetone with primary amino groups has been known for some time[69] and is discussed in further detail below within the topic of reductive alkylation. The reaction of ribulose 1,5-bisphosphate carboxylase/oxygenase with pyridoxal-5′-phosphate has been studied by Paech and Tolbert.[70] Pyridoxal-5′-phosphate inactivated the enzyme with or without reduction with $NaBH_4$. This reaction was performed in 0.1 *M* Bicine (*N,N*-(2-hydroxyethyl) glycine), 0.010 *M* $MgCl_2$, 0.2 m*M* EDTA, 0.001 *M* dithiothreitol. The reaction demonstrated an optimum at pH 8.4. Spectral studies showed the formation of a species absorbing at 432 nm. As is characteristic for the Schiff base derivative, this peak disappears on reduction to yield a species with an optimum at 325 nm ($\Delta\varepsilon$ = 4800 M^{-1} cm^{-1}). This supports the suggestion that the loss of activity observed on reaction with pyridoxal-5′-phosphate is due to the formation of a Schiff base which can be reduced with $NaBH_4$ to form a stable derivative, as opposed to the formation of a 2-azolidine ring with a second nucleophile as has been observed by other investigators.[71–73] Jones and Priest[74] have investigated the modification of apo-serine hydroxymethyltransferase with pyridoxal phosphate and the subsequent use of the enzyme-bound pyridoxal phosphate as a structural probe. Cortijo and co-workers[75] have suggested the use of the ratio of absorbance at 415 and 335 nm of enzyme-bound pyridoxal phosphate as an indication of the polarity of the medium. Cake and co-workers[76] have demonstrated that modification of activated hepatic glucorticoid receptor with pyridoxal-5′-phosphate obviated the binding of the receptor to DNA. Greatly reduced inhibition was seen with pyridoxamine-5′-phosphate, pyridoxamine, or pyridoxine. Inhibition could be reversed by gel filtration or treatment with dithiothreitol, while treatment with $NaBH_4$ resulted in irreversible inhibition of DNA binding. These investigators used 0.2 *M* borate, 0.25 *M* sucrose, 0.003 *M* $MgCl_2$ (pH 8.0) as the solvent for reaction with pyridoxal-5′-phosphate. Slebe and Martinez-Carrion[77] have introduced the use of phosphopyridoxal trifluoroethyl amine as a probe for pyridoxal phosphate binding sites in enzymes. Nishigori and Toft[78] explored the reaction of pyridoxal-5′-phosphate with the avian progesterone receptor. Reaction with pyridoxal-5′-phosphate was performed in 0.02 *M* barbital, 10% (v/v) glycerol, 0.005 m*M* dithiothreitol, 0.010 *M* KCl, pH 8.0. The modification was stabilized by $NaBH_4$. It is of interest that these investigators noted that the modification was readily reversed in Tris buffer unless stabilized by $NaBH_4$. Sugiyama and Mukohata[79] observed that modification with pyridoxal-5′-phosphate of the lysine residue in chloroplast coupling factor using 0.020 *M* Tricine, 0.001 *M* EDTA, 0.010 *M* $MgCl_2$, pH 8.0 resulted in complete inactivation of the ATPase activity. Peters and co-workers[80] reported on the inactivation of the ATPase activity in a bacterial coupling factor by reaction with pyridoxal-5′-phosphate. The modification was performed in 0.050 *M* morpholinosulfonic acid, pH 7.5. The inhibition was readily reversed by dilution or by 0.01 *M* lysine and was, as expected, stabilized by $NaBH_4$. Gould and Engel[81] reported on the reaction of mouse testicular lactate dehydrogenase with pyridoxal-5′-phosphate in 0.050 *M* sodium pyrophosphate, pH 8.7 at 25°C. This reaction resulted in the inactivation of the dehydrogenase activity. The inactivation was reversed by cysteine and stabilized by $NaBH_4$. These investigators reported that the observed absorption coefficient at 325 nm may be decreased as much as 50% with protein-bound pyridoxal phosphate. Thus, estimation of the number of lysine residues modified using the absorption coefficient obtained with model compounds might provide only a minimum value. Ogawa and Fujioka[82] studied the reaction of pyridoxal-5′-phosphate with saccharopine dehydrogenase in 0.1 *M* potassium phosphate, pH 6.8 at ambient temperature in the dark. Both spectral analysis and tritium incorporation from sodium borohydride reduction were consistent with the modification of one lysine residue per mole of enzyme being responsible for the loss of enzyme activity. A value of 1×10^4 M^{-1} cm^{-1} for the extinction coefficient at 325 nm[83] was used in this study. The concentrations of pyridoxal and pyridoxal-5′-phosphate were determined spectrophotometrically in 0.1 *M* NaOH using an extinction coefficient of 5.8×10^3 M^{-1} cm^{-1} at 300 nm and 6.6×10^3 M^{-1} cm^{-1} at 388 nm, respectively.[84] Amine compounds have the potential to interfere in the reaction of pyridoxal-5′-phosphate with proteins. Moldoon and Cidlowski[85] demonstrate that 0.1 *M* Tris, pH 7.4, markedly interfered with the modification of rat uterine estrogen receptor with pyridoxal-5′-phosphate. These investigators also noted that, as in the other studies, 0.05 *M* lysine would block the modification reaction and could also reverse the modification if the Schiff base had not been reduced. Stock solutions of pyridoxal phosphate were prepared in 0.01 *M* NaOH to avoid acid decomposition. The importance of local environmental factors in the specificity of modification by pyridoxal phosphate is emphasized by Ohsawa and Gualerzi.[86] These investigators examined the modification of *Escherichia coli* initiation factor by pyridoxal phosphate in 0.020 *M* triethanolamine, 0.03 *M* KCl, pH 7.8. In the course of the studies, it was observed that pyridoxal phosphate will not react with poly(AUG). These investigators also reported the preparation of N^6-pyridoxal lysine by reaction of pyridoxal phosphate with polylysine in 0.01 *M* sodium phosphate, pH 7.2 at 37°C followed by reduction with $NaBH_4$. The reduction was terminated by the addition of acetic acid. Acid hydrolysis (6 *N* HCl, 110°C, 22 h) yielded N^6-pyridoxal-L-lysine. Bürger and Görisch[87] reported the inactivation of histidinol dehydrogenase upon reaction with pyridoxal phosphate in 0.02 *M* Tris, pH 7.6. This modification could be reversed by dialysis unless the putative Schiff base was stabilized by reduction with NaH_4 (*n*-octyl alcohol added to prevent foaming). These investigators used a $\Delta\varepsilon$ for ε-amino pyridoxal lysine of 1×10^4 M^{-1} cm^{-1} at 325 nm. Recent applications of pyridoxal-5′-phosphate modification has been used to study hydroxymethylbilane synthetase,[88] DNA polymerase I,[89] and rabbit glycogen synthase isozymes.[90] A novel affinity label (pyridoxal-5′-diphospho-5′-ad-

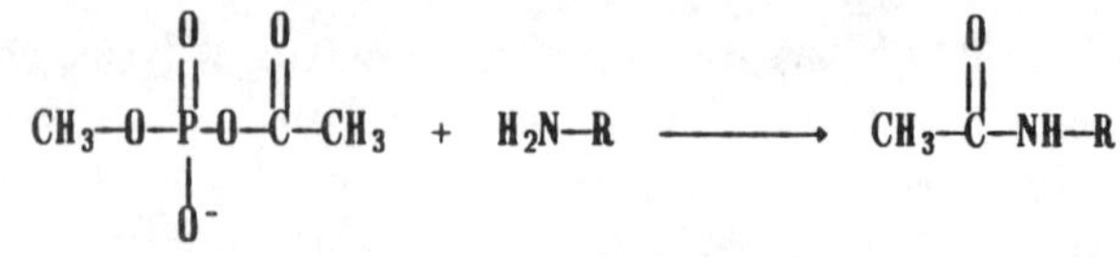

FIGURE 11. The structure of methyl acetyl phosphate, a site-specific reagent for the modification of lysine residues in proteins.

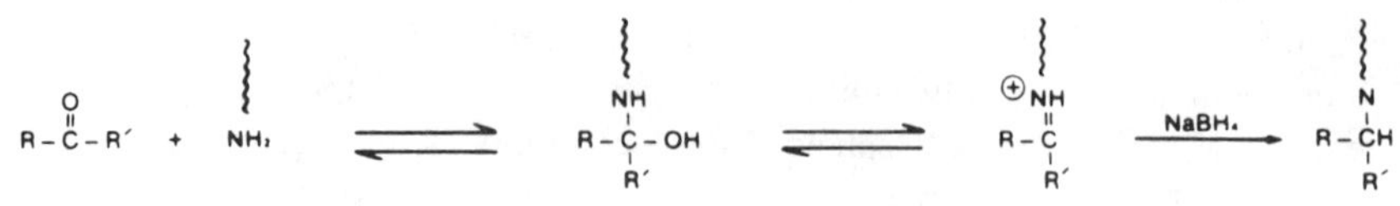

FIGURE 12. The reductive alkylation of amino groups in proteins.

enosine) utilizing pyridoxal-5′-phosphate chemistry has been used to study the adenine nucleotide binding sites in yeast hexokinase.[91]

A substantial portion of the specificity of pyridoxal-5′-phosphate in protein modification arises from electrostatic interaction(s) via the phosphate group with positively charged groups (i.e., arginine) on the protein surface. A conceptually related compound is methyl acetyl phosphate (Figure 11). The reagent was originally developed as an affinity label for D-3-hydroxybutyrate dehydrogenase.[92] Manning and co-workers have examined the chemistry of the reaction of methyl acetyl phosphate with hemoglobin in some detail.[93,94] It appears to be an affinity label for the 2,3-diphosphoglycerate binding site.[93] More recent work suggests that this reagent may be a useful probe for other anion binding sites in proteins.[94]

The modification of primary amines in proteins by reductive alkylation has proved to be a useful reaction (Figure 12). This reaction has the advantage that the basic charge properties of the modified residue are preserved. The early work on this modification has been reviewed by Means.[95] Both monosubstituted and disubstituted derivatives can be prepared depending upon reaction conditions and the nature of the carbonyl compound.

Rice and co-workers[96] reported the stabilization of trypsin by reductive methylation. This reaction utilized formaldehyde/sodium borohydride in 0.2 *M* sodium borate, pH 9.2 in the cold. Unsubstituted amino groups were present after the reaction as demonstrated by titration with trinitrobenzenesulfonic acid. The amino-terminal isoleucine residue was not modified under these conditions. Morris and co-workers[97] investigated the reductive methylation of monellin. The modification was performed in 0.2 *M* sodium borate, pH 8.0 with 1 m*M* monellin (11 m*M*) with respect to primary amino groups in the cold. Sodium borohydride was added to give a final concentration of 0.5 mg/ml, and 1 to 5 µl of 6 to 8 *M* formaldehyde was added per milliliter of solution. Tritiated formaldehyde was used to establish the extent of modification. One of the problems with the use of formaldehyde in this reaction is the presence of paraformaldehyde. Chen and Benoiton[98] obviated this difficulty by the *in situ* generation of formaldehyde from methanol.

The introduction of sodium cyanoborohydride as a reducing agent for this reaction represented a real advance. Sodium cyanoborohydride is stable in aqueous solution at pH 7.0. Unlike sodium borohydride which can reduce aldehydes and disulfide bonds, sodium cyanoborohydride only reduces the Schiff base formed in the initial process of reductive alkylation. The radiolabeling of proteins using ^{14}C-formaldehyde and sodium cyanoborohydride has been reported.[99] The modification was performed in 0.04 *M* phosphate, pH 7.0 at 25°C. The modification can be performed equally well at 0°C, but, as would be expected, it takes a longer period of time; there is no effect on the extent of the modification. In this regard, these authors estimated that the same extent of modification obtained in 1 h at 37°C could be achieved in 4 to 6 h at 25°C or 24 h at 0°C. Although the majority of experiments in this study were performed in phosphate buffer at pH 7.0, equivalent results can be obtained in Tris or HEPES (*N*-2-hydroxyethylpiperazine-*N*′-2-ethanesulfonic acid) buffer at pH 7.0. A greater extent of modification was observed with sodium cyanoborohydride at pH 7.0 than with sodium borohydride at pH 9.0. Jentoft and Dearborn have studied the use of sodium cyanoborohydride in some detail.[100] In particular, the preparation of sodium cyanoborohydride is critical, and most, if not all, commercial preparations require recrystallization prior to use. This reflects the presence of impurities which limit the extent of the reductive alkylation (see below). Recrystallization is accomplished by dissolving 11 g of sodium cyanoborohydride in 25 ml acetonitrile. Insoluble material is removed by centrifugation. Crystallization is accomplished by the addition of 150 ml methylene chloride and allowing it to stand overnight at 4°C. The recrystallized sodium cyanoborohydride is collected by filtration and stored in a vacuum desiccator. A fresh solution of reagent is prepared daily. Using [^{14}C]-formaldehyde and sodium cyanoborohydride, the major product is ε-methylated lysine, with minor incorporation of radiolabel into arginine and histidine. Optimal reductive methylation was obtained at pH values greater than 8.0 during a short-term (10 min) incubation. The effect of pH is much less pronounced at longer periods of incubation (1 to 2 h), with optimal reductive methylation occurring between pH 7.0 and 8.0. These investigators also noted that Tris, β-mercaptoethanol, dithiothreitol, ammonium ions (as ammonium sulfate), and guanidine (5 *M*) inhibited the reductive alkylation of albumin by formaldehyde and sodium cyanoborohydride in 0.1 *M* HEPES, pH 7.5.

Reductive methylation with ^{3}C-enriched formaldehyde has been used to introduce an NMR probe for the study of protein conformation.[101] A similar approach has been developed using deuterated acetone.[102]

The effect of carbonyl compounds of different size on the extent of reductive alkylation has been examined by Fretheim and co-workers.[103] The extent of modification is more a reflection of the type of alkylating agent and reaction conditions than an intrinsic property of the protein under study. For example, nearly 100% disubstitution can be obtained with formaldehyde and approximately 35% disubstitution with *n*-butanol, while only monosubstitution can be obtained with acetone, cyclopentanone, cyclohexanone, and benzaldehyde. While most of the products of reductive alkylation retained solubility, the reaction products obtained with cyclohexanone and benzaldehyde tended to precipitate. Examination of the reductive alkylation of ovomucoid, lysozyme, and ovotransferrin with different aldehydes suggests that such modification occurs without major conformational change as judged by circular dichroism measurements.[104] The same study also examined the stability of the modified proteins by scanning differential calorimetry. The extensive modification of amino groups decreases thermal stability. The destabilization effect increases with increasing size (and hydrophobicity) of the modifying aldehyde.

In another study, the reversible reductive alkylation of proteins has been examined.[105] Both glycolaldehyde and acetol will react with the primary amino groups in proteins to yield derivatives which can be cleaved with periodate under mild basic condition to yield the free amine. Sodium cyanoborohydride is much more effective in the range of pH 6.0 to 8.0, while sodium borohydride is more effective under more alkaline conditions. Treatment of 30.0 mg lysozyme in 6.0 ml 0.2 *M* sodium borate, pH 9.0 with 60 mg glycoaldehyde and 10 mg sodium borohydride at ambient temperature resulted in 60% 2-hydroxyethylation. Treatment of 20 mg ovomucoid in 2.0 ml 0.2 *M* sodium borate, pH 9.0 with 10% acetol and 30 mg sodium borohydride (added in portions) resulted in 55% hydroxyisopropylation. In both situations, the reaction was terminated by adjustment of the pH to 5 with glacial acetic acid. The extent of modification was determined either by titration with trinitrobenzenesulfonic acid and/or by amino acid analysis after acid hydrolysis. Periodate oxidation could be accomplished with 0.015 *M* sodium periodate, pH 7.9 for 30 min at ambient temperature.

The use of [^{13}C]-formaldehyde in the reductive alkylation of ribonuclease has been reported.[106] In a subsequent study,[107] Jentoft and Dearborn characterized the inhibition by cyanide of reductive alkylation with sodium cyanoborohydride (Figure 13). This is of some importance since cyanide is a product of reductive alkylation with sodium cyanoborohydride. Inhibition by cyanide can be blocked by nickel (II) or cobalt (III). The observation that nickel (II) can preclude the inhibition of reductive alkylation by cyanide was shown to obviate the previously observed necessity for recrystallization of the sodium cyanoborohydride. Additional studies on the development of reagents alternative to sodium borohydride have been reported from other laboratories. Geoghegan and co-workers[108] compared sodium cyanoborohydride, dimethylamine borane, and trimethylamine borane (Figure 14) with respect to effectiveness in reductive alkylation. Reduction at disulfide bonds was not observed with any of the three reagents. Dimethylamine borane was only slightly less effective than sodium cyanoborohydride, while trimethylamine borane was much less effective (Figure 15). This decrease in effectiveness in reductive alkylation is balanced by the absence of toxic byproducts such as cyanide evolving during the reaction.

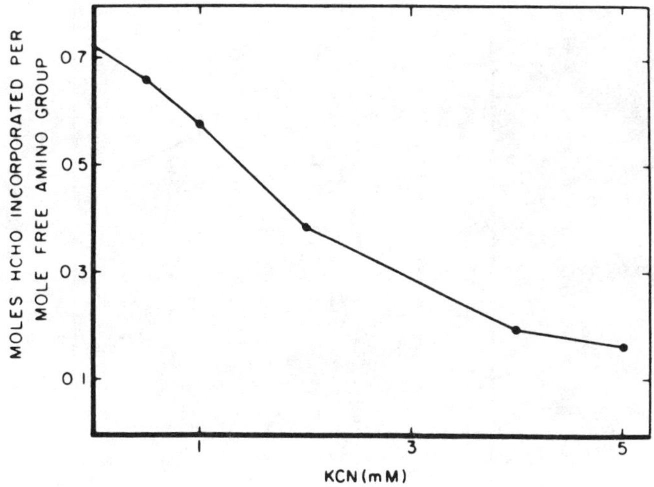

FIGURE 13. The effect of the addition of potassium cyanide (KCN) on the reductive methylation of albumin. Reaction mixtures contained 1.04 mg/ml bovine serum albumin, 2 m*M* [^{14}C]-formaldehyde, 10 m*M* sodium cyanoborohydride in 0.05 *M* HEPES, pH 7.5 with varying amounts of KCN as indicated. The extent of modification was determined by measuring the amount of radiolabel incorporation. (From Jentoft, N. and Dearborn, D. G., *Anal. Biochem.,* 106, 186, 1980. With permission.)

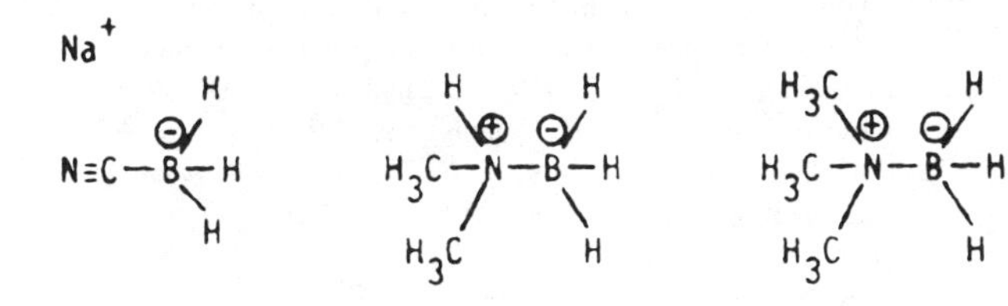

FIGURE 14. The structure of sodium cyanoborohydride, dimethylamine borane, and trimethylamine borane. (From Geoghegan, K. F., Cabacungan, J. C., Dixon, H. B. F., and Feeney, R. E., *Int. J. Pept. Protein Res.,* 17, 345, 1981. With permission.)

Figure 6 compares dimethylamine borane and trimethylamine borane in the reductive methylation of turkey ovomucoid in 0.2 *M* sodium phosphate, pH 7.0. Quantitative reductive methylation (equal to or greater than one methyl group per lysyl residue) is achieved at 10 m*M* formaldehyde with dimethylamine borane and at 50 m*M* formaldehyde with trimethylamine borane. It should be noted that a similar extent of modification is obtained with 5 m*M* formaldehyde using sodium cyanoborohydride. In a subsequent study,[109] this laboratory reported the successful use of pyridine borane in the reductive alkylation of proteins. Wu and

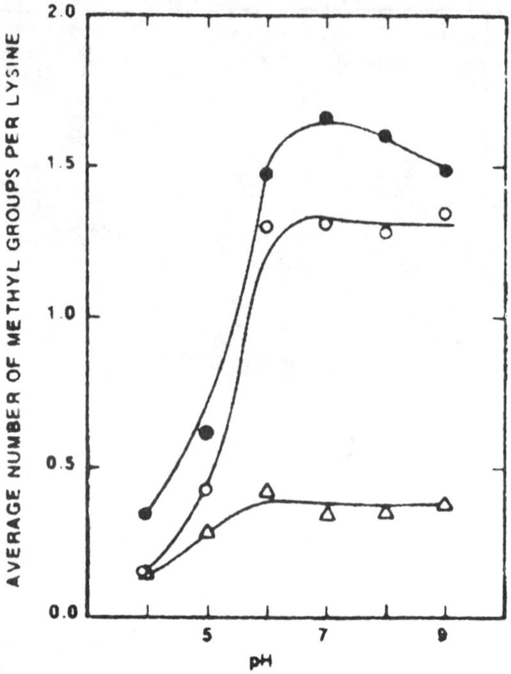

FIGURE 15. The effect of pH on the reductive methylation of turkey ovomucoid in the presence of various reducing agents. The concentration of turkey ovomucoid was 5 mg/ml, and the concentration of formaldehyde was 20 m*M* in the presence of either sodium cyanoborohydride (●), 15 m*M*, dimethylamine borane (○), 15 m*M*, or trimethylaminoborane (△), 15 m*M*. (From Geoghegan, K. F., Cabacungan, J. C., Dixon, H. B. F., and Feeney, R. E., *Int. J. Pept. Protein Res.*, 17, 345, 1981. With permission.)

Means[110] have used reductive alkylation with a nonpolar aldehyde (dodecylaldehyde) to subsequently prepare insoluble proteins by binding of the modified protein to octyl-Sepharose.

The reaction of glyceraldehyde with carbonmonoxyhemoglobin S has been explored by Acharya and Manning.[111] This reaction was performed with 0.010 *M* glyceraldehyde in phosphate-buffered saline, pH 7.4, and the resultant Schiff bases were stabilized by reduction with sodium borohydride. Using radiolabeled glyceraldehyde, these investigators were able to obtain support for the concept that there is selectivity in the reaction of sugar aldehydes with hemoglobin. The reaction product between glyceraldehyde and hemoglobin S did have stability properties without reduction that were not consistent with only Schiff base products. These investigators suggested that the glyceraldehyde-hemoglobin Schiff base could undergo an Amadori rearrangement (Figure 17) to form a stable ketoamine adduct which could be reduced with sodium

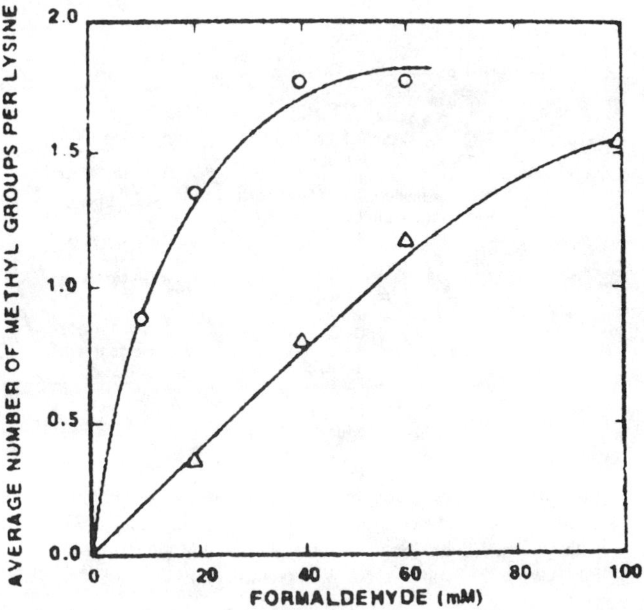

FIGURE 16. The effect of formaldehyde concentration on the reductive methylation of turkey ovomucoid. The reaction was performed in 0.2 *M* sodium phosphate, pH 7.0 with a turkey ovomucoid concentration of 5 mg/ml. The reducing agents were used at a concentration of 15 m*M* and included dimethylamine borane (○) and trimethylamine borane (△) at 22°C. The reducing agents were dissolved in methanol at a concentration of 150 m*M* and diluted 1:10 in the reaction mixture such that the final concentration of methanol was 10% (v/v). (From Geoghegan, K. F., Cabacungan, J. C., Dixon, H. B. F., and Feeney, R. E., *Int. J. Pept. Protein Res.*, 17, 345, 1981. With permission.)

borohydride to form a product identical to that obtained by direct reduction of the Schiff base. In a subsequent study, these investigators did show that the glyceraldehyde-hemoglobin S Schiff base could rearrange to form a ketamine via an Amadori rearrangement.[112] These investigators were able to use reaction with phenylhydrazine to detect the protein-bound ketamine adduct as shown in Figure 18.

Another class of aldehydes that reacts with protein to give interesting products are simple monosaccharides, which exist in solution in enol and keto forms (Figure 19). Wilson[113] showed that bovine pancreatic ribonuclease dimer would react with lactose in the presence of sodium cyanoborohydride to yield an active derivative that shows selectivity in uptake by the liver during *in vivo* experiments. The modification of ribonuclease dimer was performed in 0.2 *M* potassium phosphate, pH 7.4 (phosphate buffer was used to protect lysine-41 from modification) at 37°C for 5 days with lactose and sodium cyanoborohydride. Under these conditions, 80% of the amino groups were modified. Bunn and Higgins[114] have explored the reaction of monosaccharides

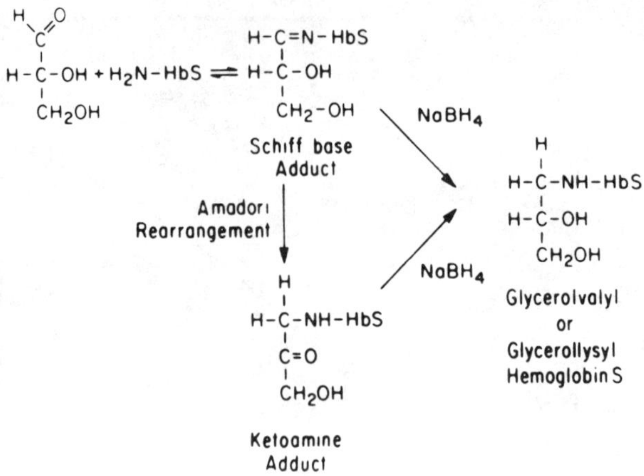

FIGURE 17. Schematic representation of the formation of glycerovaline or glycerollysine on the reaction of hemoglobin S with glyceraldehyde. (From Acharya, A. S. and Manning, J. M., *J. Biol. Chem.*, 255, 1406, 1980. With permission.)

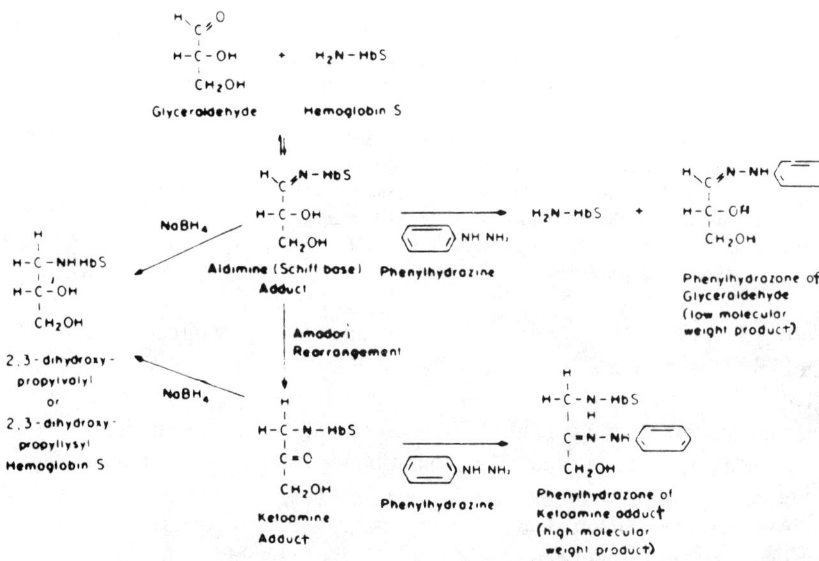

FIGURE 18. The reaction of the glyceraldehyde-hemoglobin adduct with either phenylhydrazine or sodium borohydride. (From Acharya, A. S. and Manning, J. M., *J. Biol. Chem.*, 255, 1406, 1980. With permission.)

with protein amino groups in the presence of sodium cyanoborohydride in some detail. These investigators studied the reaction of hemoglobin with various monosaccharides in Krebs-Ringer phosphate buffer, pH 7.3 (Figure 20). The extent of modification was determined using tritiated sodium

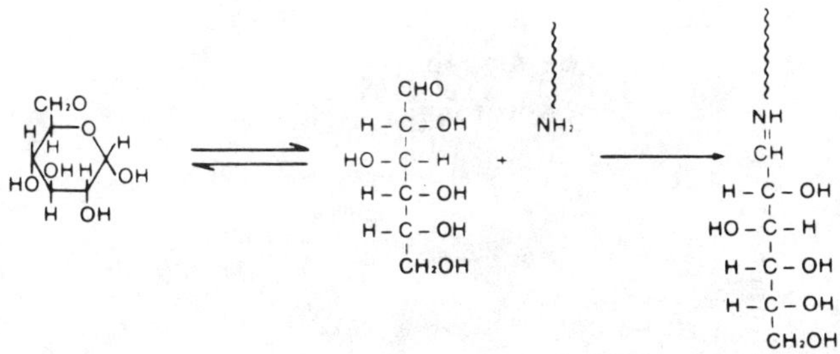

FIGURE 19. A scheme for the reaction of a monosaccharide with a primary amino group.

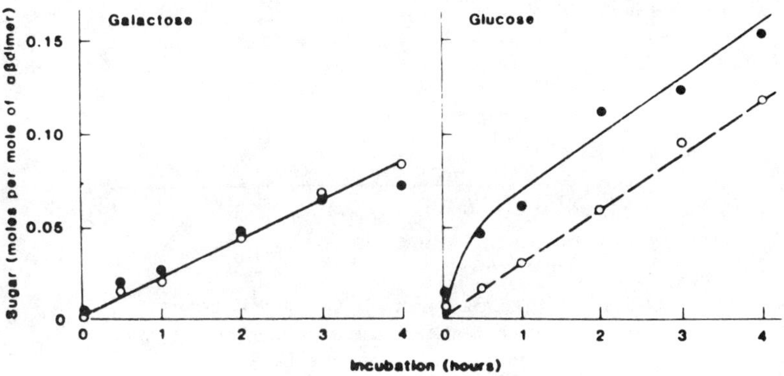

FIGURE 20. The measurement of the rate of condensation of monosaccharides with hemoglobin. The extent of reaction was measured either by incubation with unlabeled sugar followed by reduction of the aldimine linkage with tritiated sodium cyanoborohydride (open circles) or by incubation of the [^{14}C]-labeled sugar with hemoglobin followed by reduction with unlabeled sodium cyanoborohydride (closed circles). The left panel shows the rate of reaction with 42 m*M* D-galactose (k_1 = 1.9 × 10^{-3} mM^{-1} h^{-1}). The right panel shows the reaction with 42 m*M* D-glucose (k_1 = 0.6 × 10^{-3} mM^{-1} h^{-1}). The initial rapid rate of incorporation of D-[^{14}C] glucose can be explained by the small amount of rapidly reacting impurity remaining in the preparation. (From Bunn, H. F. and Higgins, R. J., *Science,* 213, 222, 1981. With permission.)

cyanoborohydride. The rate of modification was demonstrated to be a direct function of the amount of each sugar in the carbonyl (or keto) form (Figure 21). Thus, the k_1 (×10^{-3} mM^{-1} h^{-1}) for D-glucose is 0.6 with 0.002% in the carbonyl form, while the k_1 (×10^{-3} mM^{-1} h^{-1}) for D-ribose is 10.0 with 0.05% in the carbonyl form.

The reaction of 2,4,6-trinitrobenzenesulfonic acid (TNBS) with amino groups has been of value in studying the function and reactivity of the ε-amino groups of lysyl residues in proteins.[115–117] The reaction of TNBS with the primary amino groups in proteins is shown in Figure 22. The modification of amino groups with TNBS is easy to monitor by spectral analysis. In the presence of an excess of

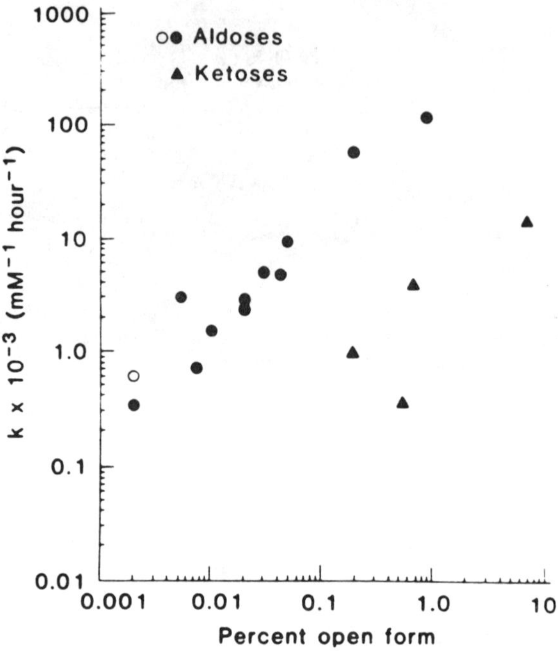

FIGURE 21. The relation between the rate of condensation of monosaccharide with hemoglobin and the equilibrium between the open and ring structures of the monosaccharide (l_1). The open circle is for glucose ($k_1 = 0.6 \times 10^{-3}$ mM^{-1} h^{-1}). The closed circles represent data for other aldoses, and the closed triangles are for ketoses. (From Bunn, H. F. and Higgins, P. J., *Science,* 213, 222, 1981. With permission.)

sulfite, absorbance at 420 nm is the most sensitive index, having $\varepsilon = 2.0 \times 10^4$ M^{-1} cm^{-1}. Absorbance at 420 nm is dependent upon the ability of the reaction product to form a complex with sulfite. It has proved convenient in our laboratories to use the fact that the spectrum of a trinitrobenzyl amino compound has an isosbestic point at 367 nm with $\varepsilon = 1.05 \times 10^4$ M^{-1} cm^{-1}. As suggested by Fields,[118] we recrystallize TNBS from 2.0 *M* HCl prior to use. We generally perform the modifications in phosphate buffer (pH 6.0 to 9.0). The derivatives of α- and ε-amino groups have similar spectra with the exception that α-amino derivatives have a slightly higher extinction coefficient at 420 nm ($\varepsilon = 2.20 \times 10^4$ M^{-1} cm^{-1}) than ε-amino groups ($\varepsilon = 1.92 \times 10^4$ M^{-1} cm^{-1}). Both of these derivatives have much higher extinction coefficients than the derivative obtained by reaction of TNBS with cysteinyl residues ($\varepsilon = 2.25 \times 10^3$ M^{-1} cm^{-1}). The α- and ε-amino derivatives can be differentiated by their stability to acid or base hydrolysis. The α-amino derivatives are unstable to acid hydrolysis (8 h at 110°C) or base hydrolysis.[119]

Frieden and co-workers have explored the reaction of trinitrobenzenesulfonic acid with bovine liver glutamate dehydrogenase.[120,121] In these studies, the

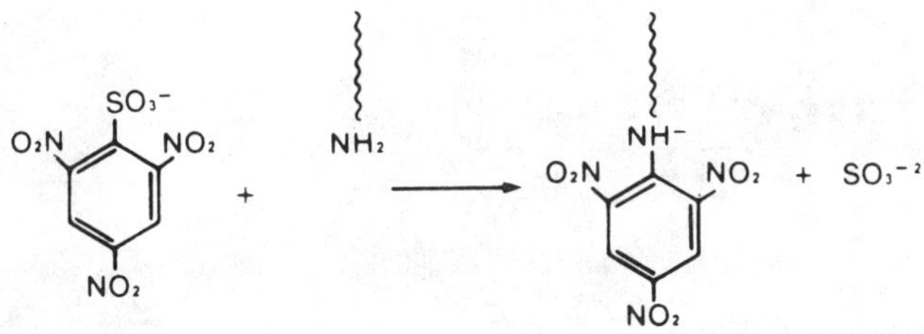

FIGURE 22. The reaction of 2,4,6-trinitrobenzenesulfonic acid with primary amines in proteins.

modification was performed in 0.04 *M* potassium phosphate, pH 8.0. Under these reaction conditions, the cysteinyl residues were not modified. The preparative reactions were terminated by reaction with β-mercaptoethanol. It is of interest that under certain conditions (with reduced coenzyme), glutamate dehydrogenase catalyzed the conversion of TNBS to trinitrobenzene.[122]

The reaction of TNBS with simple amines and hydroxide ions has been studied in some detail by Means and co-workers.[123] The reaction of TNBS with hydroxide is first order with respect to both trinitrobenzenesulfonate and hydroxide ions. Reaction with amines was considered in some detail. In general, reactivity of trinitrobenzenesulfonate with amines increases with increasing basicity except that secondary amines and *t*-alkylamines are comparatively unreactive. The specific binding of trinitrobenzenesulfonate to proteins must be considered in the study of the reaction of this compound with proteins. Only amines with a pKa >8.7 follow a simple rate law. These investigators presented the following considerations regarding the reaction of trinitrobenzenesulfonic acid with proteins. Reactivity is a sensitive measure of the basicity of an amino group. Adjacent charged groups have an influence on the rate of reaction with an increase observed with a positively charged group and a decrease with a negatively charged group. Proximity to surface hydrophobic regions which can bind TNBS can increase the observed reactivity of a particular amino group.

The reaction of TNBS with ammonium has also been investigated by Whitaker and co-workers.[124] This reaction was performed in tetraborate buffer and 1 μ*M* sulfite. The rate of the reaction was determined by following the increase in absorbance at 420 nm ($\varepsilon = 2.02 \times 10^4\ M^{-1}\ cm^{-1}$). The rate of reaction with ammonium ($k = 0.128\ min^{-1}$) was slower than that with the average amine in a protein ($k = 0.907\ min^{-1}$ for enterotoxin). The reaction with ammonium does, however, provide a sensitive assay for ammonia (as low as 6 nmol) with a precision of 1 to 2%.

Flügge and Heldt have explored the labeling of a specific membrane component with TNBS[125] and pyridoxal-5′-phosphate.[126] The modification of the phosphate translocation protein in spinach chloroplasts with TNBS was performed in 0.050 *M* HEPES, 0.33 *M* sorbitol, 0.001 *M* $MgCl_2$, 0.001 *M* $MnCl_2$, 0.002 *M* EDTA, pH 7.6 at 4°C for periods of time up to 15 min, at which point

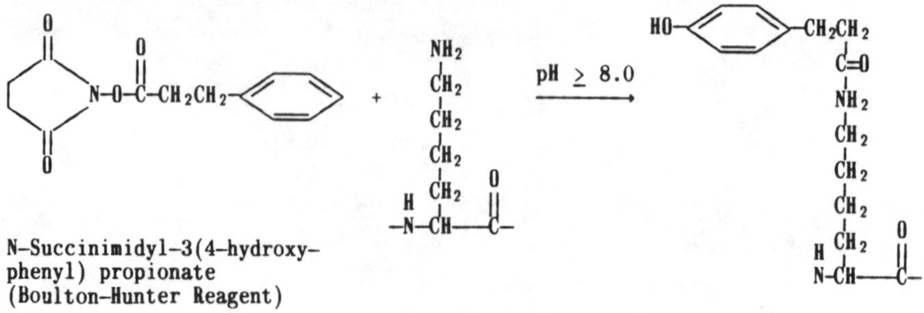

FIGURE 23. The structure of *N*-succinimidyl-3-(4-hydroxyphenyl) propionate (Boulton-Hunter reagent) and its reaction with lysine residues in proteins.

tritiated sodium borohydride was added to both terminate the reaction and radiolabel the trinitrophenyl derivatives.[127] It is possible to label components on the surface of membranes with TNBS, as the sulfonate moiety does not permit membrane penetration. The same is true for pyridoxal-5′-phosphate.

The use of trinitrobenzenesulfonate in the selective modification of membrane surface components has been explored by Salem and co-workers.[128] This study involved the modification of intact cells with the TNBS (dissolved in methyl alcohol) diluted to a 1% methanolic solution. As mentioned above, trinitrobenzenesulfonate does not pass across (or into) membranes, being more hydrophilic than, for example, fluorodinitrobenzene.

Haniu et al.[129] have examined the reaction of lysine residues in NAD(P)H:quinone reductase with TNBS as compared to the reaction of tyrosine residues with *p*-nitrobenzenesulfonyl fluoride. Isolation and characterization of the peptides containing the modified residues showed that the modified tyrosyl residues are in hydrophobic regions of the protein, while the modified lysine residues are in hydrophilic regions.

With the exception of the Boulton-Hunter reagent[130] (Figure 23), the use of *N*-hydroxysuccinimide ester derivatives to modify lysine residues has been somewhat restricted to cross-linking reagents as described in Chapter 15. However, the specificity demonstrated by this chemistry provides considerable potential for the introduction of structural probes and other unique functional groups into proteins. Yem et al.[131] have used *N*-hydroxysuccinimide chemistry to introduce biotin into recombinant interleukin-1-β (Figure 24). This is a fascinating technology with substantial promise.[132,133]

It is possible to selectively modify the α-amino groups of proteins by chemical transamination with glyoxylate (Figure 25) at a slightly acid pH.[134,135] This modification has been applied to *Euglena* cytochrome C-552. This reaction was performed in 2.0 *M* sodium acetate, 0.10 *M* acetic acid, 0.005 *M* nickel sulfate, 0.2 *M* sodium glyoxylate and resulted in the complete loss of the amino-terminal residue. Snake venom phospholipase A_2 has been subjected to chemical transamination.[135] This reaction was performed in 2.0 *M* sodium acetate, 0.4 *M* acetic acid, 0.010 *M* cupric ions, 0.1 *M* glyoxylic acid, pH 5.5.

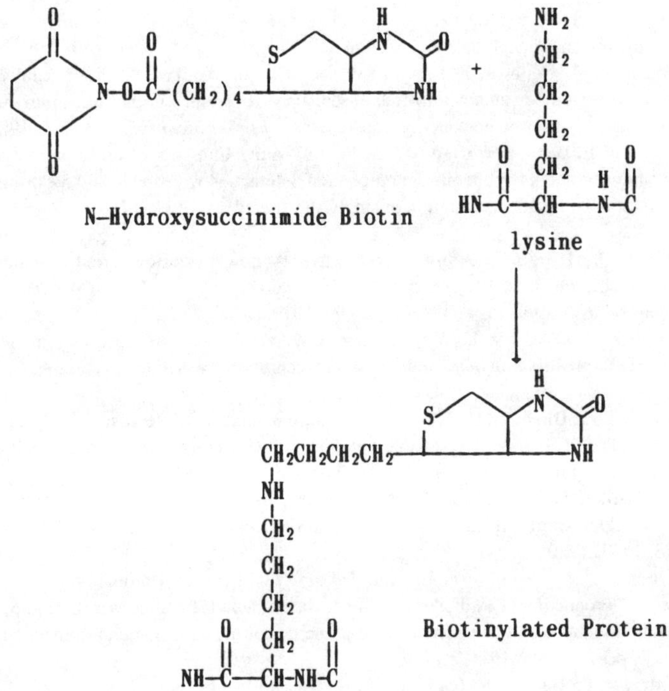

FIGURE 24. Incorporation of biotin into protein via reaction with a *N*-hydroxysuccinimide derivative.

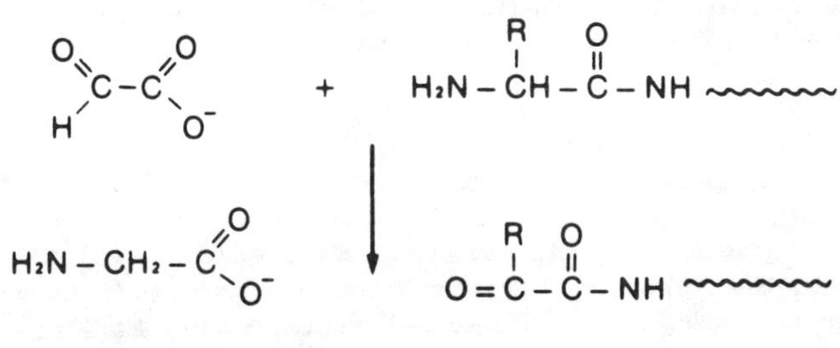

FIGURE 25. The modification of α-amino groups in proteins with glyoxylate.

REFERENCES

1. **Gurd, F. R. N.,** Carboxymethylation, *Meth. Enzymol.,* 11, 532, 1967.
2. **Sanger, F. and Tuppy, H.,** The amino acid sequence in the phenylalanyl chain of insulin. I. The identification of lower peptides from partial hydrolysates, *Biochem. J.,* 49, 463, 1951.
3. **Carty, R. P. and Hirs, C. H. W.,** Modification of bovine pancreatic ribonuclease A with 4-sulfonyloxy-2-nitrofluorobenzene, *J. Biol. Chem.,* 243, 5254, 1968.

4. **Franklin, J. G. and Leslie, J.,** Some enzymatic properties of trypsin after reaction with 1-dimethylaminonaphthalene-5-sulfonyl chloride, *Can. J. Biochem.*, 49, 516, 1971.
5. **Wagner, R., Podestá, F. E., González, D. H., and Andreo, C. S.,** Proximity between fluorescent probes attached to four essential lysyl residues in phospho*enol*pyruvate carboxylase — a resonance energy transfer study, *Eur. J. Biochem.*, 173, 561, 1988.
6. **Brautigan, D. L., Ferguson-Miller, S., and Margoliuash, E.,** Definition of cytochrome c binding domains by chemical modification. I. Reaction with 4-chloro-3,5-dinitrobenzoate and chromatographic separation of singly substituted derivatives, *J. Biol. Chem.*, 253, 130, 1978.
7. **Bello, J., Iijima, H., and Kartha, G.,** A new arylating agent, 2-carboxy-4,6-dinitrochlorobenzene. Reaction with model compounds and bovine pancreatic ribonuclease, *Int. J. Pept. Protein Res.*, 14, 199, 1979.
8. **Hall, J., Zha, X., Yu, L., Yu, C.-A., and Millett, F.,** Role of specific lysine residues in the reaction of *Rhodobacter sphaeroides* cytochrome c_2 with the cytochrome bc_1 complex, *Biochemistry*, 28, 2568, 1989.
9. **Long, J. E., Durham, B., Okamura, M., and Millett, F.,** Role of specific lysine residues in binding cytochrome c_2 to the *Rhodobacter sphaeroides* reaction center in optimal orientation for rapid electron transfer, *Biochemistry*, 28, 6970, 1989.
10. **Hiratsuka, T. and Uchida, K.,** Lysyl residues of cardiac myosin accessible to labeling with a fluorescent reagent, *N*-methyl-2-anilino-6-naphthalenesulfonyl chloride, *J. Biochem.*, 88, 1437, 1980.
11. **Onondera, M., Shiokawa, H., and Takagi, T.,** Flourescent probes for antibody active sites. I. Production of antibodies specific to the N-methyl-2-anilinonaphthalene-6-sulfonate group in rabbits and some fluorescent properties of the hapten bound to the antibodies, *J. Biochem.*, 79, 195, 1976.
12. **Cory, R. P., Becker, R. R., Rosenbluth, R., and Isenberg, I.,** Synthesis and fluorescent properties of some *N*-methyl-2-anilino-6-naphthalensulfonyl derivatives, *J. Am. Chem. Soc.*, 90, 1643, 1968.
13. **Haugland, R. P.,** *Molecular Probes. Handbook of Fluorescent Probes and Research Chemicals*, Molecular Probes, Inc., Eugene, OR, 1989, 37.
14. **Tuls, J., Geren, L., and Millett, F.,** Fluorescein isothiocyanate specifically modifies lysine 338 of cytochrome P-450_{scc} and inhibits adrenodoxin binding, *J. Biol. Chem.*, 264, 16421, 1989.
15. **Miki, M.,** Interaction of Lys-61 labeled actin with myosin subfragment-1 and the regulatory proteins, *J. Biochem. (Tokyo)*, 106, 651, 1989.
16. **Bellelli, A., Ippoliti, R., Brunori, M., Kam, Z., Benveniste, M., Emmanuel, F., Turpin, E., Alfsen, A., and Fency, J. P.,** Binding and internalization of ricin labelled with fluorescein isothiocyanate, *Biochem. Biophys. Res. Commun.*, 169, 602, 1990.
17. **Devanathan, S., Dahl, T. A., Midden, W. R., and Neckers, D. C.,** Readily available fluorescein isothiocyanate-conjugated antibodies can be easily converted into targeted phototoxic agents for antibacterial, antiviral, and anticancer therapy, *Proc. Natl. Acad. Sci. U.S.A.*, 87, 2980, 1990.
18. **Turner, D. C. and Brand, L.,** Quantitative estimation of protein binding site polarity. Fluorescence of *N*-arylaminonaphthalenesulfonates, *Biochemistry*, 7, 3381, 1968.
19. **Welches, W. R. and Baldwin, T. O.,** Active center studies on bacterial luciferase: modification of the enzyme with 2,4-dinitrofluorobenzene, *Biochemistry*, 20, 512, 1981.
20. **Shapiro, R., Fox, E. A., and Riordan, J. F.,** Role of lysines in human angiogenin: chemical modification and site-directed mutagenesis, *Biochemistry*, 28, 1726, 1989.
21. **Riordan, J. F. and Vallee, B. L.,** Acetylation, *Meth. Enzymol.*, 11, 565, 1967.
22. **Merle, M., Lefevre, G., Staub, J. F., Raulais, D., and Milhaud, G.,** Acylation of porcine and bovine calcitonin: effects on hypocalcemic activity in the rat, *Biochem. Biophys. Res. Commun.*, 79, 1071, 1977.
23. **Aviram, I.,** The role of lysines in *Euglena* cytochrome C-552. Chemical modification studies, *Arch. Biochem. Biophys.*, 181, 199, 1977.

24. **Karibian, D., Jones, C., Gertler, A., Dorrington, K. J., and Hofmann, T.,** On the reaction of acetic and maleic anhydrides with elastase. Evidence for a role of the NH_2-terminal valine, *Biochemistry,* 13, 2891, 1974.
25. **Kaplan, H., Stevenson, K. J., and Hartley, B. S.,** Competitive labeling, a method for determining the reactivity of individual groups in proteins. The amino groups of porcine elastin, *Biochem. J.,* 124, 289, 1971.
26. **Bosshard, H. R., Koch, G. L. E., and Hartley, B. S.,** The aminoacyl tRNA synthetase-tRNA complex: detection by differential labeling of lysine residues involved in complex formation, *J. Mol. Biol.,* 119, 125, 1978.
27. **Richardson, R. H. and Brew, K.,** Lactose synthase. An investigation of the interaction site of alpha-lactalbumin for galactosyltransferase by differential kinetic labeling, *J. Biol. Chem.,* 255, 3377, 1980.
28. **Rieder, R. and Bosshard, H. R.,** The cytochrome c oxidase binding site on cytochrome c. Differential chemical modification of lysine residues in free and oxidase-bound cytochrome c, *J. Biol. Chem.,* 253, 6045, 1978.
29. **Hitchcock, S. E., Zimmerman, C. J., and Smalley, C.,** Study of the structure of troponin-T by measuring the relative reactivities of lysines with acetic anhydride, *J. Mol. Biol.,* 147, 125, 1981.
30. **Hitchcock, S. E.,** Study of the structure of troponin-C by measuring the relative reactivities of lysines with acetic anhydride, *J. Mol. Biol.,* 147, 153, 1981.
31. **Hitchcock-De Gregori, S. E.,** Study of the structure of troponin-I by measuring the relative reactivities of lysine with acetic anhydride, *J. Biol. Chem.,* 257, 7372, 1982.
32. **Giedroc, D. P., Sinha, S. K., Brew, K., and Puett, D.,** Differential trace labeling of calmodulin: investigation of binding sites and conformational states by individual lysine reactivities. Effects of beta-endorphin, trifluoroperazine, and ethylene glycol bis(beta-aminoethyl ether)-*N,N,N'N'*-tetraacetic acid, *J. Biol. Chem.,* 260, 13406, 1985.
33. **Wei, Q., Jackson, A. E., Pervaiz, S., Carrawy, K. L., III, Lee, E. Y. C., Puett, D., and Brew, K.,** Effects of interactions of with calcineurin of the reactivities of calmodulin lysines, *J. Biol. Chem.,* 263, 19541, 1988.
34. **Winkler, M. A., Fried, V. A., Merat, D. L., and Cheung, W. Y.,** Differential reactivities of lysines in calmodulin complexed to phosphatase, *J. Biol. Chem.,* 262, 15466, 1987.
35. **Hitchcock-De Gregori, S. E., Lewis, S. F., and Mistrik, M.,** Lysine reactivities of tropomyosin complexed with troponin, *Arch. Biochem. Biophys.,* 264, 410, 1988.
36. **Staudenmayer, N., Smith, M. B., Smith, H. T., Spies, F. K., Jr., and Millett, F.,** An enzyme kinetics and ^{19}F nuclear magnetic resonance study of selectively trifluoroacetylated cytochrome c derivatives, *Biochemistry,* 15, 3198, 1976.
37. **Smith, M. B., Stonehuerner, J., Ahmed, A. J., Staudenmayer, N., and Millett, F.,** Use of specific trifluoroacetylation of lysine residues in cytochrome C to study the reaction with cytochrome b_5, cytochrome c_1 and cytochrome oxidase, *Biochim. Biophys. Acta,* 592, 303, 1980.
38. **Webb, M., Stonehuerner, J., and Millett, F.,** The use of specific lysine modifications to locate the reaction site of cytochrome C with sulfite oxidase, *Biochim. Biophys. Acta,* 593, 290, 1980.
39. **Ahmed, A. J. and Millett, F.,** Use of specific lysine modifications to identify the site of reaction between cytochrome C and ferricyanide, *J. Biol. Chem.,* 256, 1611, 1981.
40. **Smith, M. B. and Millett, F.,** A ^{19}F nuclear magnetic resonance study of the interaction between cytochrome C and cytochrome C peroxidase, *Biochim. Biophys. Acta,* 626, 64, 1980.
41. **Klotz, I. M.,** Succinylation, *Meth. Enzymol.,* 11, 576, 1967.
42. **Meighen, E. A., Nicolim, M. Z., and Hustings, J. W.,** Hybridization of bacterial luciferase with a variant produced by chemical modification, *Biochemistry,* 10, 4062, 1971.
43. **Shetty, K. J. and Rao, M. S. N.,** Effect of succinylation on the oligomeric structure of arachin, *Int. J. Pept. Protein Res.,* 11, 305, 1978.
44. **Shiao, D. D. F., Lumry, R., and Rajender, S.,** Modification of protein properties by change in charge. Succinylated chymotrypsinogen, *Eur. J. Biochem.,* 29, 377, 1972.

45. **Morton, R. C. and Gerber, G. E.,** Water solubilization of membrane proteins. Extensive derivatization with a novel polar derivatizing reagent, *J. Biol. Chem.,* 263, 7989, 1988.
46. **Atassi, M. Z. and Habeeb, A. F. S. A.,** Reaction of protein with citraconic anhydride, *Meth. Enzymol.,* 25, 546, 1972.
47. **Habeeb, A. F. S. A. and Atassi, M. Z.,** Enzymic and immunochemical properties of lysozyme. Evaluation of several amino group reversible blocking reagents, *Biochemistry,* 9, 4939, 1970.
48. **Shetty, J. K. and Kinsella, J. E.,** Ready separation of proteins from nucleoprotein complexes by reversible modification of lysine residues, *Biochem. J.,* 191, 269, 1980.
49. **Weisgraber, K. H., Innerarity, T. L., and Mahley, R. W.,** Role of the lysine residues of plasma lipoproteins in high affinity binding to cell surface receptors on human fibroblasts, *J. Biol. Chem.,* 253, 9053, 1978.
50. **Mahley, R. W., Weisgraber, K. H., Innerarity, T. L., and Windmueller, H. G.,** Accelerated clearance of low-density and high-density lipoproteins and retarded clearance of E apoprotein-containing lipoproteins from the plasma of rats after modification of lysine residues, *Proc. Natl. Acad. Sci. U.S.A.,* 76, 1746, 1979.
51. **Urabe, I., Yamamoto, M., Yamada, Y., and Okada, H.,** Effect of hydrophobicity of acyl groups on the activity and stability of acylated thermolysin, *Biochim. Biophys. Acta,* 524, 435, 1978.
52. **Howlett, G. J. and Wardrop, A. J.,** Dissociation and reconstitution of human erythrocyte membrane proteins using 3,4,5,6-tetrahydrophthalic anhydride, *Arch. Biochem. Biophys.,* 188, 429, 1978.
53. **Stark, G. R., Stein, W. H., and Moore, S.,** Reaction of the cyanate present in aqueous urea with amino acids and proteins, *J. Biol. Chem.,* 235, 3177, 1960.
54. **Stark, G. R. and Smyth, D. G.,** The use of cyanate for the determination of NH_2-terminal residues in proteins, *J. Biol. Chem.,* 238, 214, 1963.
55. **Stark, G. R.,** Modification of proteins with cyanate, *Meth. Enzymol.,* 25, 579, 1972.
56. **Shaw, D. C., Stein, W. H., and Moore, S.,** Inactivation of chymotrypsin by cyanate, *J. Biol. Chem.,* 239, PC 671, 1964.
57. **Cerami, A. and Manning, J. M.,** Potassium cyanate as an inhibitor of the sickling of erythrocytes *in vitro, Proc. Natl. Acad. Sci. U.S.A.,* 68, 1180, 1971.
58. **Lee, C. K. and Manning, J. M.,** Kinetics of the carbamylation of the amino groups of sickle cell hemoglobin by cyanate, *J. Biol. Chem.,* 248, 5861, 1973.
59. **Njikam, N., Jones, W. M., Nigen, A. M., Gillette, P. N., Williams, R. C., Jr., and Manning, J. M.,** Carbamylation of the chains of hemoglobin S by cyanate *in vitro* and *in vivo, J. Biol. Chem.,* 248, 8052, 1973.
60. **Nigen, A. M., Bass, B. D., and Manning, J. M.,** Reactivity of cyanate with valine-1 (α) of hemoglobin. A probe of conformational change and anion binding, *J. Biol. Chem.,* 251, 7638, 1976.
61. **Plapp, B. V., Moore, S., and Stein, W. H.,** Activity of bovine pancreatic deoxyribonuclease A with modified amino groups, *J. Biol. Chem.,* 246, 939, 1971.
62. **Sekiguchi, T., Oshiro, S., Goingo, E. M., and Nosoh, Y.,** Chemical modification of ε-amino groups in glutamine synthetase from *Bacillus stearothermophilus* with ethyl acetimidate, *J. Biochem.,* 85, 75, 1979.
63. **Browne, D. J. and Kent, S. B. H.,** Formation of non-amidine products in the reaction of primary amines with imido esters, *Biochem. Biophys. Res. Commun.,* 67, 126, 1975.
64. **Monneron, A. and d'Alayer, J.,** Effects of imido-esters on membrane-bound adenylate cyclase, *FEBS Lett.,* 122, 241, 1980.
65. **Plapp, B. V.,** Enhancement of the activity of horse liver alcohol dehydrogenase by modification of amino groups at the active sites, *J. Biol. Chem.,* 245, 1727, 1970.
66. **Dunathan, H. C.,** Stereochemical aspects of pyridoxal phosphate catalysis, *Adv. Enzymol.,* 35, 79, 1971.
67. **Shapiro, S., Enser, M., Pugh, E., and Horecker, B. L.,** The effect of pyridoxal phosphate on rabbit muscle aldolase, *Arch. Biochem. Biophys.,* 128, 554, 1968.

68. **Schnackerz, K. D. and Noltmann, E. A.,** Pyridoxal-5′-phosphate as a site-specific protein reagent for a catalytically critical lysine residue in rabbit muscle phosphoglucose isomerase, *Biochemistry,* 10, 4837, 1971.
69. **Havran, R. T. and du Vigneaud, V.,** The structure of acetone-lysine vasopressin as established through its synthesis from the acetone derivative of S-benzyl-L-cysteinyl-L-tyrosine, *J. Am. Chem. Soc.,* 91, 2696, 1969.
70. **Paech, C. and Tolbert, N. E.,** Active site studies of ribulose-1,5-bisphosphate carboxylase/oxygenase with pyridoxal-5′-phosphate, *J. Biol. Chem.,* 253, 7864, 1978.
71. **Kent, A. B., Krebs, E. G., and Fischer, E. H.,** Properties of crystalline phosphorylase b, *J. Biol. Chem.,* 232, 549, 1958.
72. **Wimmer, M. J., Mo, T., Sawyers, D. L., and Harrison, J. H.,** Biphasic inactivation of porcine heart mitochondrial malate dehydrogenase by pyridoxal-5′-phosphate, *J. Biol. Chem.,* 250, 710, 1975.
73. **Bleile, D. M., Jameson, J. L., and Harrison, J. H.,** Inactivation of porcine heart cytoplasmic malate dehydrogenase by pyridoxal-5′-phosphate, *J. Biol. Chem.,* 251, 6304, 1976.
74. **Jones, C. W., III and Priest, D. G.,** Interaction of pyridoxal-5′-phosphate with apo-serine hydroxymethyltransferase, *Biochim. Biophys. Acta,* 526, 369, 1978.
75. **Cortijo, M., Jimenez, J. S., and Lior, J.,** Criteria to recognize the structure and micropolarity of pyridoxal-5′-phosphate binding sites in proteins, *Biochem. J.,* 171, 497, 1978.
76. **Cake, M. A., DiSorbo, D. M., and Litwack, G.,** Effect of pyridoxal phosphate on the DNA binding site of activated hepatic glucocorticoid receptor, *J. Biol. Chem.,* 253, 4886, 1978.
77. **Slebe, J. C. and Martinez-Carrion, M.,** Selective chemical modification and ^{19}F NMR in the assignment of a pK value to the active site lysyl residue in aspartate transaminase, *J. Biol. Chem.,* 253, 2093, 1978.
78. **Nishigori, H. and Toft, D.,** Chemical modification of the avian progesterone receptor by pyridoxal-5′-phosphate, *J. Biol. Chem.,* 254, 9155, 1979.
79. **Sugiyama, Y. and Mukohata, Y.,** Modification of one lysine by pyridoxal phosphate completely inactivates chloroplast coupling factor 1 ATPase, *FEBS Lett.,* 98, 276, 1979.
80. **Peters, H., Risi, S., and Dose, K.,** Evidence for essential primary amino groups in a bacterial coupling factor F_1 ATPase, *Biochem. Biophys. Res. Commun.,* 97, 1215, 1980.
81. **Gould, K. G. and Engel, P. C.,** Modification of mouse testicular lactate dehydrogenase by pyridoxal 5′-phosphate, *Biochem. J.,* 191, 365, 1980.
82. **Ogawa, H. and Fujioka, M.,** The reaction of pyridoxal-5′-phosphate with an essential lysine residue of saccharopine dehydrogenase (L-lysine-forming), *J. Biol. Chem.,* 255, 7420, 1980.
83. **Forrey, A. W., Olsgaard, R. B., Nolan, C., and Fischer, E. H.,** Synthesis and properties of α- and ε-pyridoxyl lysines and their phosphorylated derivatives, *Biochimie,* 53, 269, 1971.
84. **Sober, H. A.,** *Handbook of Biochemistry,* 2nd ed., The Chemical Rubber Company, Cleveland, OH, 1970.
85. **Moldoon, T. G. and Cidlowski, J. A.,** Specific modifications of rat uterine estrogen receptor by pyridoxal-5′-phosphate, *J. Biol. Chem.,* 2, 55, 3100, 1980.
86. **Ohsawa, H., and Gualerzi, C.,** Structure-function relationship in *Escherichia coli* inhibition factors. Identification of a lysine residue in the ribosomal binding site of initiation factor by site-specific chemical modification with pyrodixal phosphate, *J. Biol. Chem.,* 256, 4905, 1981.
87. **Bürger, E. and Görisch, H.,** Evidence for an essential lysine at the active site of L-histidinol:NAD$^+$ oxidoreductase, a bifunctional dehydrogenase, *Eur. J. Biochem.,* 118, 125, 1981.
88. **Miller, A. D., Packman, L. C., Hart, G. J., Alefounder, P. R., Abell, C., and Battersby, A. R.,** Evidence that pyridoxal phosphate modification of lysine residues (Lys-55 and Lys-59) causes inactivation of hydroxymethylbilane synthase (porphobilinogen deaminase), *Biochem. J.,* 262, 119, 1989.

89. **Basu, S., Basu, A., and Modak, M. J.,** Pyridoxal 5′-phosphate mediated inactivation of *Escherichia coli* DNA polymerase I: identification of lysine-635 as an essential residue for the processive mode of DNA synthesis, *Biochemistry,* 27, 6710, 1988.
90. **Mahrenholz, A. M., Wang, Y., and Roach, P. J.,** Catalytic site of rabbit glycogen synthase isozymes. Identification of an active site lysine close to the amino terminus of the subunit, *J. Biol. Chem.,* 263, 10561, 1988.
91. **Tamura, J. K., LaDine, J. R., and Cross, R. L.,** The adenine nucleotide binding site on yeast hexokinase PII. Affinity labeling of Lys-111 by pyridoxal 5′-diphospho-5′-adenosine, *J. Biol. Chem.,* 263, 7907, 1988.
92. **Kluger, R.,** Methyl acetyl phosphate. A small anionic acetylating agent, *J. Org. Chem.,* 45, 2733, 1980.
93. **Ueno, H., Pospischil, M. A., Manning, J. M., and Kluger, R.,** Site-specific modification of hemoglobin by methyl acetyl phosphate, *Arch. Biochem. Biophys.,* 244, 795, 1986.
94. **Ueno, H., Pospischil, M. A., and Manning, J. M.,** Methyl acetyl phosphate as a covalent probe for anion-binding sites in human and bovine hemoglobins, *J. Biol. Chem.,* 264, 12344, 1989.
95. **Means, G. E.,** Reductive alkylation of amino groups, *Meth. Enzymol.,* 47, 469, 1977.
96. **Rice, R. H., Means, G. E., and Brown, W. D.,** Stabilization of bovine trypsin by reductive methylation. *Biochim. Biophys. Acta,* 492, 316, 1977.
97. **Morris, R. W., Cagan, R. H., Martenson, R. E., and Deibler, G.,** Methylation of the lysine residues of monellin, *Proc. Soc. Exp. Biol. Med.,* 157, 194, 1978.
98. **Chen, F. M. F. and Benoiton, N. L.,** Reductive *N,N*-dimethylation of amino acid and peptide derivatives using methanol as the carbonyl source, *Can. J. Biochem.,* 56, 150, 1978.
99. **Dottavio-Martin, D. and Ravel, J. M.,** Radiolabeling of proteins by reductive alkylation with [^{14}C] formaldehyde and sodium cyanoborohydride, *Anal. Biochem.,* 87, 562, 1978.
100. **Jentoft, N. and Dearborn, D. G.,** Labeling of proteins by reductive methylation using sodium cyanoborohydride, *J. Biol. Chem.,* 254, 4359, 1979.
101. **Dick, L. R., Geraldes, C. F. G. C., Sherry, A. D., Gray, C. W., and Gray, D. M.,** ^{13}C NMR of methylated lysines of fd gene 5 protein: evidence for a conformational change involving lysine 24 upon binding of a negatively charged lanthanide chelate, *Biochemistry,* 28, 7896, 1989.
102. **Brown, E. M., Pfeffer, P. E., Kumosinski, T. F., and Greenberg, R.,** Accessibility and mobility of lysine residues in β-lactoglobulin, *Biochemistry,* 27, 5601, 1988.
103. **Fretheim, K., Iwai, S., and Feeney, R. F.,** Extensive modification of protein amino groups by reductive addition of different sized substituents, *Int. J. Pept. Protein Res.,* 14, 451, 1979.
104. **Fretheim, K., Edelandsdal, B., and Harbitz, O.,** Effect of alkylation with different size substituents on the conformation of ovomucoid, lysozyme and ovotransferrin, *Int. J. Pept. Protein Res.,* 25, 601, 1985.
105. **Geoghegan, K. F., Ybarra, D. M., and Feeney, R. E.,** Reversible reductive alkylation of amino groups in proteins, *Biochemistry,* 18, 5392, 1979.
106. **Jentoft, J. E., Jentoft, N., Gerken, T. A., and Dearborn, D. G.,** ^{13}C NMR studies of ribonuclease A methylated with [^{13}C] formaldehyde, *J. Biol. Chem.,* 254, 4366, 1979.
107. **Jentoft, N. and Dearborn, D. G.,** Protein labeling by reductive methylation with sodium cyanoborohydride effect of cyanide and metal ions on the reaction, *Anal. Biochem.,* 106, 186, 1980.
108. **Geoghegan, K. F., Cabacungan, J. C., Dixon, H. B. F., and Feeney, R. E.,** Alternative reducing agents for reductive methylation of amino groups in proteins, *Int. J. Pept. Protein Res.,* 17, 345, 1981.
109. **Cabacungan, J. C., Ahmed, A. J., and Feeney, R. E.,** Amine boranes as alternative reducing agents for reductive alkylation of proteins, *Anal. Biochem.,* 124, 272, 1982.
110. **Wu, H.-L. and Means, G. E.,** Immobilization of proteins by reductive alkylation with hydrophobic aldehydes, *Biotechnol. Bioeng.,* 23, 855, 1981.

111. **Acharya, A. S. and Manning, J. M.,** Reactivity of the amino groups of carbonmonoxyhemoglobin S with glyceraldehyde, *J. Biol. Chem.,* 255, 1406, 1980.
112. **Acharya, A. S. and Manning, J. M.,** Amadori rearrangement of glyceraldehyde-hemoglobin Schiff base adducts. A new procedure for the determination of ketoamine adducts in proteins, *J. Biol. Chem.,* 255, 7218, 1980.
113. **Wilson, G.,** Effect of reductive lactosamination on the hepatic uptake of bovine pancreatic ribonuclease A dimer, *J. Biol. Chem.,* 253, 2070, 1978.
114. **Bunn, H. F. and Higgins, P. J.,** Reaction of monosaccharides with proteins: possible evolutionary significance, *Science,* 213, 222, 1981.
115. **Goldfarb, A. R.,** A kinetic study of the reactions of amino acids and peptides with trinitrobenzenesulfonic acid, *Biochemistry,* 5, 2570, 1966.
116. **Goldfarb, A. R.,** Heterogeneity of amino groups in proteins. I. Human serum albumin, *Biochemistry,* 5, 2574, 1966.
117. **Habeeb, A. F. S. A.,** Determination of free amino groups in proteins by trinitrobenzenesulfonic acid, *Anal. Biochem.,* 14, 328, 1966.
118. **Fields, R.,** The rapid determination of amino groups with TNBS, *Meth. Enzymol.,* 25, 464, 1972.
119. **Kotaki, A. and Satake, K.,** Acid and alkaline degradation of the TNP-amino acids and peptides, *J. Biochem.,* 56, 299, 1964.
120. **Coffee, C. J., Bradshaw, R. A., Goldin, B. R., and Frieden, C.,** Identification of the sites of modification of bovine liver glutamate dehydrogenase reacted with trinitrobenzenesulfonate, *Biochemistry,* 10, 3516, 1971.
121. **Goldin, B. R. and Frieden, C.,** Effects of trinitrophenylation of specific lysyl residues on the catalytic, regulatory and molecular properties of bovine liver glutamate dehydrogenase, *Biochemistry,* 10, 3527, 1971.
122. **Bates, D. J., Goldin, B. R., and Frieden, C.,** A new reaction of glutamate dehydrogenase: the enzyme-catalyzed formation of trinitrobenzene from TNBS in the presence of reduced coenzyme, *Biochem. Biophys. Res. Commun.,* 39, 502, 1970.
123. **Means, G. E., Congdon, W. I., and Bender, M. L.,** Reactions of 2,4,6-trinitrobenzenesulfonate ion with amines and hydroxide ion, *Biochemistry,* 11, 3564, 1972.
124. **Whitaker, J. R., Granum, P. E., and Aasen, G.,** Reaction of ammonia with trinitrobenzene sulfonic acid, *Anal. Biochem.,* 108, 72, 1980.
125. **Flügge, U. I. and Heldt, H. W.,** Specific labelling of the active site of the phosphate translocator in spinach chloroplasts by 2,4,6-trinitrobenzene sulfonate, *Biochem. Biophys. Res. Commun.,* 84, 37, 1978.
126. **Flügge, U. I. and Heldt, H. W.,** Specific labelling of a protein involved in phosphate transport of chloroplasts by pyridoxal-5′-phosphate, *FEBS Lett.,* 82, 29, 1977.
127. **Parrott, C. L. and Shifrin, S.,** A spectrophotometric study of the reaction of borohydride with trinitrophenyl derivatives of amino acids and proteins, *Biochim. Biophys. Acta,* 491, 114, 1977.
128. **Salem, N., Jr., Lauter, C. J., and Trams, E. G.,** Selective chemical modification of plasma membrane ectoenzymes, *Biochim. Biophys. Acta,* 641, 366, 1981.
129. **Haniu, M., Yuan, H., Chen, S., Iyanagi, T., Lee, T. D., and Shively, J. E.,** Structure-function relationship of NAD(P)H:quinone reductase: characterization of NH_2-terminal blocking group and essential tyrosine and lysine residues, *Biochemistry,* 27, 6877, 1988.
130. **Boulton, A. E. and Hunter, W. M.,** The labelling of proteins to high specific radioactivities by conjugation to a ^{125}I-containing acylating agent, *Biochem. J.,* 133, 529, 1973.
131. **Yem, A. W., Zurcher-Neely, H. A., Richard, K. A., Staite, N. D., Heinrikson, R. L., and Deibel, M. R., Jr.,** Biotinylation of reactive amino groups in native recombinant human interleukin-1β, *J. Biol. Chem.,* 264, 17691, 1989.
132. **Wilchek, M. and Bayer, E.,** Biotin-containing reagents, *Meth. Enzymol.,* 184, 123, 1990.
133. **Bayer, E. and Wilchek, M.,** Protein Biotinylation, *Meth. Enzymol.,* 184, 148, 1990.
134. **Dixon, H. B. F. and Fields, R.,** Specific modification of NH_2-terminal residues by transamination, *Meth. Enzymol.,* 25, 409, 1972.

135. **Verheij, H. M., Egmond, M. R., and de Haas, G. H.,** Chemical modification of the α-amino group in snake venom phospholipases A_2. A comparison of the interaction of pancreatic and venom phospholipases with lipid-water interfaces, *Biochemistry,* 20, 94, 1981.

Chapter 11

THE MODIFICATION OF ARGININE

The group-specific modification of arginine is relatively easy to achieve; it is somewhat more difficult to obtain the site-specific modification of arginine. Arginine residues are usually involved in binding sites rather than catalytic sites. Thus, modification of arginine rarely results in the total loss of activity. As a result, it is critical to take residual activity into consideration in determining the kinetics of inactivation. The author has found the approach developed by Levy et al.[1] quite useful. This approach has been used in a number of studies on the site-specific modification of proteins.[2–8]

Present approaches to the site-specific modification of arginyl residues in proteins used three reagents: phenylglyoxal (and derivatives such as *p*-hydroxyphenylgloxal),[9] 2,3-butanedione,[10] and 1,2-cyclohexanedione.[11] While other reagents for the site-specific modification of arginine have been described, their current use is negligible. The reader is referred to an earlier work for a more complete listing of reagents.[12]

The determination of the extent of arginine modification is generally determined by amino acid analysis after acid hydrolysis, but conditions generally need to be modified to prevent loss of the arginine derivative.[11] The Sakaguchi reaction has been used, after acid hydrolysis, to determine the extent of arginine modification by 2,3-butanedione.[13] A fluorometric method for the determination of arginine using 9,10-phenanthrenequinone[14] has been described. This method is some 1000-fold more sensitive than the Sakaguchi reaction, but some concern remains concerning the absolute accuracy of the reagent for determination of arginine in peptide linkage. This is also true of the other reagents.

The use of phenylglyoxal (Figure 1) was developed by Takahashi[9] and has since been applied to the study of the role of arginyl residues in proteins as shown in Table 1. The use of this reagent has been somewhat limited in the last 6 years, but there has been interest in modifying residues involved in ion transport. Unfortunately, many of these studies fail to recognize that phenylglyoxal, like glyoxal, will react with α-amino groups at a significant rate.[9] Polymerization was noted in a sample incubated for 21 h. The amino-terminal lysine residue was rapidly modified under these conditions. The possible effect of light on the reaction of phenylglyoxal with arginine, as has been reported for 2,3-butanedione,[15–17] has not been studied. As noted by Takahashi, the stoichiometry of the reaction involves the reaction of 2 mol of phenylglyoxal with 1 mol of arginine (Figure 1). The [^{14}C]-labeled reagent can be easily prepared.[9,18] A facile modification of the original Riley and Gray[18] method which omits the vacuum distillation step has been reported by Schloss and co-workers.[19] Radiolabeled acetophenone was added to an equal amount (on the basis of weight) of selenium dioxide in dioxane-water (30:1). The mixture was refluxed for 3 h, after which solvent was removed under a stream of nitrogen. The residue was taken up in boiling water and activated charcoal was added. The hot

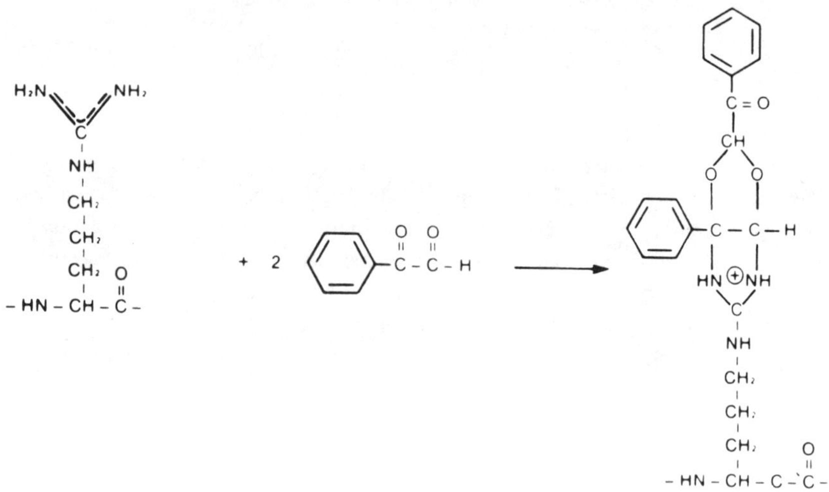

FIGURE 1. A scheme for the reaction of phenylglyoxal with arginine.

slurry was filtered through celite. The phenylglyoxal crystallized spontaneously from the filtrate on cooling.

The synthesis of phenyl [2-^{3}H]-glyoxal[20] has been reported. Borders and co-workers[21] have reported the synthesis of a chromophoric derivative, 4-hydroxy-3-nitrophenylglyoxal, which should prove quite useful in the study of arginyl residues. *p*-Hydroxylphenylglyoxal[22] has also been used as a spectrophotometric reagent for the study of this reaction.

The reaction of arginine with phenylglyoxal is greatly accelerated in bicarbonate-carbonate buffer systems.[23] Figure 2 shows the reaction of phenylglyoxal with *N*-acetylarginine, *N*-acetyllysine, and *N*-acetylcysteine in 0.083 *M* sodium bicarbonate, pH 7.5. L-Arginine reacted in the same manner, suggesting that modification of the α-amino group did not occur under these conditions. The reaction appears to be first order with respect to bicarbonate (Figure 3). The reaction of methylglyoxal with arginine is also enhanced by bicarbonate, while a similar effect is not seen with either glyoxal or 2,3-butanedione. The molecular basis for this specific buffer effect is not clear at this time nor is it known whether reaction with α-amino functional groups occurs at a different rate than with other solvent systems used for this modification of arginine with phenylglyoxal. Yamasaki and co-workers[24] reported that *p*-nitrophenylglyoxal reacts with arginine in 0.17 sodium pyrophosphate — 0.15 *M* sodium ascorbate, pH 9.0 to yield a derivative which absorbs at 475 nm. There is also reaction with histidine (the imidazole ring is critical for this reaction in that the l-methyl derivative yielded a derivative which absorbed at 475 nm, while the 3-methyl derivative did not). Free sulfhydryl groups also yielded a product with absorbance at 475 nm, but its absorbance was only 3% of that of the arginine. Branlant and co-workers[25] have used *p*-carboxyphenyl glyoxal in bicarbonate buffer at pH 8.0 to modify aldehyde reductase. Saturation kinetics were noted with the use of this reagent.

Table 1
REACTION OF PHENYLGLYOXAL WITH ARGINYL RESIDUES IN PEPTIDES AND PROTEINS

Protein	Solvent	Reagent excess[a]	Extent of modification	Ref.
Pancreatic RNase	0.1 *M* *N*-ethylmorpholine acetate, pH 8.0	—	2–3/4[b,c]	1
Porcine carboxypeptidase B	0.3 *M* borate, pH 7.9	200[d]	1[e]	2
Aspartate transcarbamylase	0.125 *M* potassium bicarbonate, pH 8.3 or 0.1 *M* *N*-ethylmorpholine, pH 8.3	—	2.2[f,g]	3
Pyruvate kinase	0.1 *M* triethanolamine, pH. 7.0	—	3/28.33[h]	4
Horse liver alcohol dehydrogenase	—	—	—	5, 6
Mitochondrial ATPase	0.097 *M* sodium borate, 0.097 *M* EDTA, pH 8.0	—	4[i]	7
Adenylate kinase	0.1 *M* triethanolamine · HCl, pH 7.0	—	—[j]	8, 9
Rhodospirillum rubrum chromatophores	0.05 *M* borate, pH 8.0	—	—[k]	10
Glutamic acid decarboxylase	0.05 *M* sodium borate[l]	—	—	11
Ribulose bisphosphate carboxylase	0.066 *M* sodium[m] bicarbonate, 0.050 M Bicine, 0.1 *M* EDTA, pH 8.0	—	2–3/35[n]	12
Yeast hexokinase	0.035 *M* Veronal, pH 7.5	—	1/18[o]	13
Propionyl CoA carboxylase	0.050 *M* borate, pH 8.0	—	—	14
β-Methylcrotonyl CoA carboxylase	0.050 *M* borate pH 8.0	—	—	14
Superoxide dismutase	0.125 *M* sodium bicarbonate, pH 8.0	—	1/4[p]	15
Myosin (subfragment 1)	0.1 *M* potassium bicarbonate, pH 8.0	—	1.7/35[q]	16
Thymidylate synthetase	0.125 *M* bicarbonate, pH 8.0[r]	—	3.6/12	17
Glutamate apodecarboxylase	0.125 *M* sodium[s] bicarbonate, pH 7.5	—	1/23[t]	18
Adenylateu kinase (yeast)	0.025 *M* HEPES, pH 7.5	—	—	19
Cardiac myosin S-1	0.1 *M* *N*-ethylmorpholine acetate, pH 7.6	—	2.8/42[v]	20
Cystathionase	0.125 *M* bicarbonate, pH 7.9	—	18/45	21
Fatty acid synthetase	0.1 *M* sodium phosphate, 0.0005 *M* dithioerythritol, 0.001 *M* EDTA, pH 7.6	—	4/106	22
Yeast inorganic pyrophosphatase	0.08 *M* *N*-ethylmorpholine acetate, pH 7.0	—	1/6	23
Porcine phospholipase A	0.125 *M* potassium bicarbonate, pH 8.5	—	1.4/4[w]	24
Superoxide dismutase[x]	0.100 *M* sodium bicarbonate, pH 8.3	50–100	0.88/4.0[y]	25, 26[z]

Table 1 (continued)
REACTION OF PHENYLGLYOXAL WITH ARGINYL RESIDUES IN PEPTIDES AND PROTEINS

Protein	Solvent	Reagent excess[a]	Extent of modification	Ref.
p-Hydroxybenzoate hydroxylase	0.050 *M* potassium phosphate, pH 8.0	250	2–3/24[aa]	27
Thymidylate synthetase	0.200 *M* *N*-ethylmorpholine, pH 7.4[cc]	65	2/12[bb]	28
Acetylcholine esterase	0.025 *M* borate, 0.005 phosphate, 0.050 *M* NaCl, pH 7.0	—	3/31[dd]	29
γ-Aminobutyrate aminotransferase	0.05 *M* Tris, pH 8.5	—	—	30
D-β-Hydroxybutyrate dehydrogenase	0.05 *M* HEPES, pH 7.5	—	—[ee]	31
Ornithine transcarboxylase	0.05 *M* Bicine, 0.1 *M* KCl, 0.0001 *M* EDTA, pH 8.05	—	—[ff]	32
Coenzyme B_{12}-dependent diol dehydrase	0.05 *M* borate, pH 8.0	—	—	33
Transketolase	0.125 *M* sodium bicarbonate, pH 7.6	—	4/34[gg]	34
ATP citrate lyase	0.050 *M* HEPES,[hh] pH 8.0	—	8.5/40	36
Malic enzyme	0.037 *M* borate,[ii] pH 7.5	—	—	37
Pyridoxamine-5′-phosphate oxidase	0.1 *M* potassium phosphate, pH 8.0, containing 5% ETOH	—[jj]	6/40	38
Ornithine transcarboxylase	0.125 *M* potassium bicarbonate, pH 8.3	—[kk]	1.5/[ll]	39
Acetate kinase	0.050 *M* triethanolamine, pH 7.6	—[mm]	—[nn]	40
Pancreatic phospholipase A_2	0.2 *M* *N*-ethylmorpholine, pH 8.0	30	1.0–1.2[oo]	41
Phosphatidylcholine transfer protein	0.1 *M* sodium bicarbonate, pH 8.0	—	4/10[pp]	42
Aldehyde reductase	0.020 *M* phosphate,[qq] pH 7.0	—	0.6/16[rr]	43
Choline acetyltransferase	0.050 *M* HEPES, pH 7.8	—	—[ss]	44
ADP-glucose synthetase	0.05 *M* potassium phosphate, 0.00025 *M* EDTA, pH 7.5	110	1[tt]	45
Pyruvate oxidase	0.1 *M* sodium phosphate, 0.010 *M* magnesium chloride, pH 7.8	—	2.5/5[uu]	46
Calcineurin	50 m*M* Tris, pH 7.5, with 0.1 *M* EDTA, 0.1 m*M* $NiCl_2$ and 0.3 m*M* $CaCl_2$	10,000[vv]	—	47
Carbon monoxide	20 m*M* sodium phosphate, pH 8.2 with 4 m*M* dithiothreitol	—	—	48
Epithelial sodium channel	pH 8.1[ww]	—	—	49

Table 1 (continued)
REACTION OF PHENYLGLYOXAL WITH ARGINYL RESIDUES IN PEPTIDES AND PROTEINS

Protein	Solvent	Reagent excess[a]	Extent of modification	Ref.
Calcium "pump"	100 m*M* *N*-ethylmorpholine, 40 m*M* KCl, 6 m*M* HEPES, 0.7 m*M* $MgCl_2$, 0.2 m*M* EGTA, pH 7.7	—	—	50
Calmodulin dependent	40 m*M* HEPES, pH 7.5 with 7% glycerol, 0.1 *M* EDTA, and 0.3 m*M* $CaCl_2$	333[xx]	—	51

[a] Reagent/protein.
[b] After 3 h at 25°C.
[c] Had modification of α-amino group and lysine residues.
[d] Reagent/arginine.
[e] After 1 h at 37°C.
[f] After 3 h at 25°C.
[g] 1.3/8 In regulatory chain.
[h] 20 Min at 37°C with 23.8 m*M* phenylglyoxal, protein 1 mg/ml.
[i] 30 Min of reaction at 30°C; the presence of efrapeptin, a low-molecular weight antibiotic which is a potent inhibitor of oxidative phosphorylation, prevented the modification of one "fast-reacting" arginyl residue.
[j] A single arginine residue is modified (Arg-97).
[k] A single site appeared to be modified with a second-order rate constant of 1.6 M^{-1} min^{-1}.
[l] pH not given; reaction at 23°C, kinetic evidence for stoichiometric inactivation.
[m] Solvent made metal-free using BioRad Chelex; reaction performed with and without $MgCl_2$.
[n] Analysis of sulfhydryl groups after phenylglyoxal modification showed no loss of cysteine. These investigations noted that modification with phenylglyoxal is apparently more specific than 2,3-butanedione.
[o] The authors claim 1:1 stoichiometry of phenylglyoxal with the arginyl residue from analysis of dependence of pseudo-first-order rate constant vs. reciprocal of reagent (phenylglyoxal concentration). Partial reactivation of modified enzyme was observed reflecting lability of modified arginine residues. Reaction also shows saturation kinetics reflecting "specific" affinity of reagent for enzyme possibly from hydrophobic interaction. These authors suggest that this phenomenon is observed with the reaction of other hydrophobic reagents with this enzyme. A similar phenomenon has been observed with trinitrobenzenesulfonic acid (see Chapter 10).
[p] Reaction was performed at 25°C for 1 h.
[q] Reaction was performed at 25°C with 3 m*M* phenylglyoxal for 3 min.
[r] Rates of enzyme inactivation were dependent upon buffer; at 5.9 m*M* phenylglyoxal, the following data were obtained, bicarbonate ($T_{1/2}$ = 6.0 min), MOPS ($T_{1/2}$ = 11.5 min), borate ($T_{1/2}$ = 34.0), and phosphate ($T_{1/2}$ = 48.0 min) at 25°C.
[s] These investigators noted a significant buffer effect on the reaction, which is more thoroughly explored in Reference 29 of Chapter 11. In this study, the following second-order rate constants were obtained with the following reagent/solvent conditions (reactions performed at 23°C): 0.69 M^{-1} min^{-1} with 2,3-butanedione 0.050 *M* borate, pH 8.0; 33.78 M^{-1} min^{-1} with glyoxal/0.125 *M* sodium bicarbonate, pH 8.0; 31.00 M^{-1} min^{-1} with methylglyoxal/0.125 *M* sodium bicarbonate, pH 8.0; and 107.68 M^{-1} min^{-1} with phenylglyoxal 0.125 *M* sodium bicarbonate, pH 8.0.
[t] 300-Fold excess of reagent, 0.083 *M* sodium bicarbonate, pH 8.1, 7 min, 23°C.

Table 1 (continued)

[u] See more complete discussion of this work in Table 2. 2,3-Butanedione or 1,2-cyclohexanedione appeared to be more effective than phenylglyoxal in this system.

[v] 6 Min, 22°C, 50% loss of activity.

[w] Determined at 99% inactivation (25°C) of phospholipase activity (release of fatty acid from egg yolk in water with 3 m*M* $CaCl_2$ and 1.4 m*M* sodium deoxycholate). These investigators (see Reference 24, Table 1) did examine the possibility of amino-terminal alanine modification; no loss of alanine was observed with 75% inactivation (0.9 mol Arg modified per mole of protein), while enzyme samples with a greater extent of inactivation did have some loss of amino-terminal alanine (quantity not given). These investigators did examine the pH dependence of enzyme inactivation by phenylglyoxal (presumably, a direct measure of the rate of arginine modification) and reported the following second-order rate constants (M^{-1} min^{-1}): pH 6.5, 0.3; pH 7.5, 1.5; pH 8.5, 3.3; and pH 9.5, 3.9. These investigators also showed that phenylglyoxal ($T_{1/2}$ = 1 min) was more effective than 2,3-butanedione ($T_{1/2}$ = 20 min) and 1,2-cyclohexanedione ($T_{1/2}$ = 120 min).

[x] Cu, Zn superoxide dismutase from *Saccharomyces cerevisiae.*

[y] Determined at 80% loss of enzymatic activity using reaction of the modified enzyme with 9,10-phenanthrenequinone. This value corresponded to that determined by the incorporation of radiolabeled phenylglyoxal assuming 2:1 adduct. Amino acid analysis with samples prepared using normal hydrolytic conditions (6 *N* HCl, 110°C, 20 h) suggested only approximately 50% of this extent of arginine modification. When thioglycolic acid was included during the hydrolysis, values for the extent of arginine modification approached those determined by the fluorescence technique and radiolabel incorporation.

[z] The study is an extension of the observation reported in Reference 25 and uses reaction with 4-hydroxy-3-nitrophenylglyoxal, a chromophoric derivative of phenylglyoxal, to identify the specific arginine residue modified. It is of some interest that the rate of reaction with this derivative is approximately sixfold less than that with the parent phenylglyoxal.

[aa] Reaction was performed at 25°C for 60 to 120 min. Loss of lysine residues was not observed under these reaction conditions. Amino acid analysis (hydrolysis in 6 *N* HCl, 110°C, 24 h) correlated well with radiolabeled phenylglyoxal incorporation assuming 2:1 stoichiometry (i.e., amino acid analysis gave 3.6 mol Arg lost per mole of enzyme, while 7.54 mol radiolabel was incorporated).

[bb] The presence of substrate, 2′-deoxyuridylate, prevents the modification of 1 mol of arginine per mole of enzyme. It is noted that these results differ from those reported in Reference 17. There were differences in solvent conditions. It is not clear why this would account for the differences observed in these two studies. It is noted that the investigators in Reference 17 obtained similar stoichiometry with 2,3-butanedione.

[cc] These investigators (see Reference 28, Table 1) examined the reaction at pH 7.4 (rate of inactivation of 32 M^{-1} min^{-1}). An approximate 100-fold increase in the rate of inactivation.

[dd] The modification with phenylglyoxal is associated with an approximate 15% loss of enzyme activity. Treatment with 2,3-butanedione under similar reaction conditions results in the modification of approximately one more mole of arginine per mole enzyme with an approximate 75% loss of catalytic activity.

[ee] Stoichiometry was not established but the data are consistent with the loss of activity resulting from the modification of a single arginine residue. Submitochondrial vesicles were used as the source of enzyme in these studies. A second-order rate constant of 1.03 M^{-1} min^{-1} was obtained from the reaction with phenylglyoxal. A value of 0.8 M^{-1} min^{-1} was obtained for reaction with 1,2-cyclohexanedione (0.050 *M* borate, pH 7.5) while a value of 4.6 M^{-1} min^{-1} was obtained for 2,3-butanedione in the borate buffer system.

[ff] See more complete discussion of this study under Table 2. For inactivation by phenylglyoxal, a second-order rate constant of 56 M^{-1} min^{-1} was obtained at pH 8.04. The reactions were performed in the dark.

Table 1 (continued)

gg Analysis of Tsou plots[38] indicates at least two classes of residues react at different rates.

hh Most studies were performed in this solvent at 30°C with a second-order rate constant of 0.33 M^{-1} s^{-1}. The rate was reduced in potassium phosphate (k = 0.25 M^{-1} s^{-1}) and borate (k = 0.078 M^{-1} s^{-1}).

ii Under these conditions at 24°C, a second-order rate constant of k = 7.08 M^{-1} min^{-1}, assuming that the rate of inactivation is directly related to the modification of arginine. With 2,3-butanedione in 0.048 *M* borate, a second-order rate constant of k = 5.4 M^{-1} min^{-1} is compared to 1.69 M^{-1} min^{-1} with methylglyoxal and 0.032 M^{-1} min^{-1} with 2,4-pentanedione.

jj The rate of inactivation at 25°C for the apoenzyme was determined to be 3.7 and 11.1 M^{-1} min^{-1} for the holoenzyme.

kk A second-order rate constant of k = 4.6 M^{-1} min^{-1} at 25°C was obtained under these conditions.

ll Based on incorporation of radiolabeled phenylglyoxal, 1.5 arginine residues are modified per 35,000 chain after 3 h of reaction. There are likely different classes of reactive arginyl residues, where the more reactive group(s) are directly associated with catalytic activity.

mm Saturation kinetics are observed with phenylglyoxal, suggesting the formation of an enzyme-inhibitor complex prior to reaction with an arginine residue(s).

nn With 95% loss of catalytic activity, there is 94% modification of arginine.

oo See Reference 24 for somewhat differing results. This study shows that this level of arginine modification is associated with 80% loss of amino-terminal alanine. It was necessary to protect the α-amino group of the amino-terminal alanine with a *t*-butyloxycarbonyl group to avoid modification under these reaction conditions. The use of radiolabeled cyclohexandedione established Arg-6 as the primary site of modification.

pp Reaction was performed for 30 min at 25°C. Extent of modification based on radiolabel incorporation and amino acid analysis.

qq For reaction at 30°C, a second-order rate constant of k = 2.6 M^{-1} min^{-1}, assuming that the loss of activity seen with phenylglyoxal directly reflects the loss of an arginine residue(s).

rr Determined from both amino acid analysis and radiolabel incorporation.

ss Phenylglyoxal was much more effective than 2,3-butanedione or camphorquinone-10-sulfonic acid.

tt Assuming 2:1 stoichiometry of phenylglyoxal to arginine; reaction at 25°C. Phenylglyoxal is much more effective than 1,2-cyclohexanedione (twofold molar excess of 1,2-cyclohexanedione had $T_{1/2}$ = 24 min).

uu From radiolabel incorporation assuming 2:1 stoichiometry. There are clearly at least two classes of reactive arginine residue. When the reaction is performed at pH 6.0, inactivation with phenylglyoxal can be partially reversed on dilution in pH 6.0 buffer.

vv Inactivation rate constant of 1.5 M^{-1} min^{-1} at pH 7.5/30°C.

ww Reaction was performed with an undefined quantity of Tris buffer. The inactivation reaction was markedly increased by the presence of sodium ions.

xx Inactivation rate constant of 132 M^{-1} min^{-1} at pH 7.5

References for Table 1

1. **Takahashi, K.,** The reaction of phenylglyoxal with arginine residues in proteins, *J. Biol. Chem.,* 243, 6171, 1968.
2. **Werber, M. M. and Sokolovsky, M.,** Chemical evidence for a functional arginine residue in carboxypeptidase B, *Biochem. Biophys. Res. Commun.,* 48, 384, 1972.
3. **Kantrowitz, E. R. and Lipscomb, W. N.,** Functionally important arginine residues of aspartate transcarbamylase, *J. Biol. Chem.,* 252, 2873, 1977.
4. **Berghäuser, J.,** Modifizierung von argininresten in pyruvat-kinase, *Hoppe-Seyler's Physiol. Chem.,* 358, 1565, 1977.

References for Table 1 (continued)

5. **Lange, L. G., III, Riordan, J. F., and Vallee, B. L.,** Functional argininyl residues as NADH binding sites of alcohol dehydrogenases, *Biochemistry,* 13, 4361, 1974.
6. **Jörnvall, H., Lange, L. G., III, Riordan, J. F., and Vallee, B. L.,** Identification of a reactive arginyl residue in horse liver alcohol dehydrogenase, *Biochem. Biophys. Res. Commun.,* 77, 73, 1977.
7. **Kohlbrenner, W. E. and Cross, R. L.,** Efrapeptin prevents modification by phenylglyoxal of an essential arginyl residue in mitochondrial adenosine triphosphatase, *J. Biol. Chem.,* 253, 7609, 1978.
8. **Berghäuser, J.,** A reactive arginine in adenylate kinase, *Biochim. Biophys. Acta,* 397, 370, 1975.
9. **Berghäuser, J. and Schirmer, R. H.,** Properties of adenylate kinase after modification of Arg-97 by phenylglyoxal, *Biochim. Biophys. Acta,* 537, 428, 1978.
10. **Vallejos, R. H., Lescano, W. I. M., and Lucero, H. A.,** Involvement of an essential arginyl residue in the coupling activity of *Rhodospirillum rubrum* chromatophores, *Arch. Biochem. Biophys.,* 190, 578, 1978.
11. **Tunnicliff, G. and Ngo, T. T.,** Functional role of arginine residues in glutamic acid decarboxylase from brain and bacteria, *Experientia,* 34, 989, 1978.
12. **Schloss, J. V., Norton, I. L., Stringer, C. D., and Hartman, F. C.,** Inactivation of ribulosebisphosphate carboxylase by modification of arginyl residues with phenylgloxal, *Biochemistry,* 17, 5626, 1978.
13. **Philips, M., Pho, D. B., and Pradel, L.-A.,** An essential arginyl residue in yeast hexokinase, *Biochim. Biophys. Acta,* 566, 296, 1979.
14. **Wolf, B., Kalousek, F., and Rosenberg, L. E.,** Essential arginine residues in the active sites of propionyl CoA carboxylase and beta-methylcrotonyl CoA carboxylase, *Enzyme,* 24, 302, 1979.
15. **Malinowski, D. P. and Fridovich, I.,** Chemical modification of arginine at the active site of the bovine erythrocyte superoxide dismutase, *Biochemistry,* 18, 5909, 1979.
16. **Mornet, D., Pantel, P., Audemard, E., and Kassab, R.,** Involvement of an arginyl residue in the catalytic activity of myosin heads, *Eur. J. Biochem.,* 100, 421, 1979.
17. **Cipollo, K. L. and Dunlap, R. B.,** Essential arginyl residues in thymidylate synthetase from amethopterin-resistant *Lactobacillus casei, Biochemistry,* 18, 5537, 1979.
18. **Cheung, S.-T. and Fonda, M. L.,** Kinetics of the inactivation of *Escherichia coli* glutamate apodecarboxylase by phenylglyoxal, *Arch. Biochem. Biophys.,* 198, 541, 1979.
19. **Varimo, K. and Londesborough, J.,** Evidence for essential arginine in yeast adenylate cyclase, *FEBS Lett.,* 106, 153, 1979.
20. **Morkin, E., Flink, I. L., and Banerjee, S. K.,** Phenylglyoxal modification of cardiac myosin S-1. Evidence for essential arginine residues at the active site, *J. Biol. Chem.,* 254, 12647, 1979.
21. **Portemer, C., Pierre, Y., Loriette, C., and Chatagner, F.,** Number of arginine residues in the substrate binding sites of rat liver cystathionase, *FEBS Lett.,* 108, 419, 1979.
22. **Poulose, A. J. and Kolattukudy, P. E.,** Presence of one essential arginine that specifically binds the 2′-phosphate of NADPH on each of the ketoacyl reductase and enoyl reductase active sites of fatty acid synthetase, *Arch. Biochem. Biophys.,* 199, 457, 1980.
23. **Bond, M. W., Chiu, N. Y., and Cooperman, B. S.,** Identification of an arginine residue important for enzymatic activity within the covalent structure of yeast inorganic pyrophosphatase, *Biochemistry,* 19, 94, 1980.
24. **Vensel, L. A. and Kantrowitz, E. R.,** An essential arginine residue in porcine phospholipase A_2, *J. Biol. Chem.,* 255, 7306, 1980.
25. **Borders, C. L., Jr. and Johansen, J. T.,** Essential arginyl residues in Cu, Zn superoxide dismutase from *Saccharomyces cerevisiae, Carlsberg Res. Commun.,* 45, 185, 1980.
26. **Borders, C. L., Jr. and Johansen, J. T.,** Identification of Arg-143 as the essential arginyl residue in yeast Cu, Zn superoxide dismutase by the use of a chromophoric arginine reagent, *Biochem. Biophys. Res. Commun.,* 96, 1071, 1980.

References for Table 1 (continued)

27. **Shoun, H., Beppu, T., and Arima, K.,** An essential arginine residue at the substrate-binding site of *p*-hydroxybenzoate hydroxylase, *J. Biol. Chem.*, 255, 9319, 1980.
28. **Belfort, M., Maley, G. F., and Maley, F.,** A single functional arginyl residue involved in the catalysis promoted by *Lactobacillus casei* thymidylate synthetase, *Arch. Biochem. Biophys.*, 204, 340, 1980.
29. **Müllner, H. and Sund, H.,** Essential arginine residue in acetylcholinesterase from *Torpedo californica*, *FEBS Lett.*, 119, 283, 1980.
30. **Tunnicliff, G.,** Essential arginine residues at the pyridoxal phosphate binding site of brain α-aminobutyrate aminotransferase, *Biochem. Biophys. Res. Commun.*, 97, 160, 1980.
31. **El Kebbaj, M. S., Latruffe, N., and Gaudemer, Y.,** Presence of an essential arginine residue in D-β-hydroxybutyrate dehydrogenase from mitochondrial inner membrane, *Biochem. Biophys. Res. Commun.*, 96, 1569, 1980.
32. **Marshall, M. and Cohen, P. P.,** Evidence for an exceptionally reactive arginyl residue at the binding site for carbamyl phosphate in bovine ornithine transcarbamylase, *J. Biol. Chem.*, 255, 7301, 1980.
33. **Kuno, S., Toraya, T., and Fukui, S.,** Coenzyme B_{12}-dependent diol dehydrase: chemical modification with 2,3-butanedione and phenylglyoxal, *Arch. Biochem. Biophys.*, 205, 240, 1980.
34. **Kremer, A. B., Egan, R. M., and Sable, H. Z.,** The active site of transketolase. Two arginine residues are essential for activity, *J. Biol. Chem.*, 255, 2405, 1980.
35. Deleted in proof.
36. **Ramakrishna, S. and Benjamin, W. B.,** Evidence for an essential arginine residue at the active site of ATP citrate lyase from rat liver, *Biochem. J.*, 195, 735, 1981.
37. **Chang, G.-G. and Huang, T.-M.,** Modification of essential arginine residues of pigeon liver malic enzyme, *Biochim. Biophys. Acta*, 660, 341, 1981.
38. **Choi, J.-D. and McCormick, D. B.,** Roles of arginyl residues in pyridoxamine-5′-phosphate oxidase from rabbit liver, *Biochemistry*, 20, 5722, 1981.
39. **Fortin, A. F., Hauber, J. M., and Kantrowitz, E. R.,** Comparison of the essential arginine residue in *Escherichia coli* ornithine and aspartate transcarbamylases, *Biochim. Biophys. Acta*, 662, 8, 1981.
40. **Wong, S. S. and Wong, L.-J.,** Evidence for an essential arginine residue at the active site of *Escherichia coli* acetate kinase, *Biochim. Biophys. Acta*, 660, 142, 1981.
41. **Fleer, E. A. M., Puijk, W. C., Slotboom, A. J., and DeHaas, G. H.,** Modification of arginine residues in porcine pancreatic phospholipase A_2, *Eur. J. Biochem.*, 116, 277, 1981.
42. **Akeroyd, R., Lange, L. G., Westerman, J., and Wirtz, K. W. A.,** Modification of the phosphatidylcholine-transfer protein from bovine liver with butanedione and phenylglyoxal. Evidence for one essential arginine residue, *Eur. J. Biochem.*, 121, 77, 1981.
43. **Branlant, G., Tritsch, D., and Biellmann, J.-F.,** Evidence for the presence of anion-recognition sites in pig-liver aldehyde reductase. Modification by phenylglyoxal and *p*-carboxyphenyl glyoxal of an arginyl residue located close to the substrate-binding site, *Eur. J. Biochem.*, 116, 505, 1981.
44. **Mautner, H. G., Pakyla, A. A., and Merrill, R. E.,** Evidence for presence of an arginine residue in the coenzyme A binding site of choline acetyltransferase, *Proc. Natl. Acad. Sci. U.S.A.*, 78, 7449, 1981.
45. **Carlson, C. A. and Preiss, J.,** Involvement of arginine residues in the allosteric activation of *Escherichia coli* ADP-glucose synthetase, *Biochemistry*, 21, 1929, 1982.
46. **Koland, J. G., O'Brien, T. A., and Gennis, R. B.,** Role of arginine in the binding of thiamin pyrophosphate to *Escherichia coli* pyruvate oxidase, *Biochemistry*, 21, 2656, 1982.
47. **King, M. M. and Heiny, L. P.,** Chemical modification of the calmodulin-stimulated phosphatase, calcineurin, by phenylglyoxal, *J. Biol. Chem.*, 262, 10658, 1987.

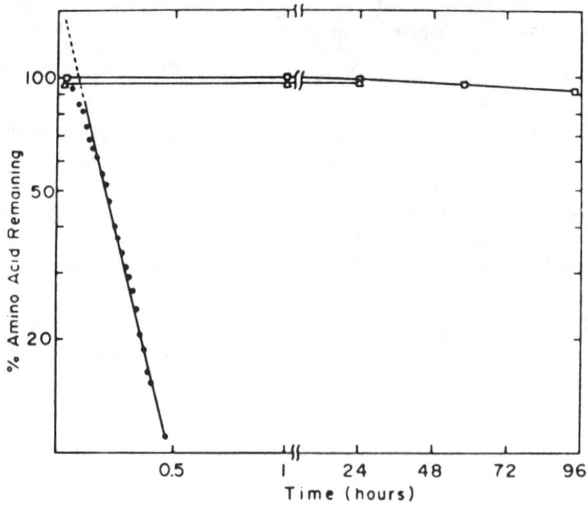

FIGURE 2. The modification of amino acid derivatives with phenylglyoxal in bicarbonate buffer. Shown is the time course for the reaction of *N*-acetylarginine (●), *N*-acetyllysine (□), and *N*-acetylcysteine (△) with phenylglyoxal in 83 m*M* bicarbonate buffer, pH 7.5 at 23°C. The reaction with *N*-acetylarginine was monitored by the increase in absorbance at 340 nm. The amounts of unreacted *N*-acetyllysine and *N*-acetylcysteine were determined by reaction with 2,4,6-trinitrobenzenesulfonic acid. (From Cheung, S.-T. and Fonda, M. L., *Biochem. Biophys. Res. Commun.*, 90, 940, 1979. With permission.)

References for Table 1 (continued)

48. **Shanmugasundaram, T., Kumar, G. K., Shenoy, B. C., and Wood, H. G.,** Chemical modification of the functional arginine residues of carbon monoxide dehydrogenase from *Clostridium thermoaceticum, Biochemistry,* 28, 7112, 1989.
49. **Garty, H., Yeger, O., and Asher, C.,** Sodium-dependent inhibition of the epithelial sodium channel by an arginyl-specific reagent, *J. Biol. Chem.*, 263, 5550, 1988.
50. **Missiaen, L., Raeymaekers, L., Droogmans, G., Wuytack, F., and Casteels, R.,** Role of arginine residues in the stimulation of the smooth-muscle plasma-membrane Ca^{2+} pump by negatively charged phospholipids, *Biochem. J.*, 264, 609, 1989.
51. **King, M. M.,** Conformation-sensitive modification of the type II calmodulin-dependent protein kinase by phenylglyoxal, *J. Biol. Chem.*, 263, 4754, 1988.

Eun[26] has examined the effect of borate on the reaction of arginine with phenylglyoxal and *p*-hydroxyphenylglyoxal. The base buffer of these studies was 0.1 *M* sodium pyrophosphate, pH 9.0. Spectroscopy was used to follow the rate of arginine modification. The rate of modification of either free arginine or *N*-acetyl-L-arginine with phenylglyoxal was 10 to 15 times faster than that of *p*-hydroxyphenylglyoxal in the base buffer system. The inclusion of sodium borate (10 to 50 m*M*) markedly increased the rate of the reaction (approximately 20-fold) of *p*-hydroxyphenylglyoxal with either arginine or *N*-acetyl-L-arginine, while there was only a slight enhancement of the phenylglyoxal reaction. In a related study,[27] the effect of phenylglyoxal on sodium-channel gating in frog myelinated nerve was compared with that of *p*-hydroxyphenylglyoxal or *p*-nitrophenylglyoxal. Both *p*-hydroxyphe-

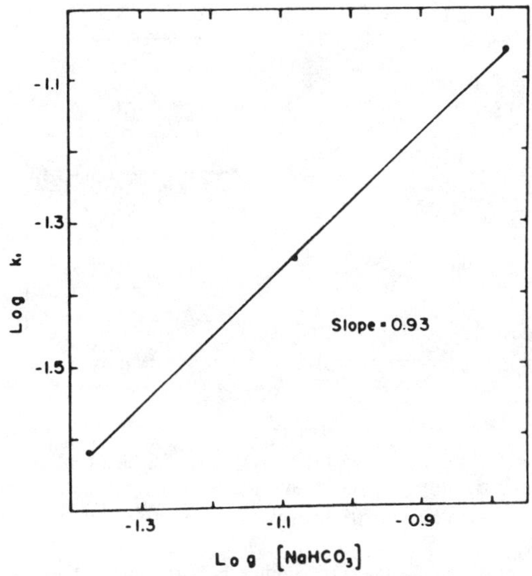

FIGURE 3. The effect of bicarbonate concentration on the rate of reaction of arginine with phenylglyoxal. Shown is a plot of logarithm apparent first-order rate constants vs. logarithm bicarbonate concentrations. The reaction mixtures contained 5 m*M* *N*-acetylarginine and 25 m*M* phenylglyoxal in sodium bicarbonate, pH 7.5 at 25°C. The absorbance at 340 nm was recorded, and the rate constants were obtained from the slopes of the plots of ln $(A_\infty - A_t)$ vs. time. The slope of the line obtained in this figure is 0.93, suggesting that the reaction is first order with respect to bicarbonate. (From Cheung, S.-T. and Fonda, M. L., *Biochem. Biophys. Res. Commun.*, 90, 940, 1979. With permission.)

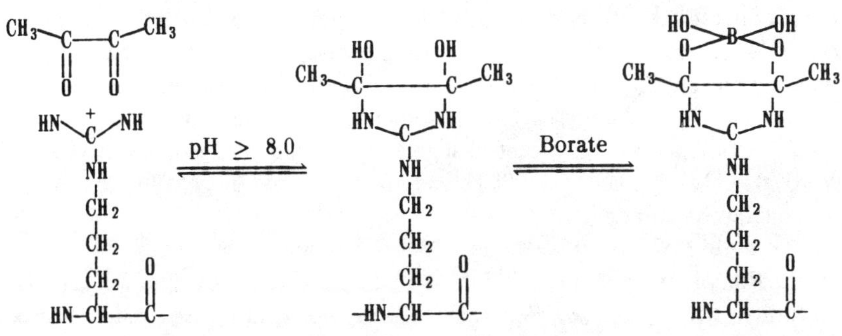

FIGURE 4. A scheme for the reaction of arginine with 2,3-butanedione.

nylglyoxal and *p*-nitrophenylglyoxal had less effect than phenylglyoxal in reduced sodium current. The results are discussed in terms of the differences in hydrophobicity of the reagents, but it is clear that the intrinsic difference in reagent effectiveness described by Eun may be responsible, in part, for the observed differences.

2,3-Butanedione (Figure 4) is the second well-characterized reagent for the selective modification of arginyl residues in proteins. Yankeelov and co-workers

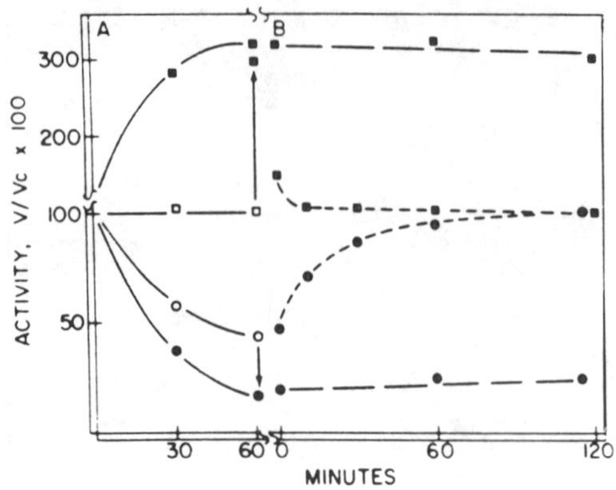

FIGURE 5. The modification of carboxypeptidase A with 2,3-butanedione in borate buffer. (A) Changes in esterase (□, ■) and peptidase (○, ●) activities on modification of carboxypeptidase A (0.15 m*M*) with 2,3-butanedione in 0.05 *M* borate — 1 *M* NaCl, pH 7.5 (9 m*M* reagent, closed symbols), or in 0.02 *M* Veronal — 1.0 *M* NaCl, pH 7.5 (75 m*M* reagent, open symbols) at 20°C. The changes in the activity immediately on addition of borate after 1 h to the sample reacted in Veronal buffer are indicated by the arrows. (B) Changes in activities of the samples reacted in borate buffer subsequent to gel filtration through Bio-Gel P-4 equilibrated either with 0.05 *M* borate — 1.0 *M* NaCl, pH 7.5 (———), or with 0.02 *M* Veronal — 1.0 *M* NaCl, pH 7.5 (- - - -). (From Riordan, J. F., *Biochemistry,* 12, 3915, 1973. With permission.)

introduced the use of this reagent.[10,28] There were problems with the specificity of the reaction (c.f. Reference 28) and the time required for modification until the observation of Riordan[29] that borate had a significant effect on the nature of the reaction of 2,3-butanedione with arginyl residues in proteins. Figure 5 shows the effect of borate (0.05 *M* borate, 1.0 *M* NaCl, pH 7.5) on the changes in biological activity occurring on the reaction of carboxypeptidase A (0.15 m*M*) with 2,3-butanedione (freshly distilled). Note that, in particular, the enhancement of esterase activity in the presence of butanedione is dependent on the presence of borate buffer, as no significant change is seen with butanedione in 0.02 *M* Veronal — 1.0 *M* NaCl, pH 7.5. The removal of borate by gel filtration results in the recovery of activity.

The ability of 2,3-butanedione to act as a photosensitizing agent for the destruction of amino acids and proteins in the presence of oxygen was emphasized in work by Fliss and Viswanatha.[15] Figure 6 shows the destruction of certain amino acids in the presence of 2,3-butanedione and oxygen at pH 6.0 (phosphate) at 36°C upon irradiation at 350 to 375 nm ("Blak-Lite" UV-Lamp, 100-W bulb, 20 cm from sample contained in a quartz cuvette). As would be expected from consideration of early photooxidation work, tryptophan and histidine are lost most rapidly with methionine; cystine and tyrosine are lost at a much slower rate. Loss is not seen on irradiation in the absence of 2,3-butanedione (open symbols). Azide (10 m*M*), a singlet oxygen scavenger, greatly reduces the rate of loss of amino acids. The absence of oxygen also greatly reduces the rate of loss of sensitive amino acids.

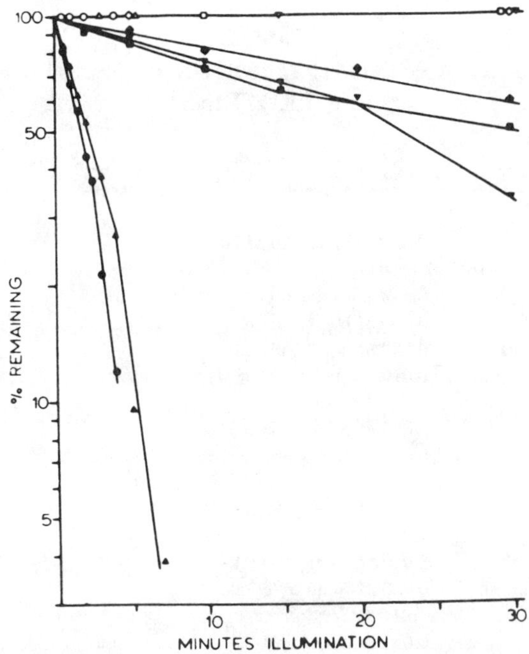

FIGURE 6. 2,3-Butanedione-sensitized destruction of α-amino acids. ●, Tryptophan (99.5 μ*M*); ■, tyrosine (333 μ*M*); ▲, histidine (92 μ*M*); ▼, methionine (1000 m*M*); and ◆, cystine (33.3 μ*M*) in the presence of 2,3-butanedione (9580 μ*M*) and continuous oxygenation were irradiated at pH 6.0 (irradiation was performed in quartz cuvettes 20 cm from a "Blak-Lite" UV light source, Canlab catalog No. L6093–1, equipped with a 100-W lamp emitting light almost exclusively in the range of 350 to 375 nm) at 36°C. Open symbols represent preparations of amino acids (at the same concentrations as the experiments described above) irradiated in the absence of 2,3-butanedione. The above experiments used freshly distilled monomer preparation of 2,3-butanedione. (From Fliss, H. and Viswanatha, T., *Can. J. Biochem.*, 57, 1267, 1979. With permission.)

These observations have been confirmed and extended by other laboratories.[16,17] An examination of recent studies using 2,3-butanedione to modify arginyl residues in proteins is presented in Table 2.

The use of 1,2-cyclohexanedione under very basic conditions to modify arginyl residues was demonstrated in 1967.[30] However, it was not until Patthy and Smith[11] reported on the reaction of 1,2-cyclohexanedione in borate with arginyl residues in proteins that the use of this reagent became practical. These investigators reported that 1,2-cyclohexanedione reacted with arginyl residues in 0.2 *M* borate, pH 9.0. At alkaline pH, reaction of 1,2-cyclohexanedione with arginine (Figure 7) forms N^5-(4-oxo-1,3-diazaspiro[4,4]non-2-yliodene)-L-ornithine (CHD-arginine), a reaction which cannot be reversed. Between pH 7.0 and 9.0 a compound is formed from arginine and 1,2-cyclohexanedione, N^7-N^8-(1,2-dihydroxycyclohex-1,2-ylene)-L-arginine (DHCH-arginine). This compound is stabilized by the presence of borate and is unstable in the presence of buffers such as Tris. This compound

Table 2
USE OF 2,3-BUTANEDIONE TO MODIFY ARGINYL RESIDUES IN PROTEINS

Protein	Solvent	Reagent excess[a]	Stoichiometry	Ref.
Carboxypeptidase A	0.05 *M* borate, 1.0 *M* NaCl, pH 7.5	—[b]	2/10	1
Chymotrypsin	0.1 *M* phosphate, pH 6.0	100[c]	1/3[d]	2
Thymidylate synthetase	0.050 *M* borate, pH 8.0	—	—[e]	3
Prostatic acid phosphatase	0.050 *M* borate, pH 8.0	—	—[f]	4
Purine nucleoside phosphorylase	0.0165 *M* borate, pH 8.0	—	—[g]	5
Yeast hexokinase PII	0.050 *M* borate, pH 8.3	—	4.2/18[h]	6
Isocitrate dehydrogenase	0.05 *M* MES, pH 6.2, 20% glycerol, 0.0021 *M* $MnSO_4$	—	1.6/13.4[i]	7
Stearylcoenzyme A desaturase	0.050 *M* sodium borate, pH 8.1	2,500	2[j]	8
Superoxide dismutase	0.050 *M* borate, pH 9.0	—[k]	1.3/4[l]	9
Energy-independent transhydrogenase	0.050 *M* sodium[m] borate, pH 7.8	—	—	10
Enolase	0.050 *M* borate, pH 8.3, 0.001 *M* Mg $(OAc)_2$, 0.01 m*M* EDTA	260	3/16[n]	11
NADPH-dependent aldehyde reductase	0.050 *M* borate, pH 7.0	—[o]	1/18[p]	12
Aryl sulfatase A	0.050 *M*[q] $NaHCO_3$, pH 8.0	—	—	13
Na^+, K^+-ATPase	0.04 *M* TES, 0.02 *M* borate, pH 7.4	—	—	14
Carbamate kinase	0.005 *M* triethanolamine, 0.050 *M* borate, pH 7.5	2,000	1.2/3.0[r]	15
Thymidylate synthetase	0.050 *M* borate, 0.001 *M* EDTA, pH 8.0	1,201	2.1/12[s]	16
(K^+ + H^+)-ATPase	0.125 *M* sodium borate, pH 7.0	—	—[t]	17
Cu, Zn superoxide dismutase	0.050 *M* borate, pH 8.3	—[u]	—	18
Fatty acid synthetase	0.020 *M* borate, 0.200 *M* KCl, 0.001 *M* dithiothreitol, 0.001 m*M* EDTA, pH 7.6	—	—[v]	19
Acetylcholinesterase	0.005 *M* phosphate, 0.025 *M* borate, 0.050 *M* NaCl, pH 7.0	—	4/31[w]	20
Coenzyme B_{12}-dependent diol dehydrase	0.050 *M* borate, pH 8.5	—	—[x]	21
Ornithine transcarbamylase	0.05 *M* Bicine,[y] 0.1 m*M* EDTA, 0.1 *M* KCl, pH 7.67	—	0.88/11[z]	22
Glycogen phosphorylase	0.020 *M* sodium tetraborate, 1 m*M* EDTA, pH 7.5	—	—	23
Cytochrome c	0.05 *M* sodium bicarbonate, pH 7.5	9,900[aa]	2/2[bb]	24

Table 2 (continued)
USE OF 2,3-BUTANEDIONE TO MODIFY ARGINYL RESIDUES IN PROTEINS

Protein	Solvent	Reagent excess[a]	Stoichiometry	Ref.
Bacteriorhodopsin	0.100 *M* borate, pH 8.2	66,700	4/79[cc]	25
α-Ketoglutarate dehydrogenase	0.050 *M* sodium borate, pH 8.0	—[dd]	—[ee]	26
Acetate kinase	0.050 *M* borate, pH 8.6	—	—	27
Malic enzyme	0.045 *M* borate,[ff] pH 7.5	—	—	28
Glucose phosphate isomerase	0.05 *M* sodium borate, pH 8.7	—	7.8/30[gg]	29
Saccharopine dehydrogenase	0.08 *M* HEPES, 0.2 *M* KCl, 0.01 *M* borate, pH 8.0	—[hh]	8/38[ii]	30
Testicular hyaluronidase	0.050 *M* borate, pH 8.3	—[jj]	3.6/28	31
Glutathione reductase	0.050 *M* sodium borate, pH 8.3, 1 m*M* EDTA	20,000	5.3[kk]	32,33
Escherichia coli fibrillar adhesins	100 m*M* sodium borate with 0.9% NaCl, pH 7.5	—	—	34
Inter-α-trypsin inhibitor	10 m*M* HEPES, 175 m*M* NaCl, 100 m*M* sodium borate, pH 7.4	—	—	35
D-Glyceraldehyde-3-phosphate dehydrogenase	100 m*M* medinal[ll] buffer, pH 8.3 with 5 m*M* EDTA and 2 m*M* dithiothreitol	—	—	36

[a] Mole reagent per mole protein unless otherwise indicated.

[b] This study demonstrated that, in the presence of borate, there is essentially no difference in the reaction of 2,3-butanedione monomer and butadione trimer. It is noted that the commercially available 2,3-butanedione should be distilled immediately prior to use.

[c] This study used 2,3-butanedione trimer prepared by allowing 2,3-butanedione (40 ml) to stand with 80 g untreated Permutit under dry air (after shaking to obtain an even dispension of 2,3-butanedione in Permutit) for 4 to 6 weeks at ambient temperature. The mixture was extracted with anhydrous ether. The ether extract was taken to an oil with dry air. The oil was allowed to stand for 5 to 7 days to permit crystallization of the timer.

[d] In the absence of light, there is also some loss of lysine; there is no loss of catalytic activity. In the presence of sunlight, there was rapid inactivation of the enzyme with loss of lysine, arginine (less than in the dark), and tyrosine. With the exception of tyrosine modification, the changes in amino acid composition in the reaction exposed to light were less than those for the dark reaction despite the more significant loss of activity. Study of the wavelength dependence demonstrates that light of 300 nm is most effective. 2,3-Butanedione monomer was not effective in this photoinactivation process.

[e] Stoichiometry of reaction was not established. Inactivation was reversed by gel filtration in 0.05 *M* Tris, 0.010 *M* β-mercaptoethanol, pH 8.0.

[f] 30°C.

[g] Ambient temperature. Calf spleen enzyme had 26 Arg modified at 98% loss of activity. Reaction with arginyl residues (as judged by loss of catalytic activity) was 50% as rapid with 2,3-butanedione in borate ($T_{1/2}$ = 40.3 min) as with phenylglyoxal in Tris buffer ($T_{1/2}$ = 19.2 min).

[h] Reaction was performed at 25°C. Determined by amino acid analysis after acid hydrolysis (6 *N* HCl, 110°C, 18 h). MgATP (5 m*M*) did not protect against either modification or loss of enzymatic activity, but MgATP and glucose reduced extent of modification from 3.3 arginine residues per

Table 2 (continued)

subunit (65% inactivation) to 2.1 residues per subunit (20% inactivation). Inactivation was also observed with phenylglyoxal in 0.050 *M* Bicine, pH 8.3. Stoichiometry with this modification was not established.

i Determined by amino acid analysis. As indicated, the maximum value obtained is 1.6 residues modified out of an average of 13.4 arginyl residues per subunit.

j The modification was performed at 25°C. The presence of stearyl-CoA greatly decreased the rate and extent of inactivation by 2,3-butanedione. When the modified enzyme is taken into 0.020 Tris (acetate), 0.100 *M* NaCl, pH 8.1 by gel filtration, there is the rapid recovery of activity and the concomitant decrease in the extent of arginine modification. A similar extent of modification and loss of catalytic activity was seen with 1,2-cyclohexanedione in 0.1 *M* sodium borate, pH 8.1.

k Inactivation occurred at a rate of 10.9 M^{-1} min^{-1} under these conditions (compared to 4.0 M^{-1} min^{-1} with phenylglyoxal in bicarbonate/carbonate and 6.6 M^{-1} min^{-1} with 1,2-cyclohexanedione in 0.050 *M* borate, pH 9.0). Inactivation with 2,3-butanedione is not observed in 0.05 *M* bicarbonate/carbonate, pH 9.0 at 25°C; however, there is reduced modification of arginine (0.4 residue per subunit as compared to 1.3 residues per subunit with 77% inactivation).

l The majority of arginine modification could be reversed by the removal of reagent and borate solvent by dialysis vs. 0.05 *M* potassium phosphate, pH 7.8. Enzymic activity was also recovered as a result of the dialysis procedure. These investigators were able to obtain evidence supporting the selective modification of Arg[141] by either 2,3-butanedione, 1,2-cyclohexandione, or phenylglyoxal.

m The modification was performed at 22°C. These studies were performed with bacterial membrane preparations. Stoichiometry was not established. Analysis of the rates of inactivation suggested that inactivation was due to the modification of a single arginine residue. NADH, which stimulates the transhydrogenation of 3-acetylpyridine-NAD by NADPH, protects the enzyme from inactivation.

n The modification was performed at 25°C. The extent of modification was determined by amino acid analysis after acid hydrolysis. The extent of modification reported was obtained after 75 min of reaction concomitant with 85% loss of activity. The presence of substrate, α-phosphoglycerate, reduced the extent of modification to 2 mol arginine per subunit with only 5% loss of catalytic activity.

o A second-order rate constant of 0.0635 M^{-1} min^{-1} was obtained for the loss of enzymic activity upon reaction with 2,3-butanedione in 0.050 *M* borate, pH 7.0 at 25°C. This presumably reflects the modification of a single arginine residue (see Footnote p). The inactivation of the enzyme by 1,2-cyclohexanedione, methylglyoxal, and phenylglyoxal is compared with that by 2,3-butanedione (all at 10 m*M* in 0.05 *M* borate, pH 7.0). Butanedione is clearly most effective followed by phenylglyoxal, methylglyoxal, and 1,2-cyclohexanedione. The authors note that the enzyme under study aldehyde reductase can utilize methylglyoxal and phenylglyoxal as substrates, precluding their rigorous evaluation in this study.

p Obtained by amino acid analysis after acid hydrolysis (6 *N* HCl, 110°C, 24 h). The control preparation yielded a value of 17.8 ± 1 Arg, while the modified enzyme yielded a value of 16.7 ± 1 Arg. The presence of cofactor yielded a preparation with 17.5 ± 1 Arg.

q The reactions are reported at 25°C. Borate buffers could not be used since borate is a competitive inhibitor of the enzyme and prevents inactivation in bicarbonate buffer. Reaction with phenylglyoxal in the same solvent.

r Reaction was performed at 25°C. Stoichiometry was established by amino acid analysis after acid hydrolysis (6 *N* HCl, 100°C, 20 h). Arginine is the only amino acid modified under these reaction conditions. These values were obtained at 80% inactivation. The presence of ADP reduced activity loss to 55%, with the extent of arginine modification reduced to 0.4 to 0.5 residues.

Table 2 (continued)

s Reaction was performed at 25°C for 90 min. Stoichiometry was determined by amino acid analysis after acid hydrolysis (6 *N* HCl, 110°C, 24 h).

t The use of isolated "membrane fraction" prevented the establishment of stoichiometry in these studies. Analysis of the dependence of reaction rate on concentration of 2,3-butanedione is consistent with the modification of a single arginine residue. As expected, the stability of modification is dependent upon the presence of borate. Gel filtration into HEPES (0.125 *M*, pH 7.0) and subsequent inactivation at 37°C resulted in the recovery of a substantial amount of catalytic activity. Similar results were obtained with imidazole and Tris buffers under similar reaction conditions. This reactivation does not occur when the incubation following gel filtration is performed at 0°C instead of 37°C.

u A reaction rate with a second-order constant of k = 5.2 M^{-1} min^{-1} is obtained at 25°C. Inactivation is dependent on the presence of borate as inactivation is not observed with use of Bicine buffer. Dialysis vs. 0.025 *M* phosphate, pH 7.0 for 21 h at 4°C results in an increase in activity of 14 to 85%, while complete recovery of activity is achieved after 21 h of dialysis.

v Stoichiometry was not established for the reaction with 2,3-butanedione. As shown in Table 1, reaction with phenylglyoxal modifies approximately 4 of the 106 arginyl residues in each subunit of fatty acid synthetase. The loss of the biological activity as determined either by fatty acid synthetase activity, ketoreductase activity, or enoylreductase activity was considerably more rapid with phenylglyoxal than with 2,3-butanedione. It is noted that these reactions are performed in borate buffer for the studies with 2,3-butanedione and in phosphate buffer for the studies with phenylglyoxal (both buffers at pH 7.6 with the reactions performed at 30°C).

w Reactions were performed at 25°C. The modification of arginyl residues is associated with an approximate 70% loss of enzymatic activity. The presence of *N*-phenylpyridinium-2-aldoxine iodide reduces the extent of arginine modification by approximately 1 mol/mol of enzyme with concomitant protection of enzymatic activity. It should be noted that modification of this enzyme with phenylglyoxal results in the modification of 3 mol of arginine per mole of enzyme with 17% loss of enzymatic activity (see Table 1). It is not clear when modification of a particular arginyl residue with the two reagents is a mutually exclusive event.

x Reactions were performed at 25°C. Rigorous evaluation of the stoichiometry of the reaction is not available. Analysis of the dependence of first-order rate constants on reagent concentration (double-logarithmic relationships) is consistent with the modification of a single arginyl residue. The inactivation was reversed by 100-fold dilution into 0.05 *M* potassium phosphate, pH 8.5 at 25°C.

y The inactivation of ornithine transcarbamylase is readily reversible in this solvent; the presence of borate precludes reactivation observed on dilution of modified enzyme in solvent. A value of 179 M^{-1} min^{-1} for the second-order rate constant for reaction of 2,3-butanedione with ornithine transcarbamylase under these conditions was recorded.

z Obtained at 88% inactivation.

aa Reaction was performed at 22°C.

bb Determined by amino acid analysis. The reaction is readily reversible, even in the presence of borate.

cc Determined by amino acid analysis. Constructed Scatchard plot shows that two residues were not available for modification with 2,3-butanedione.

dd Second-order rate constant is k = 2.95 M^{-1} min^{-1} in this solvent, assuming that loss in catalytic activity is a measure of reaction with arginine.

ee Stoichiometry was not established. Kinetic analysis suggests that inactivation of catalytic activity results from the modification of a single arginine residue.

ff Modification reaction was performed at 24°C. Very little inactivation is observed if the reaction is performed in Tris buffer at the same pH. Reactivation of enzyme modified in borate buffer is observed when the inactivated enzyme is diluted in borate buffer.

Table 2 (continued)

gg The reaction was performed at 25°C for 4 h. The presence of the competitive inhibitor, 6-phosphogluconate, protected 1 mol of arginine per mole of enzyme from modification, suggesting that there is a single arginine residue critical for catalytic activity. A 20-fold increase in inhibitor concentration resulted in the modification of greater than 95% of the total arginine residues.

hh Second-order rate constant of k = 7.5 M^{-1} min^{-1} at 25°C was obtained from the analysis of reaction rate data. pH dependence study showed optimal rate of inactivation at pH 8.2.

ii Determined by amino acid analysis on 95+ % inactivated enzyme. Plotting loss of activity vs. arginine residues modified suggests that inactivation is due to the modification of a single arginine residue. Inactivation occurs with loss of sulfhydryl content.

jj Second-order rate constant of k = 13.57 M^{-1} min^{-1} obtained at 20°C. Inactivation much less rapid in 0.050 *M* HEPES, pH 8.3 ($T_{1/2}$ = 30 min in borate; 11.5 min in HEPES).

kk Reactions were performed at 30°C. Modification associated with 80 to 90% inactivation. Reaction with phenylglyoxal (0.050 *M* sodium phosphate, 1 m*M* EDTA, pH 7.6) at 2000-fold molar excess led to the modification of two arginyl residues at a level of 90% inactivation. The extent of arginine was determined by spectrophotometric analysis (increase in absorbance at 250 nm, $\Delta\varepsilon$ = 11,000 M^{-1} cm^{-1}; see Reference 33).

ll Reaction readily reversible, reflecting the absence of borate in the buffer.

References for Table 2

1. **Riordan, J. F.,** Functional arginyl residues in carboxypeptidase A. Modification with butanedione, *Biochemistry,* 12, 3915, 1973.
2. **Fliss, H., Tozer, N. M., and Viswanatha, T.,** The reaction of chymotrypsin with 2,3-butanedione trimer, *Can. J. Biochem.,* 53, 275, 1975.
3. **Cipollo, K. L. and Dunlap, R. B.,** Essential arginyl residues in thymidylate synthetase, *Biochem. Biophys. Res. Commun.,* 81, 1139, 1978.
4. **McTigue, J. J. and Van Etten, R. L.,** An essential arginine residue in human prostatic acid phosphatase, *Biochim. Biophys. Acta,* 523, 422, 1978.
5. **Jordan, F. and Wu, A.,** Inactivation of purine nucleoside phosphorylase by modification of arginine residues, *Arch. Biochem. Biophys.,* 190, 699, 1978.
6. **Borders, C. L., Jr., Cipollo, K. L., and Jordasky, J. F.,** Role of arginyl residues in yeast hexokinase PII, *Biochemistry,* 17, 2654, 1978.
7. **Hayman, S. and Colman, R. F.,** Effect of arginine modification on the catalytic activity and allosteric activation by adenosine diphosphate of the diphosphopyridine nucleotide specific isocitrate dehydrogenase of pig heart, *Biochemistry,* 17, 4161, 1978.
8. **Enoch, H. G. and Strittmatter, P.,** Role of tyrosyl and arginyl residues in rat liver microsomal stearylcoenzyme A desaturase, *Biochemistry,* 17, 4927, 1978.
9. **Malinowski, D. P. and Fridovich, I.,** Chemical modification of arginine at the active site of the bovine erythrocyte superoxide dismutase, *Biochemistry,* 18, 5909, 1979.
10. **Homyk, M. and Bragg, P. D.,** Steady-state kinetics and the inactivation by 2,3-butanedione of the energy-independent transhydrogenase of *Escherichia coli* cell membranes, *Biochim. Biophys. Acta,* 571, 201, 1979.
11. **Borders, C. L., Jr. and Zurcher, J. A.,** Rabbit muscle enolase also has essential argininyl residues, *FEBS Lett.,* 108, 415, 1979.
12. **Davidson, W. S. and Flynn, T. G.,** A functional arginine residue in NADPH-dependent aldehyde reductase from pig kidney, *J. Biol. Chem.,* 254, 3724, 1979.
13. **James, G. T.,** Essential arginine residues in human liver arylsulfatase A, *Arch. Biochem. Biophys.,* 197, 57, 1979.
14. **Grisham, C. M.,** Characterization of essential arginyl residues in sheep kidney (Na^+ + K^+)-ATPase, *Biochem. Biophys. Res. Commun.,* 88, 229, 1979.

References for Table 2 (continued)

15. **Pillai, R. P., Marshall, M., and Villafranca, J. J.,** Modification of an essential arginine of carbamate kinase, *Arch. Biochem. Biophys.,* 199, 16, 1980.
16. **Cipollo, K. L. and Dunlap, R. B.,** Essential arginyl residues in thymidylate synthetase from amethopterin-resistant *Lactobacillus casei, Biochemistry,* 18, 5537, 1979.
17. **Schrijen, J. J., Luyben, W. A. H. M., DePont, J. J. H. M., and Bonting, S. L.,** Studies on ($K^+ + H^+$)-ATPase. I. Essential arginine residue in its substrate binding center, *Biochim. Biophys. Acta,* 597, 331, 1980.
18. **Belfort, M., Maley, G. F., and Maley, F.,** A single functional arginyl residue involved in the catalysis promoted by *Lactobacillus casei* thymidylate synthetase, *Arch. Biochem. Biophys.,* 204, 340, 1980.
19. **Poulose, A. J. and Kolattukudy, P. E.,** Presence of one essential arginine that specifically binds the 2′-phosphate of NADPH on each of the ketoacyl reductase and enoyl reductase active sites of fatty acid synthetase, *Arch. Biochem. Biophys.,* 199, 457, 1980.
20. **Müllner, H. and Sund, H.,** Essential arginine residue in acetylcholinesterase from *Torpedo californica, FEBS Lett.,* 119, 283, 1980.
21. **Kuno, S., Toraya, T., and Fukui, S.,** Coenzyme B_{12}-dependent diol dehydrase: chemical modification with 2,3-butanedione and phenylglyoxal, *Arch. Biochem. Biophys.,* 205, 240, 1980.
22. **Marshall, M. and Cohen, P. P.,** Evidence for an exceptionally reactive arginyl residue at the binding site for carbamyl phosphate in bovine ornithine transcarbamylase, *J. Biol. Chem.,* 255, 7301, 1980.
23. **Dreyfus, M., Vandenbunder, B., and Buc, H.,** Mechanism of allosteric activation of glycogen phosphorylase probed by the reactivity of essential arginyl residues. Physicochemical and kinetic studies, *Biochemistry,* 19, 3634, 1980.
24. **Pande, J. and Myer, Y. P.,** The arginines of cytochrome c. The reduction-binding site for 2,3-butanedione and ascorbate, *J. Biol. Chem.,* 255, 11094, 1980.
25. **Tristram-Nagle, S. and Packer, L.,** Effects of arginine modification on the photocycle and proton pumping of bacteriorhodopsin, *Biochem. Int.,* 3, 621, 1981.
26. **Gomazkova, V. S., Stafeeva, O. A., and Severin, S. E.,** The role of arginine residues in the functioning of α-ketoglutarate dehydrogenase from pigeon breast muscle, *Biochem. Int.,* 2, 51, 1981.
27. **Wong, S. S. and Wong, L.-J.,** Evidence for an essential arginine residue at the active site of *Escherichia coli* acetate kinase, *Biochim. Biophys. Acta,* 660, 142, 1981.
28. **Chang, G.-G. and Huang, T.-M.,** Modification of essential arginine residues of pigeon liver malic enzyme, *Biochim. Biophys. Acta,* 660, 341, 1981.
29. **Lu, H. S., Talent, J. M., and Gracy, R. W.,** Chemical modification of critical catalytic residues of lysine, arginine and tryptophan in human glucose phosphate isomerase, *J. Biol. Chem.,* 256, 785, 1981.
30. **Fujioka, M. and Takata, Y.,** Role of arginine residue in saccharopine dehydrogenase (L-lysine forming) from baker's yeast, *Biochemistry,* 20, 468, 1981.
31. **Gacesa, P., Savitsky, M. J., Dodgson, K. S., and Olavesen, A. H.,** Modification of functional arginine residues in purified bovine testicular hyaluronidase with butane-2,3-dione, *Biochim. Biophys. Acta,* 661, 205, 1981.
32. **Boggaram, V. and Mannervik, B.,** Essential arginine residues in the pyridine nucleotide binding sites of glutathione reductase, *Biochim. Biophys. Acta,* 701, 119, 1982.
33. **Takahashi, K.,** Further studies on the reactions of phenylgloxal and related reagents with proteins, *J. Biochem.,* 81, 403, 1977.
34. **Jacobs, A. A. C., van Mechelen, J. R., and De Graaf, F. K.,** Effect of chemical modification on the K99 and K88ab fibrillar adhesins of *Escherichia coli, Biochim. Biophys. Acta,* 832, 148, 1985.

References for Table 2 (continued)

35. **Swaim, M. W. and Pizzo, S. V.,** Modification of the tandem reactive centres of human inter-trypsin inhibitor with butanedione and *cis*-dichlorodiamineplatinum (II), *Biochem. J.*, 254, 171, 1988.
36. **Asryants, R. A., Kuzminskaya, E. V., Tishkov, V. I., Douzhenkova, I. V., and Nagradova, N. K.,** An examination of the role of arginine residues in the functioning of D-glyceraldehyde-3-phosphate dehydrogenase, *Biochim. Biophys. Acta*, 997, 159, 1989.

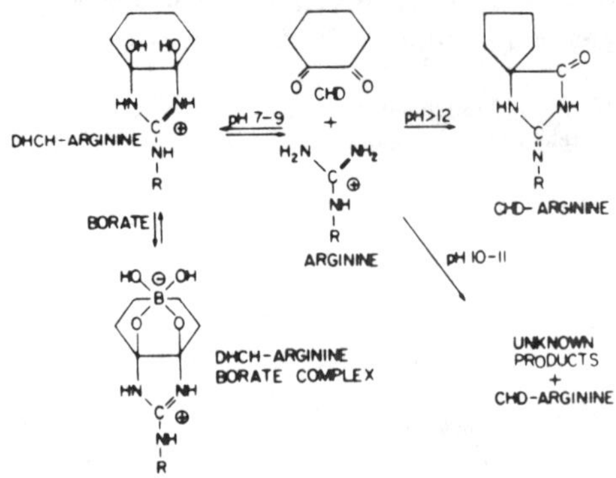

SCHEME I

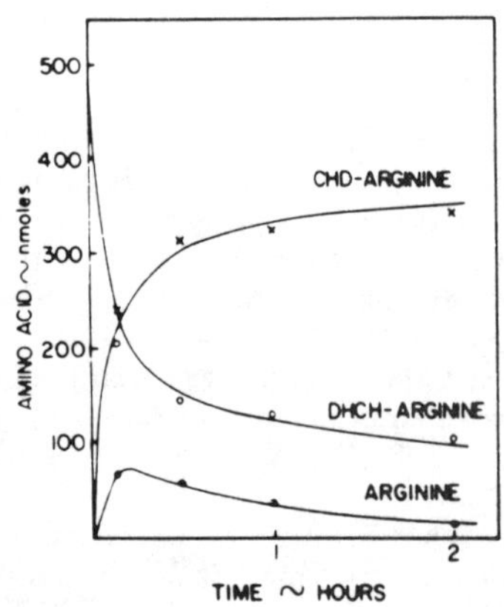

FIGURE 7. The reaction of arginine with 1,2-cyclohexanedione. Scheme I shows a representation of the reaction of 1,2-cyclohexanedione with arginine. The figure shows the conversion of DHCH-arginine to CHD-arginine in 0.5 *M* NaOH. Amino acids were determined on the amino acid analyzer. (From Patthy, L. and Smith, E. L., *J. Biol. Chem.*, 250, 557, 1975. With permission.)

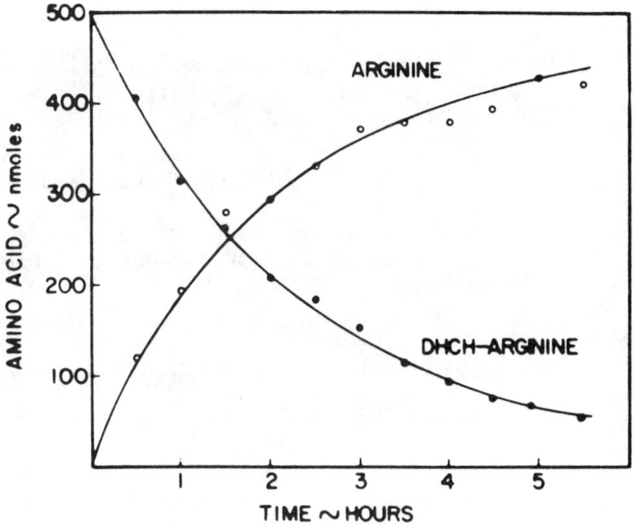

FIGURE 8. Disappearance of DHCH-arginine and formation of arginine on treatment with neutral hydroxylamine. DHCH-arginine (0.1 *M*) was incubated in 0.5 *M* hydroxylamine, pH 7.0 at 37°C. The amino acids were determined on the amino acid analyzer. Both sets of data are first order with a half time of 100 min. (From Patthy, L. and Smith, E. L., *J. Biol. Chem.*, 250, 557, 1975. With permission.)

is readily converted back to free arginine in 0.5 *M* hydroxylamine, pH 7.0 (Figure 8).

These authors have subsequently used this reagent to identify functional residues in bovine pancreatic ribonuclease A and egg white lysozyme.[31] Extent of modification of arginine residues in protein by 1,2-cyclohexanedione is generally assessed by amino acid analysis after acid hydrolysis. Under the conditions normally used for acid hydrolysis (6 *N* HCl, 110°C, 24 h), the borate-stabilized reaction product between arginine and 1,2-cyclohexanedione is unstable, and there is partial regeneration of arginine and the formation of unknown degradation products.[11] Acid hydrolysis in the presence of an excess of mercaptoacetic acid (20 μl/ml of hydrolysate) prevents the destruction of DHCD-arginine.[11] Table 3 lists some of the enzymes in which structure-function relationships have been studied by reaction with 1,2-cyclohexanedione.

Table 3
REACTION OF ARGINYL RESIDUES IN PROTEINS WITH 1,2-CYCLOHEXANEDIONE

Protein	Solvent	Reagent excess	Extent of modification	Ref.
Ribonuclease A	0.2 *M* sodium borate, pH 9.0	50,000	3/4	1
Lysozyme	0.2 *M* sodium borate, pH 9.0	50,000	11/11	1
Kunitz bovine trypsin inhibitor	0.2 *M* sodium borate, pH 9.0	—	5.5/6	2
Threonine dehydrogenase	25 μ*M* triethanolamine, 25 μ*M* sodium borate with pH 2.5 μ*M* 2-mercaptoethanol, pH 7.4	—	—[a]	3
Phosphoenolpyruvate carboxykinase	65 m*M* Tris-Cl, pH 7.4	—	—[b]	4

[a] Rate of inactivation with 1,2-cyclohexanedione is less than that observed with corresponding molar excesses of either phenylglyoxal or 2,3-butanedione.

[b] Rate constant for inactivation of 0.313 M^{-1} min^{-1}, pH 7.4 at 22°C.

References for Table 3

1. **Patthy, L. and Smith, E. L.,** Identification of functional arginine residues in ribonuclease A and lysozyme, *J. Biol. Chem.*, 250, 565, 1975.
2. **Menegatti, E., Ferroni, R., Benassi, C. A., and Rocchi, R.,** Arginine modification in Kunitz bovine trypsin inhibitor through 1,2-cyclohexanedione, *Int. J. Pept. Protein Res.*, 10, 146, 1977.
3. **Epperly, B. R. and Dekker, E. E.,** Inactivation of *Escherichia coli* L-threonine dehydrogenase by 2,3-butanedione. Evidence for a catalytically essential arginine residue, *J. Biol. Chem.*, 264, 18296, 1989.
4. **Cheng, K.-C. and Nowak, T.,** Arginine residues at the active site of avian liver phosphoenolpyruvate carboxykinase, *J. Biol. Chem.*, 264, 3317, 1989.

REFERENCES

1. **Levy, H. M., Leber, P. D., and Ryan, E. M.,** Inactivation of myosin by 2,4-dinitrophenol and protection by adenosine triphosphate and other phosphate compounds, *J. Biol. Chem.*, 238, 3654, 1963.
2. **Zanetti, G., Gozzer, C., Sacchi, G., and Curti, B.,** Modification of arginyl residues in ferrodoxin-NADP reductase from spinach leaves, *Biochim. Biophys. Acta,* 569, 127, 1979.
3. **Carrillo, N., Arana, J. L., and Vallejos, R. H.,** An essential carboxyl group at the nucleotide binding site of ferredoxin-$NADP^+$ oxidoreductase, *J. Biol. Chem.*, 256, 6823, 1981.
4. **Bhagwat, A. S. and Ramakrishna, J.,** Essential histidine residues of ribulosebisphosphate carboxylase indicated by reaction with diethylpyrocarbonate and rose bengal, *Biochim. Biophys. Acta,* 662, 181, 1981.

5. **Chang, G.-G. and Huang, T.-M.,** Modification of essential arginine residues of pigeon liver malic enzyme, *Biochim. Biophys. Acta,* 660, 341, 1981.
6. **Dunham, K. R. and Selman, B. R.,** Regulation of spinach chloroplast coupling factor 1 ATPase activity, *J. Biol. Chem.,* 256, 212, 1981.
7. **Gacesa, P., Savitzky, M. J., Dodgson, K. S., and Olavesen, A. H.,** Modification of functional arginine residues in purified bovine testicular hyaluronidase with butane-2,3-dione, *Biochim. Biophys. Acta,* 661, 205, 1981.
8. **Lundblad, R. L., Noyes, C. M., Featherstone, G. L., Harrison, J. H., and Jenzano, J. W.,** The reaction of bovine alpha-thrombin with tetranitromethane, *J. Biol. Chem.,* 263, 3729, 1988.
9. **Takahashi, K.,** The reaction of phenylglyoxal with arginine residues in proteins, *J. Biol. Chem.,* 243, 6171, 1968.
10. **Yankeelov, J. A., Jr., Mitchell, C. D., and Crawford, T. H.,** A simple trimerization of 2,3-butanedione yielding a selective reagent for the modification of arginine in proteins, *J. Am. Chem. Soc.,* 90, 1664, 1968.
11. **Patthy, L. and Smith, E. L.,** Reversible modification of arginine residues. Application to sequence studies by restriction of tryptic hydrolysis to lysine residues, *J. Biol. Chem.,* 250, 557, 1975.
12. **Lundblad, R. L.,** *Chemical Reagents for Protein Modification,* 2nd ed., CRC Press, Boca Raton, FL, 1991, 173.
13. **Enoch, H. G. and Strittmatter, P.,** Role of tyrosyl and arginyl residues in rat liver microsomal stearyl-coenzyme A desaturase, *Biochemistry,* 17, 4927, 1978.
14. **Smith, R. E. and MacQuarrie, R.,** A sensitive fluorometric method for the determination of arginine using 9,10-phenanthrenequinone, *Anal. Biochem.,* 90, 246, 1978.
15. **Fliss, H. and Viswanatha, T.,** 2,3-Butanedione as a photosensitizing agent: application to α-amino acids and α-chymotrypsin, *Can. J. Biochem.,* 57, 1267, 1979.
16. **Gripon, J.-C. and Hofmann, T.,** Inactivation of aspartyl proteinases by butane-2,3-dione. Modification of tryptophan and tyrosine residues and evidence against reaction of arginine residues, *Biochem. J.,* 193, 55, 1981.
17. **Mäkinen, K. K., Mäkinen, P.-L., Wilkes, S. H., Bayliss, M. E., and Prescott, J. M.,** Photochemical inactivation of *Aeromonas* aminopeptidase by 2,3-butanedione, *J. Biol. Chem.,* 257, 1765, 1982.
18. **Riley, H. A. and Gray, A. R.,** Phenylglyoxal, in *Organic Syntheses,* Collective Vol. 2, Blatt, A. H., Ed., John Wiley & Sons, New York, 1943, 509.
19. **Schloss, J. V., Norton, I. L., Stringer, C. D., and Hartman, F. C.,** Inactivation of ribulosebisphosphate carboxylase by modification of arginyl residues with phenylglyoxal, *Biochemistry,* 17, 5626, 1978.
20. **Augustus, B. W. and Hutchinson, D. W.,** The synthesis of phenyl[2-^{3}H]glyoxal, *Biochem. J.,* 177, 377, 1979.
21. **Borders, C. L., Jr., Pearson, L. J., McLaughlin, A. E., Gustafson, M. E., Vasiloff, J., An, F. Y., and Morgan, D. J.,** 4-Hydroxy-3-nitrophenylglyoxal. A chromophoric reagent for arginyl residues in proteins, *Biochim. Biophys. Acta,* 568, 491, 1979.
22. **Yamasaki, R. B., Vega, A., and Feeney, R. E.,** Modification of available arginine residues in proteins by *p*-hydroxyphenylglyoxal, *Anal. Biochem.,* 109, 32, 1980.
23. **Cheung, S.-T. and Fonda, M. L.,** Reaction of phenylglyoxal with arginine. The effect of buffers and pH, *Biochem. Biophys. Res. Commun.,* 90, 940, 1979.
24. **Yamasaki, R. B., Shimer, D. A., and Feeney, R. E.,** Colorimetric determination of arginine residues in proteins by *p*-nitrophenylglyoxal, *Anal. Biochem.,* 111, 220, 1981.
25. **Branlant, G., Tritsch, D., and Biellmann, J.-F.,** Evidence for the presence of anion-recognition sites in pig-liver aldehyde reductase. Modification by phenylglyoxal and *p*-carboxyphenyl glyoxal of an arginyl residue located close to the substrate-binding site, *Eur. J. Biochem.,* 116, 505, 1981.

26. **Eun, H.-M.,** Arginine modification by phenyglyoxal and (*p*-hydroxyphenyl)glyoxal: reaction rates and intermediates, *Biochem. Int.,* 17, 719, 1988.
27. **Meves, H., Rubly, N., and Stämpfli, R.,** The action of arginine-specific reagents on ionic and gating currents in frog myelinated nerve, *Biochim. Biophys. Acta,* 943, 1, 1988.
28. **Yankeelov, J. A., Jr.,** Modification of arginine in proteins by oligomers of 2,3-butanedione, *Biochemistry,* 9, 2433, 1970.
29. **Riordan, J. F.,** Functional arginyl residues in carboxypeptidase A. Modification with butanedione, *Biochemistry,* 12, 3915, 1973.
30. **Toi, K., Bynum, E., Norris, E., and Itano, H. A.,** Studies on the chemical modification of arginine. I. The reaction of 1,2-cyclohexanedione with arginine and arginyl residues of proteins, *J. Biol. Chem.,* 242, 1036, 1967.
31. **Patthy, L. and Smith, E. L.,** Identification of functional arginine residues in ribonuclease A and lysozyme, *J. Biol. Chem.,* 250, 565, 1975.

Chapter 12

CHEMICAL MODIFICATION OF TRYPTOPHAN

The specific chemical modification of tryptophan in protein is one of the more challenging problems in protein chemistry. First, the solvent conditions for providing specificity of modification are, in general, somewhat harsh. Second, there is the considerable possibility of either the concomitant or separate modification of a different amino acid residue. Third, the analysis for the determination of the exact extent of modification requires a rigorous approach combining spectral analysis and amino acid analysis[1,2] after hydrolysis in a solvent which will not destroy tryptophan.

Treatment of tryptophan with hydrogen peroxide results in the oxidation of the indole ring.[3–6] Usually, the reaction is performed at alkaline pH (1.0 *M* sodium bicarbonate, pH 8.4) with the H_2O_2 dioxane mixture prepared as described by Hachimori and co-workers.[3] The loss of tryptophan is monitored by the change in absorbance at 280 nm.[3,5,6] The difference in the molar extinction coefficient between tryptophan and the fully oxidized derivative is 3490 M^{-1} cm^{-1}.

While the reaction of *N*-bromosuccinimide (NBS) with tryptophan (Figure 1) continues to be useful a method for obtaining the site-specific chemical modification of this residue in proteins,[7–11] NBS is only rarely used today for the quantitative analysis of tryptophan in proteins. However, the data obtained from such studies can be quite valuable in the use of this reagent for the site-specific modification of tryptophanyl residues in proteins. For the quantitative determination of the tryptophan content in proteins, the basic approach is to add NBS to the protein (pH 4.0 to 5.0, usually acetate buffers) until there is no further decrease in absorbance at 280 nm. The change in the molar extinction coefficient of tryptophan on conversion to the oxindole derivative is taken to be 4×10^3 M^{-1} cm^{-1}.[12] It has been our experience that one must either perform the reaction in 8.0 *M* urea (pH adjusted to 4.0) or with the reduced, carboxymethylated derivative.[13] The spectra must be obtained as soon as possible after the addition of the *N*-bromosuccinimide since, unless the excess reagent and low molecular weight products of the reaction are rapidly removed, there is a reversal of the decrease in absorbance.[14] This is not a trivial consideration since there is at least one study[15] where there is a real difference in the extent of modification as determined by spectroscopy or amino acid analysis. The rigorous evaluation[14] of the reaction of *N*-bromosuccinimide with model tryptophanyl and tyrosyl compounds reported from Keitaro Hiromi's laboratory provides considerable insight into the problems to be encountered with the study of intact proteins. Figure 2 shows the changes in the UV spectrum of *N*-acetyltryptophan ethyl ester (ATEE) upon reaction with *N*-bromosuccinimide. These spectra were obtained within 5 min after the initiation of the reaction. At ratios of *N*-bromosuccinimide to *N*-acetyltryptophan ethyl ester of greater

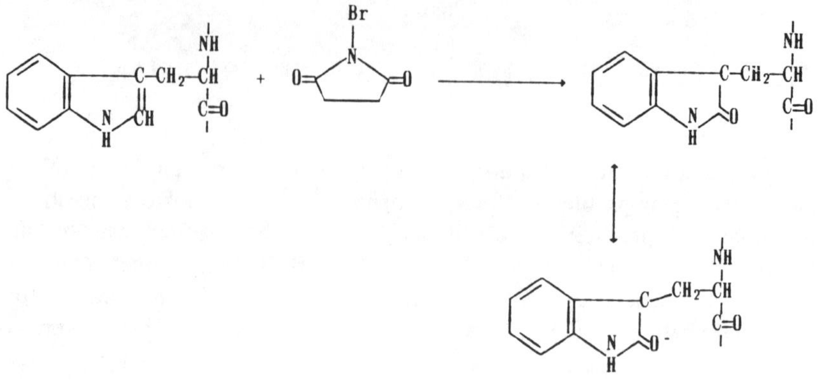

FIGURE 1. The reaction of tryptophan with *N*-bromosuccinimide.

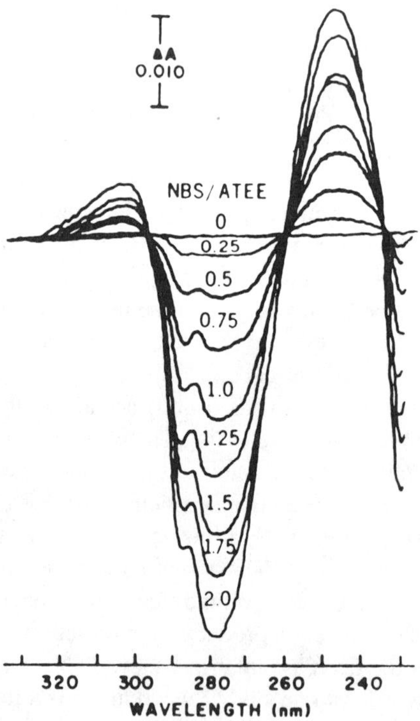

FIGURE 2. Difference UV absorption spectra of *N*-acetyltryptophan ethyl ester caused by reaction with *N*-bromosuccinimide. *N*-acetyltryptophan ethyl ester 19 μ*M*, numbers indicate the molar ratios of *N*-bromosuccinimide to *N*-acetyltryptophan ethyl ester. The reactions were performed in 0.1 *M* acetate buffer, pH 4.5 at 25°C. The difference spectra were obtained within 5 min of the start of the reaction. (From Ohnishi, M., Kawagishi, T., Abe, T., and Hiromi, K., *J. Biochem.*, 87, 273, 1980. With permission.)

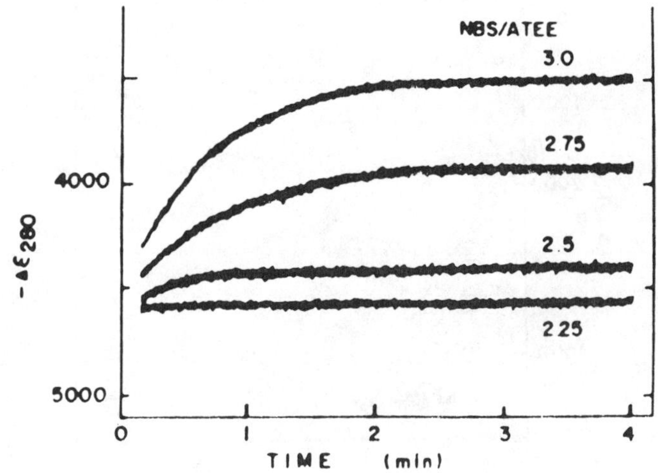

FIGURE 3. Time course of the difference absorbance change of *N*-acetyltryptophan ethyl ester at 280 nm caused by *N*-bromosuccinimide. *N*-acetyltryptophan ethyl ester, 49 μ*M;* 0.1 *M* acetate buffer, pH 4.5 at 25°C. The numbers in the figure indicate the molar ratios of *N*-bromosuccinimide to *N*-acetyltryptophan ethyl ester. (From Ohnishi, M., Kawagishi, T., Abe, T., and Hiromi, K., *J. Biochem.,* 87, 273, 1980. With permission.)

than 2, there is an apparent reversal of the decrease in absorbance at 280 nm as shown in Figure 3. Figure 4 shows the spectral changes occurring upon the reaction of *N*-bromosuccinimide with *N*-acetyltryptophan ethyl ester as a function of time and molar excess of *N*-bromosuccinimide. The maximal decrease in absorbance occurs at a ratio of *N*-bromosuccinimide to tryptophan of 2. If the data are obtained by stopped-flow spectroscopy, the molar excess of *N*-bromosuccinimide does not have an effect on the maximum decrease observed, but when the spectrum is obtained 5 min after the initiation of the reaction, there is a decrease in the observed magnitude of change in absorbance at 280 nm. The evaluation of spectral changes in a protein is further complicated by the reaction of *N*-bromosuccinimide with tyrosine. This is demonstrated in Figure 5, which shows the spectral changes occurring as a result of reaction of *N*-bromosuccinimide with *N*-acetyltyrosine ethyl ester. Here, an increase in absorbance at 280 nm can be observed. The use of this procedure for the analysis of tryptophan content in proteins has been largely supplanted by ion-exchange analysis following modified hydrolytic procedures.[1,2,15]

The primary use of the *N*-bromosuccinimide modification of proteins has been in studies on the effect of such modification on biological (catalytic) activity. In general, the modification reaction is performed in 0.1 *M* sodium acetate, pH 4 to 5. The *N*-bromosuccinimide should be recrystallized from water before use. The presence of halides such as chloride or bromide in the

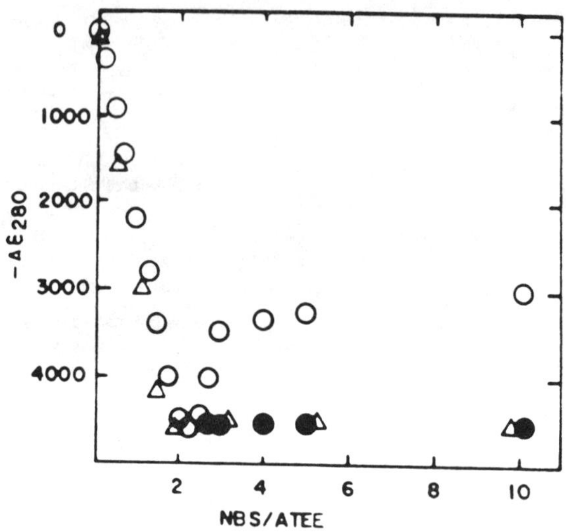

FIGURE 4. Dependence of the difference UV absorption change (decrease) at 280 nm of *N*-acetyltryptophan ethyl ester (ATEE) on time and molar excess of *N*-bromosuccinimide (NBS). The open circles show the difference at 5 min after the addition of *N*-bromosuccinimide, and the closed circles show the change at 0 min (obtained by extrapolation of the time curves). The triangles represent data obtained by the stopped-flow method. The concentration of *N*-acetyltryptophan ethyl ester was 49 μ*M* in 0.1 *M* acetate buffer, pH 4.5 at 25°C. (From Ohnishi, M., Kawagishi, T., Abe, T., and Hiromi, K., *J. Biochem.*, 87, 273, 1980. With permission.)

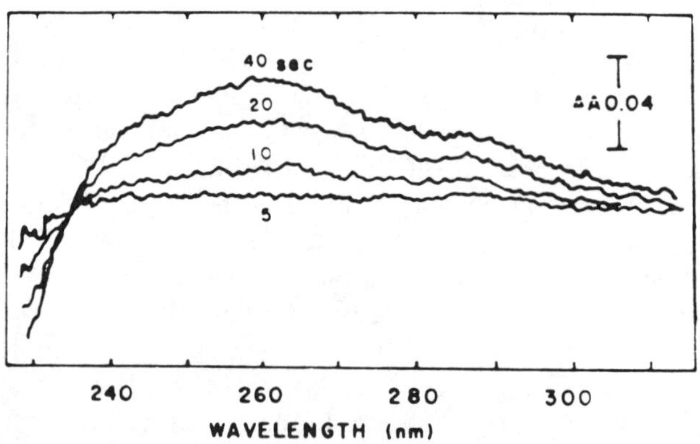

FIGURE 5. The difference UV absorption spectra of *N*-acetyltyrosine ethyl ester on reaction with *N*-bromosuccinimide. *N*-acetyltyrosine ethyl ester, 25 μ*M; N*-bromosuccinimide, 100 μ*M;* 0.1*M* acetate buffer, pH 4.5 at 25°C. The data were obtained with a stopped-flow spectrophotometer in a rapid scanning mode. The spectra were recorded at 5, 10, 20, and 40 s after the start of the reaction. A spectrum was obtained within 10 ms (scan speed). (From Ohnishi, M., Kawagishi, T., Abe, T., and Hiromi, K., *J. Biochem.*, 87, 273, 1980. With permission.)

solvent must be avoided since the addition of *N*-bromosuccinimide will oxidize these ions to the elemental form with disastrous and irreproducible effects on the proteins under study. In general, a twofold molar excess of *N*-bromosuccinimide per mole of tryptophan is necessary to achieve modification. Daniel and Trowbridge[16] found that (at pH 4.0) the reaction of *N*-bromosuccinimide with acetyl-L-tryptophan ethyl ester required 1.5 mol of *N*-bromosuccinimide per mole of the acetyl-L-tryptophan ethyl ester, while trypsinogen required 2.0 to 2.3 mol *N*-bromosuccinimide per mole of tryptophan oxidized, and trypsin required 1.5 to 2.0 mol *N*-bromosuccinimide per mole of tryptophan oxidized. At pH 4.0, only tryptophan in dihydrofolate reductase reacts with NBS, while at pH 6.0, a sulfhydryl group apparently is preferentially oxidized by the reagent prior to the reaction of tryptophan.[17] Poulos and Price have reported on the reaction of a tryptophanyl residue in bovine pancreatic DNase with *N*-bromosuccinimide.[18] Prior reaction of the DNase with another "tryptophan" reagent, 2-hydroxy-5-nitrobenzyl bromide, modified a different residue from the one modified by *N*-bromosuccinimide. These investigators used spectral analysis to determine the extent of tryptophan modification. Subsequent studies from another laboratory[15] on the modification of DNase with *N*-bromosuccinimide suggested that apparently 2 mol of tryptophan are modified per mole of enzyme at 100% inactivation with a sixfold molar excess of *N*-bromosuccinimide in 0.01 *M* $CaCl_2$ at pH 4.0. Using amino acid analysis (after hydrolysis in 6 *N* HCl containing mercaptoacetic acid, phenol, and 3-(2-aminoethyl) indole for 24 h at 110°C), these investigators showed that all three tryptophanyl residues are modified under the above experimental conditions. The study on the modification of tryptophan in galactose oxidase[19] is worth comment in that these investigators report the amino acid composition of the modified protein after hydrolysis in 3 *N* *p*-toluenesulfonic acid. There was excellent agreement between the extent of tryptophan modification as judged by direct amino acid analysis and the value observed by spectral analyses. Tryptophan is responsible for the majority of the innate fluorescence of proteins, and oxidation by *N*-bromosuccinimide obviates this property as shown in Figure 6.

There are several other facets of the use of *N*-bromosuccinimide for the modification of tryptophanyl residue in proteins which should be considered. The use of the reagent at mildly acidic pH has been mentioned above. Not only does increasing pH decrease specificity in terms of reaction with amino acid residues other than tryptophan, but there is a decrease in the modification of tryptophan. This is shown by the studies[20] on the modification of a glucoamylase from *Aspergillus saitoi*. As shown in Figure 7, there is a modest decrease in modification as the pH is increased from 4.0 to 6.0 with a dramatic decrease at pH 7.0. At pH values close to neutrality, there is the increased possibility of modification of amino residues other than tryptophan. The pH dependence of *N*-bromosuccinimide tryptophan modification may well reflect local confor-

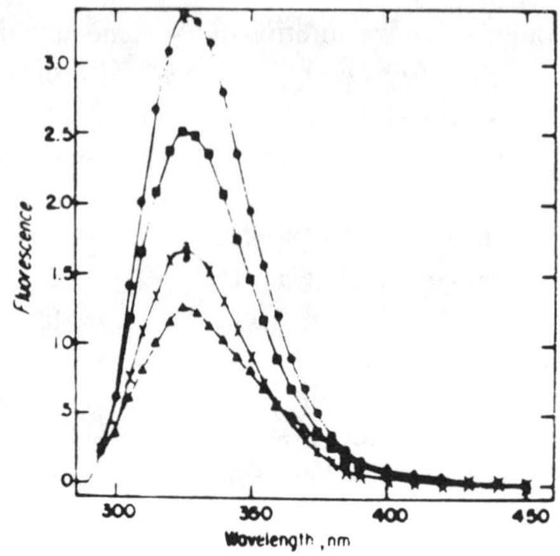

FIGURE 6. Corrected fluorescence emission spectra of deoxygenated solutions of galactose oxidase: unmodified (●), with 0.85 oxidized tryptophans (■), with 2.0 oxidized tryptophans (×), and with 3.0 oxidized tryptophans (▲). The spectra were recorded in 100 m*M* sodium acetate, pH 4.15, after the modification with *N*-bromosuccinimide was performed in 5 m*M* sodium acetate, pH 4.15. The protein concentration was 0.14 mg/ml. The error bars represent the standard deviation of time-averaged recordings. (From Kosman, D. J., Ettinger, M. J., Bereman, R. D., and Giordano, R. S., *Biochemistry,* 16, 1597, 1977. With permission.)

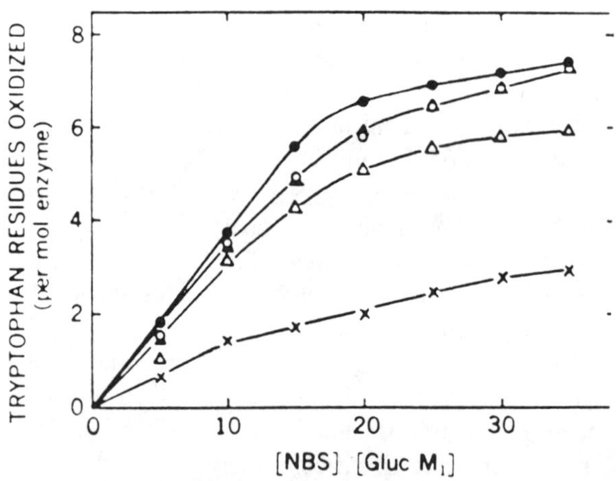

FIGURE 7. *N*-bromosuccinimide oxidation of a glucoamylase (Gluc M_1) as a function of pH. *N*-bromosuccinimide (6.6 m*M*) was added in 1- to 10-μl portions at 5 min intervals to 1 ml of 6.67 m*M* Gluc M_1 in 0.1 *M* acetate buffer at pH 4.0 (●), pH 4.5 (○), pH 5.0 (▲), pH 6.0 (△), and pH 7.0 (×). The decrease in absorbance at 280 nm was measured after each addition of *N*-bromosuccinimide at 25°C. The amount of tryptophan residues oxidized was calculated according to the method of Spande and Witkop. (From Inokuchi, N., Takahashi, T., Yoshimoto, A., and Irie, M., *J. Biochem.,* 91, 1661, 1982. With permission.)

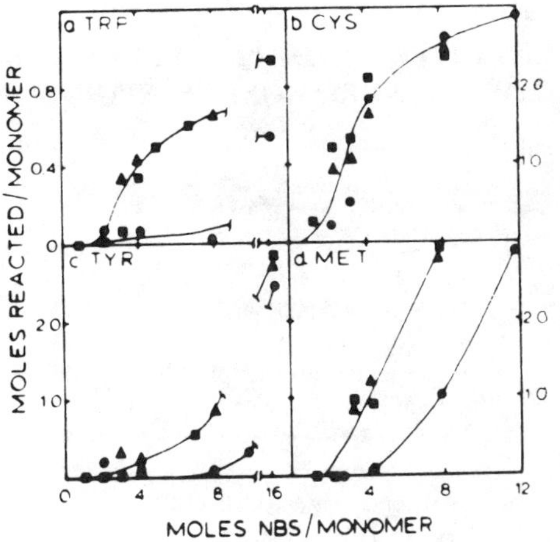

FIGURE 8. The modification of *E. coli* lac repressor protein with *N*-bromosuccinimide. Shown are the moles of amino acid reacted with *N*-bromosuccinimide per monomer of repressor protein. The modification was performed in 1.0 *M* Tris/Cl, pH 7.8 for 15 min at ambient temperature in the dark. The reactions were terminated by the addition of dithiothreitol and the modified protein preparations dialyzed vs. water for subsequent analysis. The amount of tryptophan, tyrosine, and methionine were determined by amino acid analysis after hydrolysis in methanesulfonic acid. Cysteine was determined by titration with 2-chloromercuri-4-nitrophenol in 8 *M* urea. ■——■, Repressor reacted with *N*-bromosuccinimide alone; ●——●, repressor reacted with *N*-bromosuccinimide in the presence of isopropyl-1-thio-β-D-galactoside; ▲——▲, repressor reacted with *N*-bromosuccinimide in the presence of *o*-nitrophenyl-β-D-fucoside. (From O'Gorman, R. B. and Matthews, K. S., *J. Biol. Chem.*, 252, 3565, 1977. With permission.)

mational effects rather than intrinsic chemistry. In studies[21] on the reaction on *Escherichia coli* lac repressor protein with *N*-bromosuccinimide at pH 7.8 (1.0 *M* Tris), cysteine was modified as readily as tryptophan with lesser modification of methionine and tyrosine (Figure 8). Although the great majority of *N*-bromosuccinimide modifications of proteins are performed at pH values less than 5.0 to avoid modification of other functional groups, success can be achieved under less acidic conditions. Kumar et al.[22] modified tryptophanyl residues in transcarboxylase in 0.25 *M* potassium phosphate, pH 6.5, containing 0.1 *M* dithiothreitol and 0.1 m*M* phenylmethylsulfonyl fluoride. These investigators also demonstrated that the loss of activity upon modification was not a reflection of gross conformational change by examination of the quenching of intrinsic fluorescence and changes in the susceptibility to tryptic cleavage. In general, modification should occur at a 4 to 6 molar excess (with respect to total tryptophan) of *N*-bromosuccinimide. Under most reaction conditions, the modification of tryptophanyl residues with *N*-bromosuccinimide is quite rapid. Time-dependent reactions have, however, been observed such as that reported for xylanase.[23] Reaction with 2-hydroxy-5-nitrobenzyl bromide was also time dependent under these reaction conditions. The use of *N*-bromosuccinimide in the study of proteins is summarized in Table 1.

Table 1
EXAMPLES OF THE MODIFICATION OF PROTEINS WITH *N*-BROMOSUCCINIMIDE

Protein	Solvent	Molar excess[a]	Extent of modification	Ref.
Trypsinogen	pH 7.0[b]	1–4	1–2	1
Trypsin	pH 4.0[b]	1–4	1–2	1
Dihydrofolate reductase	0.1 *M* sodium phosphate, pH 6.0	15	2.0	2
		15	2.7	2
	0.1 *M* sodium acetate, pH 4.0	12	3.8	2
	0.13 *M* sodium acetate formate, 5.3 *M* urea, pH 4.0			
Bovine pancreatic DNase	0.1 *M* sodium acetate, pH 4.0 containing 0.033 *M* $CaCl_2$	6	1.0	3
Bovine pancreatic DNase	pH 4.0, 0.010 *M* $CaCl_2$	1–6	3[c]	4[c]
Dihydrofolate reductase	0.05 *M* potassium phosphate, pH 6.5	20	2.0	5
Pyrocatechase (*Brevibacterium fuscum*)[d]	0.1 *M* phosphate, pH 7.0	—	2	6
Relaxin	0.2 *M* sodium acetate, pH 4.7	—		7
Rhodopsin	0.1 *M* Tris acetate, pH 7.4 containing 1% emulphogene	50	6	8
		100	9	8
Pig kidney amino acylase	0.1 *M* sodium acetate, pH 5.0, 1.0 *M* urea	50	6	9
Galactose oxidase	0.005 *M* sodium acetate, pH 4.15	7	2	10
Bovine thrombin	0.1 sodium acetate, pH 4.0	1	0.5	11
		2	1.1	11
Papain[e]	0.05 *M* sodium acetate, pH 4.75	6[f]	1.4	12
Lac repressor protein	1.0 *M* Tris HCl, pH 7.8	8	0.7[g]	13
α-Mannosidase (*Phaseolus vulgaris*)	1.0 *M* sodium acetate, pH 4.0	35	10	14
α-Amylase (*Bacillus subtilis*)	0.01 *M* sodium phosphate, pH 7.0	8	2	15
		50	4	15
Dihydrofolate reductase	0.015 *M* bis Tris HCl, pH 6.5			16
	0.5 *M* KCl	4	1.2	
Xylanase	50 m*M* NaOAc, pH 4.5	—	—[h]	17
Transcarboxylase	250 m*M* potassium phosphate, pH 6.5[j]	—	10/90[i]	18
Cellulase	50 m*M* NaOAc, pH 5.0	30	8/12	19
Winged bean	0.1 *M* sodium citrate, pH 6.0	10	2/4	20

[a] Reagent to protein.
[b] pH maintained at 4.0 by addition of KOH.
[c] Spectral analysis suggested that 2 mol of tryptophan oxidized, while amino acid analysis demonstrates that all three tryptophan residues modified.
[d] Thiophenylated apoenzyme [apoenzyme modified with 5,5′-dithiobis-(2-nitrobenzoic acid)].
[e] Not activated.
[f] Also modified tyrosine at this concentration.
[g] Also had substantial modification of tyrosine, cysteine, and methionine.

Table 1 (continued)
EXAMPLES OF THE MODIFICATION OF PROTEINS WITH *N*-BROMOSUCCINIMIDE

[h] There is an apparent time-dependent reaction with both *N*-bromosuccinimide and 2-hydroxy-5-nitrobenzyl bromide.

[i] Containing 0.1 m*M* dithiothreitol and 0.1 *M* phenylmethylsulfonyl fluoride. The reaction was terminated by the addition of a tenfold molar excess of *N*-acetyltryptophanamide.

[j] At maximum inactivation. A total of 40 tryptophanyl residues were available for oxidation in the native state.

References for Table 1

1. **Daniel, V. W., III and Trowbridge, C. G.,** The effect of *N*-bromosuccinimide upon trypsinogen activation and trypsin catalysis, *Arch. Biochem. Biophys.*, 134, 506, 1969.
2. **Freisheim, J. H. and Huennekens, F. M.,** Effect of *N*-bromosuccinimide on dihydrofolate reductase, *Biochemistry*, 8, 2271, 1969.
3. **Poulos, T. L. and Price, P. A.,** The identification of a tryptophan residue essential to the catalytic activity of bovine pancreatic deoxyribonuclease, *J. Biol. Chem.*, 246, 4041, 1971.
4. **Sartin, J. L., Hugli, T. E., and Liao, T.-H.,** Reactivity of the tryptophan residues in bovine pancreatic deoxyribonuclease with *N*-bromosuccinimide, *J. Biol. Chem.*, 255, 8633, 1980.
5. **Warwick, P. E., D'Souza, L., and Freisheim, J. H.,** Role of tryptophan in dihydrofolate reductase, *Biochemistry*, 11, 3775, 1972.
6. **Nagami, K.,** The participation of a tryptophan residue in the binding of ferric iron to pyrocatechase, *Biochem. Biophys. Res. Commun.*, 51, 364, 1973.
7. **Schwabe, C. and Braddon, S. A.,** Evidence for the essential tryptophan residue at the active site of relaxin, *Biochem. Biophys. Res. Commun.*, 68, 1126, 1976.
8. **Cooper, A. and Hogan, M. E.,** Reactivity of tryptophans in rhodopsin, *Biochem. Biophys. Res. Commun.*, 68, 178, 1976.
9. **Kördel, W. and Schneider, F.,** Chemical modification of two tryptophan residues abolishes the catalytic activity of aminoacylase, *Hoppe Seyler's Z. Physiol. Chem.*, 357, 1109, 1976.
10. **Kosman, D. J., Ettinger, M. J., Bereman, R. D., and Giordano, R. S.,** Role of tryptophan in the spectral and catalytic properties of the copper enzyme, galactose oxidase, *Biochemistry*, 16, 1597, 1977.
11. **Uhteg, L. C. and Lundblad, R. L.,** The modification of tryptophan in bovine thrombin, *Biochim. Biophys. Acta*, 491, 551, 1977.
12. **Glick, B. R. and Brubacher, L. S.,** The chemical and kinetic consequences of the modification of papain by *N*-bromosuccinimide, *Can. J. Biochem.*, 55, 424, 1977.
13. **O'Gorman, R. B. and Matthews, K. S.,** *N*-bromosuccinimide modification of *lac* repressor protein, *J. Biol. Chem.*, 252, 3565, 1977.
14. **Paus, E.,** The chemical modification of tryptophan residues of α-mannosidase from *Phaseolus vulgaris*, *Biochim. Biophys. Acta*, 533, 446, 1978.
15. **Fujimori, H., Ohnishi, M., and Hiromi, K.,** Tryptophan residues of saccharifying α-amylase from *Bacillus subtilis*. A kinetic discrimination of states of tryptophan residues using *N*-bromosuccinimide, *J. Biochem.*, 83, 1503, 1978.
16. **Thomson, J. W., Roberts, G. C. K., and Burgen, A. S. V.,** The effects of modification with *N*-bromosuccinimide on the binding of ligands to dihydrofolate reductase, *Biochem. J.*, 187, 501, 1980.
17. **Keskar, S. S., Srinivasan, M. C., and Deshpande, V. V.,** Chemical modification of a xylanase from a thermotolerant *Streptomyces*. Evidence for essential tryptophan and cysteine residues at the active site, *Biochem. J.*, 261, 49, 1989.
18. **Kumar, G. K., Beegen, H., and Wood, H. G.,** Involvement of tryptophans at the catalytic and subunit-binding domains of transcarboxylase, *Biochemistry*, 27, 5972, 1988.
19. **Clarke, A. J.,** Essential tryptophan residues in the function of cellulase from Schizophyllum commune, *Biochim. Biophys. Acta*, 912, 424, 1987.
20. **Higuchi, M., Inoue, K., and Iwai, K.,** A tryptophan residue is essential to the sugar-binding site of winged bean basis lectin, *Biochim. Biophys. Acta*, 829, 51, 1985.

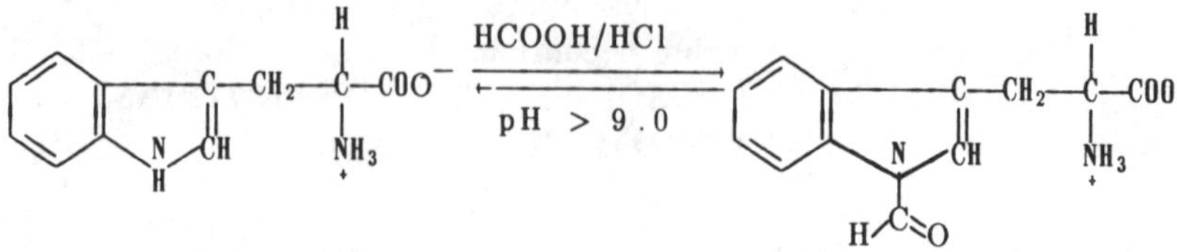

FIGURE 9. A scheme for the reversible formylation of tryptophan residues.

The reaction of *N*-bromosuccinimide with proteins can also result in the cleavage of peptide bonds at tryptophan, tyrosine, and histidine.[24] Thus, the careful investigator will also evaluate the integrity of the polypeptide chain(s) of the protein of interest. Whereas peptide bond cleavage is usually an unwanted side reaction, Feldhoff and Peters[25] have devised a procedure which has enhanced specificity for tryptophan. Their procedure uses 8.0 *M* urea, 2.0 *M* acetic acid as the solvent with a 20-fold molar excess of *N*-bromosuccinimide. Their approach offers at least two advantages: first, the protein is denatured so that all residues should be equally available, and second, the *N*-bromosuccinimide reacts with urea to yield *N*-bromourea, a less severe oxidizing agent which should have increased specificity for tryptophanyl residues. The use of *N*-chlorosuccinimide for peptide bond cleavage of tryptophanyl residues is considered a superior approach[26] (see Chapter 5). A direct comparison of *N*-bromosuccinimide and *N*-chlorosuccinimide in the modification of the single tryptophanyl residue in *Clostridium perfringes* ε-toxin has been performed.[27] Modification with *N*-bromosuccinimide (50 m*M* sodium acetate, pH 5.0) resulted in the total loss of tryptophan and marked reduction in tyrosine (71% decrease) and methionine (79% decrease). Reaction with *N*-chlorosuccinimide (50 m*M* Tris, pH 8.5) resulted in the total loss of both tryptophan and methionine, but no significant change in tyrosine. Reaction with chloroamine T (50 m*M* Tris, pH 8.5) resulted only in the loss of methionine. Activity was lost only with the modification of tryptophan. Peptide bond cleavage was not observed under these reaction conditions.

The conversion of tryptophanyl residues to 1-formyltryptophanyl residues has been reported. The reaction conditions are somewhat harsh, but the procedure is reversible (Figure 9), should prove quite useful for small peptides, and has been applied to several proteins. Coletti-Previero and co-workers[28] have successfully applied this procedure to bovine pancreatic trypsin. Trypsin was dissolved in formic acid saturated with HCl at a concentration of 2.5 mg/ml at 20°C. The formylation reaction is associated with an increase in absorbance at 298 nm[29] (Figure 10). Therefore, it is possible to follow the reaction spectrophotometrically. The reaction is judged complete when there is no further increase in absorbance at 298 nm. The above reaction with trypsin was complete after an incubation period of 1 h. The solvent was partially removed *in vacuo* over KOH pellets followed by lyophilization. The formyltryptophan derivative is unstable at alkaline pH. At pH 9.5 (pH-stat), conversion back to tryptophan is complete after 200 min incubation at 20°C. Holmgren has suc-

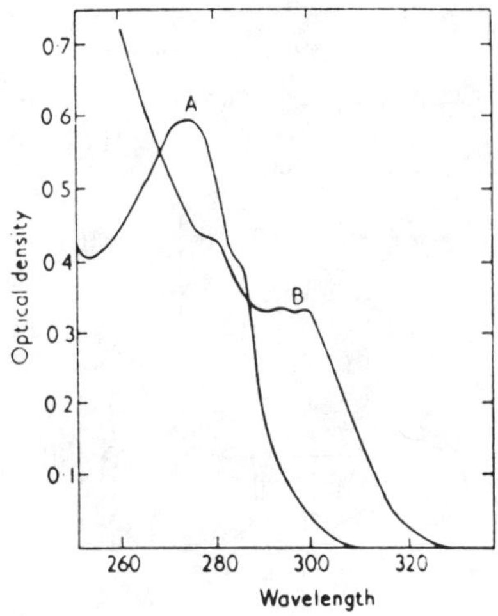

FIGURE 10. Changes in the UV absorption spectrum of trypsin occurring as a result of formylation of tryptophan residues. Shown is the spectrum of trypsin (18 μ*M* in 8.0 *M* urea, pH 4.0) before (curve A) and after (curve B) 1-formylation of tryptophanyl residues. Formylation of trypsin was accomplished by dissolving trypsin in formic acid saturated with gaseous HCl at 20°C (2.5 mg/ml). At suitable time intervals, 0.4 ml samples of the solution were diluted with 2 ml of 8 *M* urea, pH 4.0 for recording the UV spectra. When the maximum increase in absorbance at 298 nm was reached (about 60 min), the solvent was partially removed under vacuum over KOH pellets for 15 min in order to eliminate most of the HCl and the sample was subsequently lyophilized. (From Coletti-Previero, M.-A., Previero, A., and Zuckerkandl, E., *J. Mol. Biol.*, 39, 493, 1969. With permission.)

cessfully applied this procedure to thioredoxin.[30] A more recent study on *N*-formylation of tryptophanyl residues in proteins involved the study of epitopes in horse heart cytochrome c.[31] The single tryptophanyl residue was formylated with formic acid saturated with HCl. The modified protein had markedly reduced affinity for a monoclonal antibody resulting from local conformational change.

One of the most useful modification procedures for tryptophanyl residues in proteins involves the use of 2-hydroxy-5-nitrobenzyl bromide and its various derivatives (Figure 11). 2-Hydroxy-5-nitrobenzyl bromide, frequently referred to as Koshland's reagent, was introduced by Koshland and co-workers.[32,33] Barman and Koshland[34] have reported the use of 2-hydroxy-5-nitrobenzyl bromide for the quantitative determination of tryptophanyl residues in proteins. Although this approach to the quantitative determination of tryptophanyl residues in proteins has been largely replaced by the development of new methods for the hydrolysis of proteins, it can still be useful in certain instances. For this procedure, the sample is incubated for 16 to 20 h at 37°C in 1.0 ml 10 *M* urea (the urea should be recrystallized ($EtOH/H_2O$) prior to use), pH 2.7 (pH adjusted with concentrated HCl). This solution is cooled to ambient temperature, and approximately 5.0 mg of 2-hydroxy-5-nitrobenzyl bromide (in 0.1 ml

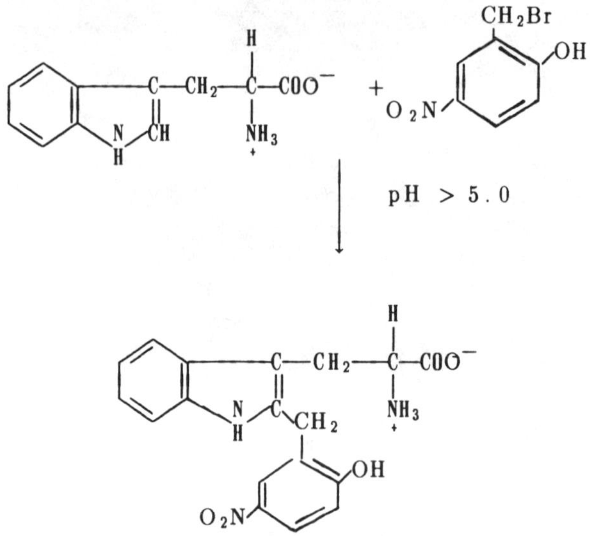

FIGURE 11. A scheme for the reaction of 2-hydroxy-5-nitrobenzyl bromide with tryptophan.

acetone) is added followed by vigorous stirring (we have found the Pierce Reacti-Vials® very useful for this purpose). Occasionally, a precipitate of 2-hydroxy-5-nitrobenzyl alcohol (the hydrolytic product of 2-hydroxy-5-nitrobenzyl bromide) forms, which can be removed by centrifugation. The labeled protein is obtained free of reagent by gel filtration. This step is generally performed under acidic conditions (e.g., 0.18 *M* acetic acid, 10% acetic acid, or 10% formic acid).* Depending upon the protein under study, it might be necessary to perform this step in 10 *M* urea (pH 2.7) to maintain the solubility of the modified protein. A portion of the modified protein is taken to a pH >12 with NaOH. The extent of incorporation is determined at 410 nm using an extinction coefficient of 18,000 M^{-1} cm^{-1}. It is necessary to determine the concentration of protein by a technique other than absorbance at 280 nm because of the modification of tryptophan. We have found it convenient to either use amino acid analysis after acid hydrolysis or the ninhydrin reaction[35] after alkaline hydrolysis.[36]

The most frequent use of 2-hydroxy-5-nitrobenzyl bromide has been in the specific modification of tryptophan in peptides and proteins (Table 2). Under appropriate reaction conditions (pH 4.0 or below), the reagent is highly specific for reaction with tryptophan. We have, on occasion, seen the modification of methionine residues under these conditions. This reagent also has the advan-

* Despite its wide use as a solvent for peptides and proteins, the use of formic acid is not recommended because of the potential of side reactions at amino functional groups; see Shively, J. E., Hawke, D., and Jones, B. N., Microsequence analysis of peptides and protein. III. Artifacts and effects of impurities on analysis, *Anal. Biochem.*, 120, 312, 1982.

Table 2
EXAMPLES OF THE MODIFICATION OF PROTEINS WITH 2-HYDROXY-5-NITROBENZYL BROMIDE

Protein	Solvent	Molar excess	Residues modified	Ref.
Pepsin	0.1 *M* NaCl[a]	300	2/4	1
Streptococcal proteinase	0.46 *M* sodium phosphate, pH 3.1	200	1.8/4	2
Pancreatic deoxyribonuclease	0.050 *M* $CaCl_2$[b]	100	1/3	3
Carbonic anhydrase	0.1 *M* phosphate, pH 6.8	100	—[c]	4
Trypsin	0.1 *M* NaCl, 0.02 *M* $CaCl_2$,[d] pH 4.2 (pH-stat)	ca. 100	1/4	5
Human chorionic somatomammotropin	0.05 *M* glycine, pH 2.8	—	—	6
Naja naja neurotoxin	0.2 *M* acetic acid[e]	40	—[f]	7
Glyceraldehyde-3-phosphate dehydrogenase	pH 6.75[g]	30[h]	1/3	8
α-Mannosidase *(Phaseolus vulgaris)*	0.1 *M* sodium acetate, pH 3.7	100	5/28	9
Thrombin	0.2 *M* acetate, pH 4.0	100	1/8	10
Laccase	pH 6.95[i]	50	0.30/6	11
	pH 4.00[i]	50	0.58/6	11
	pH 3.30[i]	110	2.39/6	11
Human serum albumin	10 *M* urea, pH 4.4	1000	1.1/1[j]	12
Xylanase	50 m*M* NaOAc, pH 5.0	—	—[k]	13
Winged bean	0.1 *M* sodium citrate, pH 3.1	100	0.9/4	14
		200	1.7/4	
		400	1.8/4	

[a] pH adjusted with 50% acetic acid.
[b] pH remained between 4.0 an 4.5 without need for buffer.
[c] Variation with respect to enzyme source.
[d] pH maintained at 4.2 by addition of NaOH (pH-stat).
[e] pH 2.7.
[f] Polymerization occurred.
[g] pH maintained at 6.75 by the addition of 0.1 *M* NaOH.
[h] Dimethyl (2-hydroxy-5-nitrobenzyl) sulfonium bromide was used in the experiments. Prior to reaction, the active site sulfhydryl was blocked by reaction with 5,5′-dithiobis-(2-nitrobenzoate).
[i] Unbuffered, pH maintained by titration with NaOH.
[j] Incorporation determined at pH 7.4 after the following relationship:[51] moles 2-hydroxy-5-nitrobenzyl bromide per mole albumin = $(A_{410} \times 69,000 \times 0.498)/13,800 \times (A_{280} - 0.167) \times A_{410}$.
[k] A time-dependent inactivation reaction was observed.

References for Table 2

1. **Dopheide, T. A. A. and Jones, W. M.,** Studies on the tryptophan residues in porcine pepsin, *J. Biol. Chem.,* 243, 3906, 1968.
2. **Robinson, G. W.,** Reaction of a specific tryptophan residue in streptococcal proteinase with 2-hydroxy-5-nitrobenzyl bromide, *J. Biol. Chem.,* 245, 4832, 1970.
3. **Poulos, T. L. and Price, P. A.,** The identification of a tryptophan residue essential to the catalytic activity of bovine pancreatic deoxyribonuclease, *J. Biol. Chem.,* 246, 4041, 1971.

References for Table 2 (continued)

4. **Lindskog, S. and Nilsson, A.,** The location of tryptophanyl groups in human and bovine carbonic anhydrases. Ultraviolet difference spectra and chemical modification, *Biochim. Biophys. Acta,* 295, 117, 1973.
5. **Imhoff, J. M., Keil-Dlouha, V., and Keil, B.,** Functional changes in bovine α- and β-trypsins caused by the substitution of tryptophan-199, *Biochimie,* 55, 521, 1973.
6. **Neri, P., Arezzini, C., Botti, R., Cocola, F., and Tarli, P.,** Modification of the tryptophanyl residue and its effect on the immunological and biological activity of human chorionic somatomammotropin, *Biochim. Biophys. Acta,* 322, 88, 1973.
7. **Karlsson, E., Eaker, D., and Drevin, H.,** Modification of the invariant tryptophan residue of two *Naja naja* neurotoxins, *Biochim. Biophys. Acta,* 328, 510, 1973.
8. **Heilman, H. D. and Pfleiderer, G.,** On the role of tryptophan residues in the mechanism of action of glyceraldehyde-3-phosphate dehydrogenase as tested by specific modification, *Biochim. Biophys. Acta,* 384, 331, 1975.
9. **Paus, E.,** The chemical modification of tryptophan residues of α-mannosidase from *Phaseolus vulgaris, Biochim. Biophys. Acta,* 533, 446, 1978.
10. **Uhteg, L. C. and Lundblad, R. L.,** The modification of tryptophan in bovine thrombin, *Biochim. Biophys. Acta,* 491, 551, 1977.
11. **Clemmer, J. D., Carr, J., Knaff, D. B., and Holwerda, R. A.,** Modification of laccase tryptophan residues with 2-hydroxy-5-nitrobenzyl bromide, *FEBS Lett.,* 91, 346, 1978.
12. **Fehske, K. J., Müller, W. E., and Wollert, U.,** The modification of the lone tryptophan residue in human serum albumin by 2-hydroxy-5-nitrobenzyl bromide. Characterization of the modified protein and the binding of L-tryptophan and benzodiazepines to the tryptophan-modified albumin, *Hoppe Seyler's Z. Physiol. Chem.,* 359, 709, 1978.
13. **Keskar, S. S., Srinivasan, M. C., and Deshpande, V. V.,** Chemical modification of a xylanase from a thermotolerant *Streptomyces.* Evidence for essential tryptophan and cysteine residues at the active site, *Biochem. J.,* 261, 49, 1989.
14. **Higuchi, M., Inoue, K., and Iwai, K.,** A tryptophan residue is essential to the sugar-binding site of winged bean basis lectin, *Biochim. Biophys. Acta,* 829, 51, 1985.

tage of being a "reporter" group in the sense that the spectrum of the hydroxynitrobenzyl derivative is sensitive to changes in the microenvironment as shown in Figure 12. This decrease observed in absorbance at 410 nm associated with an increase in absorbance at 320 nm upon the addition of dioxane is similar to that seen with acidification and reflects the increase in the pKa of the phenolic hydroxyl group. Titration curves of oxidized and reduced laccase[37] which had been modified with 2-hydroxy-5-nitrobenzyl bromide are shown in Figure 13. From this experiment, it was concluded that the tryptophanyl residues in laccase modified with 2-hydroxy-5-nitrobenzyl bromide are in an essentially aqueous microenvironment. The chemistry of the reaction of 2-hydroxy-5-nitrobenzyl bromide with tryptophan has been studied in some detail.[38] Disubstitution on the indole ring is a possibility and is usually seen as a sudden "break" in the plot of extent of modification vs. reagent excess (see Figure 14).

In our hands, the following procedure has been found useful. The protein or peptide to be modified is taken into 0.1 to 0.2 *M* sodium acetate buffer, pH 4 to 5. Reaction with other nucleophilic centers on the protein will become more

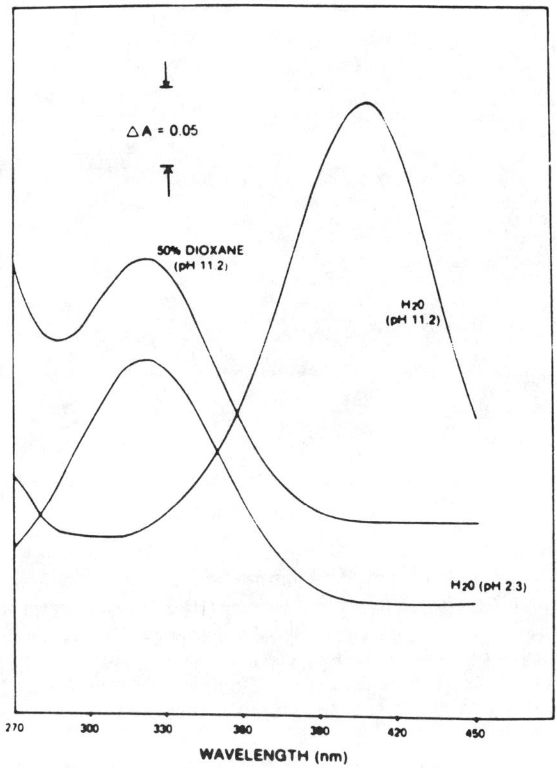

FIGURE 12. The UV absorption spectra of 2-hydroxy-5-nitrobenzyl alcohol (HNB-OH) in different solvents. The concentration of HNB-OH was 33.2 μ*M*. (From Clemmer, J. D., Carr, J., Knaff, D. B., and Holwerda, R. A., *FEBS Lett.*, 91, 346, 1978. With permission.)

of a problem as one approaches neutral pH. A 100-fold molar excess of 2-hydroxy-5-nitrobenzyl bromide (dissolved in a suitable water-miscible organic solvent such as acetone or dimethyl sulfoxide) is added in the dark. After 5 min (this time period was arbitrarily selected; the reaction can be considered to be essentially instantaneous to either modify tryptophan or undergo hydrolysis), the reaction mixture is taken by gel filtration into a solvent suitable for subsequent analysis. The extent of modification is determined under basic conditions as described above for the use of this reagent in the quantitative determination of tryptophan.

The use of 2-hydroxy-5-nitrobenzyl bromide does present problems in that the reagent is extremely sensitive to hydrolysis and is not very soluble under aqueous conditions. These difficulties are avoided, and the characteristics of the reaction are preserved by the use of the dimethyl sulfonium salt obtained

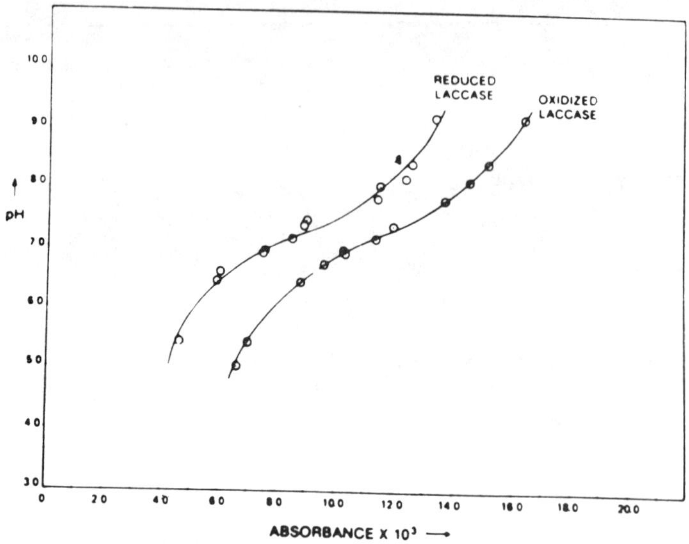

FIGURE 13. Effect of pH on the absorbance (410 nm) of two forms of laccase modified with 2-hydroxy-5-nitrobenzyl bromide (HNB-laccase). Shown are spectrophotometric titration curves for oxidized and reduced HNB-laccase (2.5 μ*M*; 0.38 mol HNB per mole of laccase). The absorbance at 410 nm is shown as a function of pH (phosphate buffers). (From Clemmer, J. D., Carr, J., Knaff, D. B., and Holwerda, R. A., *FEBS Lett.*, 91, 346, 1978. With permission.)

from the reaction of 2-hydroxy-5-nitrobenzyl bromide with dimethyl sulfide.[39] This compound is easily synthesized or can be obtained from various commercial sources. This water-soluble sulfonium salt derivative has recently been used to modify a tryptophanyl residue in rabbit skeletal myosin subfragment-1.[40] Purification of peptide-containing modified tryptophanyl residues was achieved by immunoaffinity chromatography using rabbit antibody to bovine serum albumin previously modified with dimethyl-(2-hydroxy-5-nitrobenzyl) sulfonium bromide.

Horton and Koshland[41] have also developed a clever approach for modification of hydrolytic enzymes. If 2-hydroxy-5-nitrobenzyl bromide is substituted at the phenolic hydroxyl, it is essentially unreactive as originally shown for the methoxy derivative. Horton and Young[42] prepared 2-acetoxy-5-nitrobenzyl bromide. This derivative, like the methoxy derivative, is essentially unreactive. There is considerable structural identity between 2-acetoxy-5-nitrobenzyl bromide and *p*-nitrophenyl acetate, which is a nonspecific substrate for chymotrypsin (Figure 15). α-Chymotrypsin removes the acetyl group from 2-acetoxy-5-nitrobenzyl bromide, thus generating 2-hydroxy-5-nitrobenzyl bromide at the active site which then either rapidly reacts with a neighboring nucleophile or undergoes hydrolysis. Uhteg and Lundblad[43] have used both the

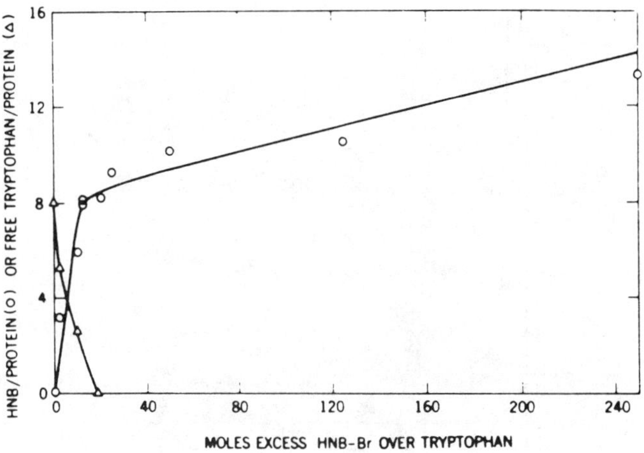

FIGURE 14. The titration of the tryptophanyl residues of carboxymethyl chymotrypsinogen with 2-hydroxy-5-nitrobenzyl bromide. Carboxymethyl chymotrypsinogen (reduced and *S*-carboxymethylated with iodoacetic acid) was incubated in 10 *M* urea, pH 2.7 for 16 to 18 h at which point 1-ml portions (5 mg of protein) were reacted with increasing amounts of 2-hydroxy-5-nitrobenzyl bromide dissolved in 0.1 ml acetone. The modified protein was separated from excess reagent by gel filtration (G-25 Sephadex) and subsequently analyzed for tryptophan (amino acid analysis after alkaline hydrolysis) and for the incorporation of the 2-hydroxy-5-nitrobenzyl group. (From Barman, T. E. and Koshland, D. E., Jr., *J. Biol. Chem.*, 242, 5771, 1967. With permission.)

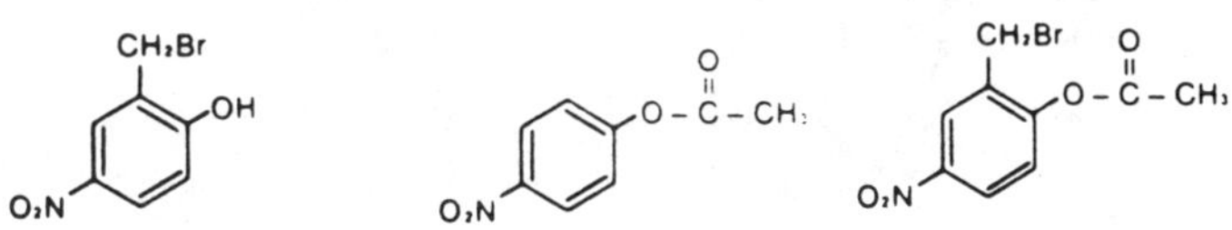

FIGURE 15. The structures of 2-hydroxy-5-nitrobenzyl bromide (left), *p*-nitrophenyl acetate (center), and 2-acetoxy-5-nitrobenzyl bromide.

acetoxy and butyroxy derivatives in the study of thrombin. A similar approach has been used in the study of papain with 2-chloromethyl-4-nitrophenyl *N*-carbobenzoxy-glycinate.[44] It has been subsequently shown that this modification occurs at a specific tryptophan residue in papain.[45]

Reagents with reaction characteristics similar to 2-hydroxy-5-nitrobenzyl bromide are the *o*-nitrophenylsulfenyl derivatives.[46] The reaction product resulting from the sulfonylation of lysozyme[47] in *o*-nitrobenzenesulfenyl chloride (2-nitrophenylsulfenyl chloride) (40-fold molar excess) pH 3.5 (0.1 *M* sodium acetate) has spectral characteristics which can be used to determine the extent of reagent incorporation (at 365 nm $\varepsilon = 4 \times 10^{-3}\ M^{-1}\ cm^{-1}$) (Figure 16). These reagents show considerable specificity for the modification of tryptophan at pH $\leq$4.0 (Figure 17). Possible side reactions with other nucleophiles

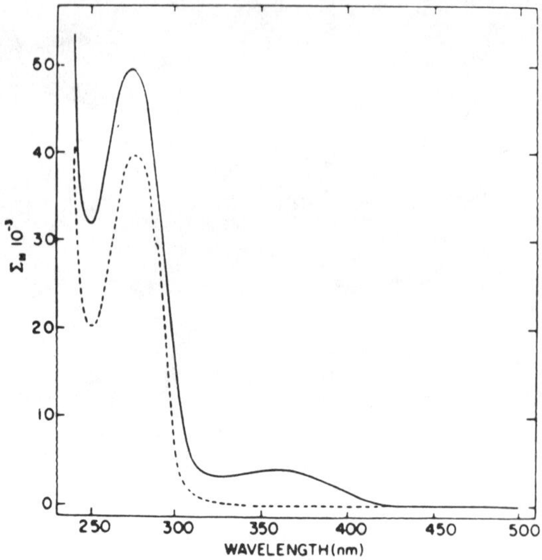

FIGURE 16. The UV absorbance spectrum of egg white lysozyme after modification of tryptophan-62 with 2-nitrophenylsulfenyl chloride (*o*-nitrobenzenesulfenyl chloride). Shown is the UV absorption spectrum of 1-NPS-lysozyme (—) and native lysozyme (- - -). The measurements were performed in water at pH 7.0. (From Shechter, Y., Burstein, Y., and Patchornik, A., *Biochemistry*, 11, 653, 1972. With permission.)

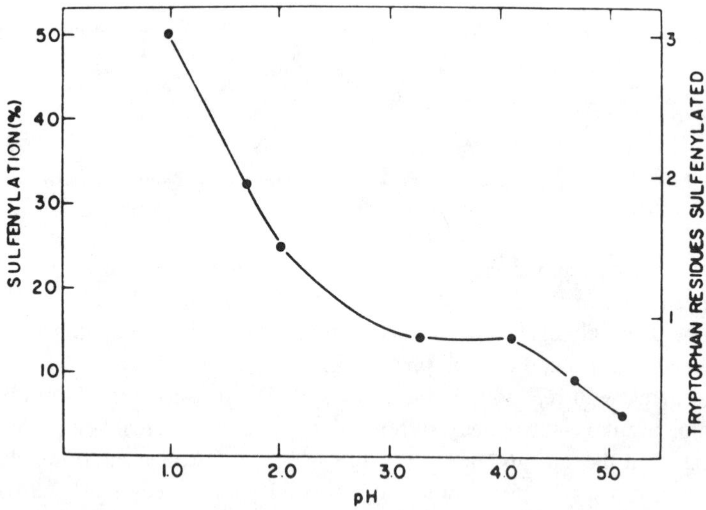

FIGURE 17. The extent of sulfenylation of tryptophanyl residues in egg white lysozyme by 2-nitrophenylsulfenyl chloride as a function of the pH. Sulfenylation was carried out at protein concentration of 0.5 μmol in 1 ml of 0.1 *M* buffered solutions (HCl-KCl at pH 1 to 2; sodium acetate at pH 3 to 5) with 20 μmol of 2-nitrophenylsulfenyl chloride for 5 h. (From Shechter, Y., Burstein, Y., and Patchornik, A., *Biochemistry,* 11, 653, 1972. With permission.)

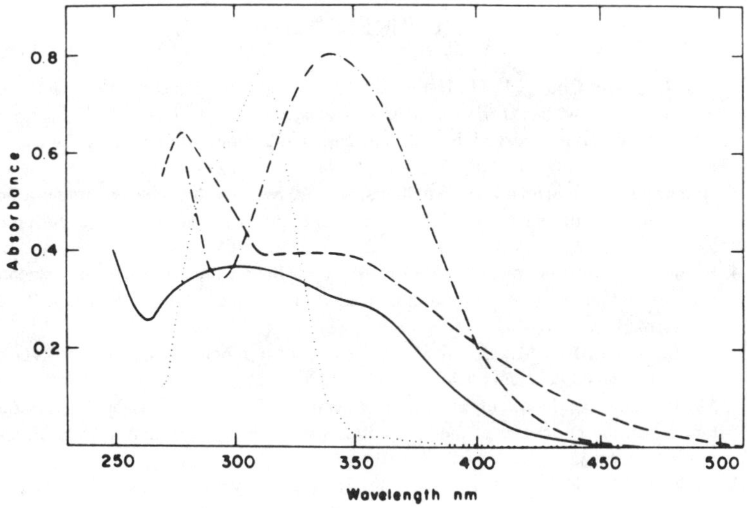

FIGURE 18. UV absorption spectra of the derivatives formed by the thiolysis of 2-(2,4-dinitrophenylsulfenyl) tryptophan. Shown are the UV absorption spectra of 2,4-dinitrophenylsulfenyltryptophan (- - - -), 2-thioltryptophan (····), ditryptophanyl-2-disulfide (——), and *S*-2,4-dinitrophenyl-2-mercaptoethanol (—·—). The spectral studies were performed in 0.1 *M* ammonium bicarbonate. (From Wilchek, M. and Miron, T., *Biochem. Biophys. Res. Commun.*, 47, 1015, 1972. With permission.)

such as amino groups need to be considered. In the case of human chorionic somatomammotropin and human pituitary growth hormone,[48] reaction with *o*-nitrophenylsulfenyl chloride (2-nitrophenylsulfenyl chloride) was achieved in 50% acetic acid, but not in 0.1 sodium acetate, pH 4.0. Wilchek and Miron[49] have reported on the reaction of 2,4-dinitrophenylsulfenyl chloride with tryptophan in peptides and protein, and subsequent conversion of the modified tryptophan to 2-thiotryptophan by reaction with β-mercaptoethanol at pH 8.0. The thiolysis of the modified tryptophan is responsible for changes in the spectral properties of the derivative (Figure 18). The characteristics of the modified tryptophan have resulted in the development of a facile purification scheme for peptides containing the modified tryptophan residues.[50,51] Mollier et al.[52] examined the reaction of *o*-nitrophenylsulfenyl chloride (2-nitrophenylsulfenyl chloride) with notexin (a phospholipase obtained from *Notechis scutalatus scutalatus* venom which contains two tryptophanyl residues). Reactions with 2-nitrophenylsulfenyl chloride (twofold molar excess) in 50% (v/v) acetic acid resulted in two derivative proteins on HPLC analysis. One derivative contained two modified tryptophanyl residues (20 and 110), while the other derivative was modified only at position 20.

REFERENCES

1. **Liu, T.-Y. and Chang, Y. H.,** Hydrolysis of proteins with *p*-toluenesulfonic acid. Determination of tryptophan, *J. Biol. Chem.*, 246, 2842, 1971.
2. **Simpson, R. J., Neuberger, M. R., and Liu, T.-Y.,** Complete amino acid analysis of proteins from a single hydrolysate, *J. Biol. Chem.*, 251, 1936, 1976.
3. **Hachimori, Y., Horinishi, H., Kurihara, K., and Shibata, K.,** States of amino residues in proteins. V. Different reactivities with H_2O_2 of tryptophan residues in lysozyme, proteinases and zymogens, *Biochim. Biophys. Acta*, 93, 346, 1964.
4. **Kotoku, I., Matsushima, A., Bando, M., and Inada, Y.,** Tyrosine and tryptophan residues and amino groups in thrombin related to enzymic activities, *Biochim. Biophys. Acta*, 214, 490, 1970.
5. **Sanda, A. and Irie, M.,** Chemical modification of tryptophan residues in ribonuclease from a *Rhizopus* sp., *J. Biochem.*, 87, 1079, 1980.
6. **Matsushima, A., Takiuchi, H., Saito, Y., and Inada, Y.,** Significance of tryptophan residues in the D-domain of the fibrin molecule in fibrin polymer formation, *Biochim. Biophys. Acta*, 625, 230, 1980.
7. **Turk, T., Macek, P., and Gubensek, F.,** The role of tryptophan in structural and functional properties of equinatoxin II, *Biochim. Biophys. Acta*, 1119, 1, 1992.
8. **Hsieh, P.-C., Shenoy, B. C., Haase, F. C., Jentoft, J. E., and Phillips, N. F. B.,** Involvement of tryptophan(s) at the active site of polyphosphate/ATP glucokinase from Mycobacterium tuberculosis, *Biochemistry*, 32, 6243, 1993.
9. **Kaneko, S.-I., Ichiba, T., Hirano, N., and Hachimori, A.,** Modification of a single tryptophan of the inorganic pyrophosphatase from thermophilic bacterium PS-3: possible involvement in its substrate binding, *Biochim. Biophys. Acta*, 1077, 281, 1991.
10. **Stevens, W. K. and Nesheim, M. E.,** Structural changes in the protease domain of prothrombin upon activation as assessed by *N*-bromosuccinimide modification of tryptophan residues in prethrombin-2 and thrombin, *Biochemistry*, 32, 2787, 1993.
11. **Nakazawa, M. and Manabe, K.,** Differential exposure of tryptophan residues in the red and far-red light absorbing forms of phytochrome, as revealed by chemical modification, *Plant Cell Physiol.*, 34, 1097, 1993.
12. **Green, N. M.,** Avidin. III. The nature of the biotin binding site, *Biochem. J.*, 89, 599, 1963.
13. **Crestfield, A. M., Moore, S., and Stein, W. H.,** The preparation and enzymatic hydrolysis of reduced and *S*-carboxymethylated proteins, *J. Biol. Chem.*, 238, 622, 1963.
14. **Ohnishi, M., Kawagishi, T., Abe, T., and Hiromi, K.,** Stopped-flow studies on the chemical modification with *N*-bromosuccinimide of model compounds of tryptophan residues, *J. Biochem.*, 87, 273, 1980.
15. **Sartin, J. L., Hugli, T. E., and Liao, T.-H.,** Reactivity of the tryptophan residues in bovine pancreatic deoxyribonuclease with *N*-bromosuccinimide, *J. Biol. Chem.*, 255, 8633, 1980.
16. **Daniel, V. W., III and Trowbridge, C. G.,** The effect of *N*-bromosuccinimide upon trypsinogen activation and trypsin catalysis, *Arch. Biochem. Biophys.*, 134, 506, 1969.
17. **Freisheim, J. H. and Huennekens, F. M.,** Effect of *N*-bromosuccinimide on dihydrofolate reductase, *Biochemistry*, 8, 2271, 1969.
18. **Poulos, T. L. and Price, P. A.,** The identification of a tryptophan residue essential to the catalytic activity of bovine pancreatic deoxyribonuclease, *J. Biol. Chem.*, 246, 4041, 1971.
19. **Kosman, D. J., Ettinger, M. J., Bereman, R. D., and Giordano, R. S.,** Role of tryptophan in the spectral and catalytic properties of the copper enzyme, galactose oxidase, *Biochemistry*, 16, 1597, 1977.
20. **Inokuchi, N., Takahashi, T., Yoshimoto, A., and Irie, M.,** *N*-Bromosuccinimide oxidation of a glucoamylase from *Aspergillus saitoi*, *J. Biochem.*, 91, 1661, 1982.
21. **O'Gorman, R. B. and Matthews, K. S.,** *N*-Bromosuccinimide modification of lac repressor protein, *J. Biol. Chem.*, 252, 3565, 1977.

22. **Kumar, G. K., Beegen, H., and Wood, H. G.,** Involvement of tryptophans at the catalytic and subunit-binding domains of transcarboxylase, *Biochemistry,* 27, 5972, 1988.
23. **Keskar, S. S., Srinivasan, M. C., and Deshpande, V. V.,** Chemical modification of a xylanase from a thermotolerant *Streptomyces.* Evidence for essential tryptophan and cysteine residues at the active site, *Biochem. J.,* 261, 49, 1989.
24. **Ramachandran, L. K. and Witkop, B.,** *N*-Bromosuccinimide cleavage of peptides, *Meth. Enzymol.,* 11, 283, 1967.
25. **Feldhoff, R. C. and Peters, T., Jr.,** Determination of the number and relative position of tryptophan residues in various albumins, *Biochem. J.,* 159, 529, 1976.
26. **Shechter, Y., Patchornik, A., and Burstein, Y.,** Selective chemical cleavage of tryptophanyl peptide bonds by oxidative chlorination with *N*-chlorosuccinimide, *Biochemistry,* 15, 5071, 1976.
27. **Sakurai, J. and Nagahama, M.,** Role of one tryptophan residue in the lethal activity of Clostridium perfringens epsilon toxin, *Biochem. Biophys. Res. Commun.,* 128, 760, 1985.
28. **Coletti-Previero, M.-A., Previero, A., and Zuckerkandl, E.,** Separation of the proteolytic and esteratic activities of trypsin by reversible structural modifications, *J. Mol. Biol.,* 39, 493, 1969.
29. **Previero, A., Coletti-Previero, M.-A., and Cavadore, J. C.,** A reversible chemical modification of the tryptophan residue, *Biochim. Biophys. Acta,* 147, 453, 1967.
30. **Holmgren, A.,** Reversible chemical modification of the tryptophan residues of thioredoxin from *Eschericia coli* B., *Eur. J. Biochem.,* 26, 528, 1972.
31. **Cooper, H. M., Jemmerson, R., Hunt, D. F., Griffin, P. R., Yates, J. R., III, Shabanowitz, J., Zhu, N.-Z., and Paterson, Y.,** Site-directed chemical modification of horse cytochrome c results in changes in antigenicity due to local and long-range conformation perturbations, *J. Biol. Chem.,* 262, 11591, 1987.
32. **Koshland, D. E., Jr., Karkhanis, Y. D., and Latham, H. G.,** An environmentally-sensitive reagent with selectivity for the tryptophan residue in proteins, *J. Am. Chem. Soc.,* 86, 1448, 1964.
33. **Horton, H. R. and Koshland, D. E., Jr.,** A highly reactive colored reagent with selectivity for the tryptophanyl residue in proteins, 2-hydroxy-5-nitrobenzyl bromide, *J. Am. Chem. Soc.,* 87, 1126, 1965.
34. **Barman, T. E. and Koshland, D. E., Jr.,** A colorimetric procedure for the quantitative determination of tryptophan residues in proteins, *J. Biol. Chem.,* 242, 5771, 1967.
35. **Moore, S.,** Amino acid analysis: aqueous dimethyl sulfoxide as solvent for the ninhydrin reaction, *J. Biol. Chem.,* 243, 6281, 1968.
36. **Fruchter, R. G. and Crestfield, A. M.,** Preparation and properties of two active forms of ribonuclease dimer, *J. Biol. Chem.,* 240, 3868, 1965.
37. **Clemmer, J. D., Carr, J., Knaff, D. B., and Holwerda, R. A.,** Modification of laccase tryptophan residues with 2-hydroxy-5-nitrobenzyl bromide, *FEBS Lett.,* 91, 346, 1978.
38. **Loudon, G. M. and Koshland, D. E., Jr.,** The chemistry of a reporter group: 2-hydroxy-5-nitrobenzyl bromide, *J. Biol. Chem.,* 245, 2247, 1970.
39. **Horton, H. R. and Tucker, W. P.,** Dimethyl (2-hydroxy-5-nitrobenzyl) sulfonium salts. Water-soluble environmentally sensitive protein reagents, *J. Biol. Chem.,* 245, 3397, 1970.
40. **Peyser, Y. M., Muhlrad, A., and Werber, M. M.,** Tryptophan-130 is the most reactive tryptophan residue in rabbit skeletal myosin subfragment-1, *FEBS Lett.,* 259, 346, 1990.
41. **Horton, H. R. and Koshland, D. E., Jr.,** Reactions with reactive alkyl halides, *Meth. Enzymol.,* 11, 556, 1967.
42. **Horton, H. R. and Young, G.,** 2-Acetoxy-5-nitrobenzyl chloride. A reagent designed to introduce a reporter group near the active site of chymotrypsin, *Biochim. Biophys. Acta,* 194, 272, 1969.
43. **Uhteg, L. C. and Lundblad, R. L.,** The modification of tryptophan in bovine thrombin, *Biochim. Biophys. Acta,* 491, 551, 1977.

44. **Mole, J. E. and Horton, H. R.,** A kinetic analysis of the enhanced catalytic efficiency of papain modified by 2-hydroxy-5-nitrobenzylation, *Biochemistry,* 12, 5285, 1973.
45. **Chang, S.-M. T. and Horton, H. R.,** Structure of papain modified by reaction with 2-chloromethyl-4-nitrophenyl *N*-carbobenzoxylglycinate, *Biochemistry,* 18, 1559, 1979.
46. **Fontana, A. and Scoffone, E.,** Sulfenyl halides as modifying reagents for polypeptides and proteins, *Meth. Enzymol.,* 25, 482, 1972.
47. **Shechter, Y., Burstein, Y., and Patchornik, A.,** Sulfenylation of tryptophan-62 in hen egg-white lysozyme, *Biochemistry,* 11, 653, 1972.
48. **Bewley, T. A., Kawauchi, H., and Li, C. H.,** Comparative studies of the single tryptophan residue in human chorionic somatomammotropin and human pituitary growth hormone, *Biochemistry,* 11, 4179, 1972.
49. **Wilchek, M. and Miron, T.,** The conversion of tryptophan to 2-thioltryptophan in peptides and proteins, *Biochem. Biophys. Res. Commun.,* 47, 1015, 1972.
50. **Chersi, A. and Zito, R.,** Isolation of tryptophan-containing peptides by adsorption chromatography, *Anal. Biochem.,* 73, 471, 1976.
51. **Rubinstein, M., Schechter, Y., and Patchornik, A.,** Covalent chromatography — the isolation of tryptophanyl containing peptides by novel polymeric reagents, *Biochem. Biophys. Res. Commun.,* 70, 1257, 1976.
52. **Mollier, P., Chwetzoff, S., Bouet, F., Harvey, A. L., and Ménez, A.,** Tryptophan 110, a residue involved in the toxic activity but not in the enzymatic activity of notexin, *Eur. J. Biochem.,* 185, 263, 1989.

Chapter 13

THE MODIFICATION OF TYROSINE

The modification of tyrosyl residues in proteins is comparatively easy to achieve. As with most of the other amino acid residues, it is relatively difficult to obtain site-specific modification. However, the ability to utilize tyrosine modification as means to introduce spectral probes into proteins has been quite useful. The material in this chapter focuses on the modification of tyrosine with aromatic diazonium compounds, iodine, *N*-acetylimidazole, and tetranitromethane. The reader is recommended to several recent papers for the current application of these reactions.[1–6]

Diazonium salts readily couple with proteins to form colored derivatives with interesting spectral properties.[7–9] Reaction with diazonium salts is accomplished at alkaline pH (pH 8 to 9, bicarbonate/carbonate or borate buffers). It is relatively difficult to obtain specific residue class modification with the aromatic diazonium salts, but tyrosine, lysine, and histidine are rapidly modified.[10,11] The reaction of tyrosyl residues with diazotized arsanilic acid is shown in Figure 1. The reaction of chymotrypsinogen A with diazotized arsanilic acid has been investigated.[12] Diazotization of arsanilic acid is accomplished by treatment of *p*-arsanilic acid with nitrous acid (0.55 m*M* sodium nitrite in 0.15 *M* HCl at 0°C). After adjustment of the pH to 5.5 with NaOH, the reagent is diluted to a final concentration of 0.02 *M*. Reaction with chymotrypsinogen is accomplished in 0.5 *M* sodium bicarbonate buffer, pH 8.5 with a 20-fold excess of reagent at 0°C. The reaction is terminated by the addition of a sufficient quantity of aqueous phenol (0.1 *M*) to react with excess reagent. The extent of the formation of monoazotyrosyl and monoazohistidyl derivatives is determined by spectral analysis.[10,11] The extent of reagent incorporation is determined by atomic absorption analysis for arsenic. Tyrosine (~1.0 mol/mol) and lysine (~4 mol/mol) were the only amino acid residues modified to any significant extent under these reaction conditions. The arsaniloazo functional group provides a spectral probe that can be used to study conformational change in proteins. In this particular study, there was a substantial change in the circular dichroism spectrum during the activation of the modified chymotrypsinogen preparation by trypsin.

The reaction of α-chymotrypsin with three diazonium salt derivatives (analogs) of *N*-acetyl-D-phenylanine methyl ester[13] has been studied. The corresponding aromatic amine was converted to the diazonium salt by the action of nitrous acid (sodium nitrite per 0.6 *M* HCl at 0°C) and, after neutralization (NaOH) and dilution with 0.2 *M* sodium borate, pH 8.4, was used immediately for the modification of α-chymotrypsin (diazonium salt at a tenfold molar excess) in 0.2 *M* sodium borate, pH 8.4 at ambient temperature for 1 h. The reaction was terminated by gel filtration (G-25 Sephadex) in 0.001 *M* HCl.

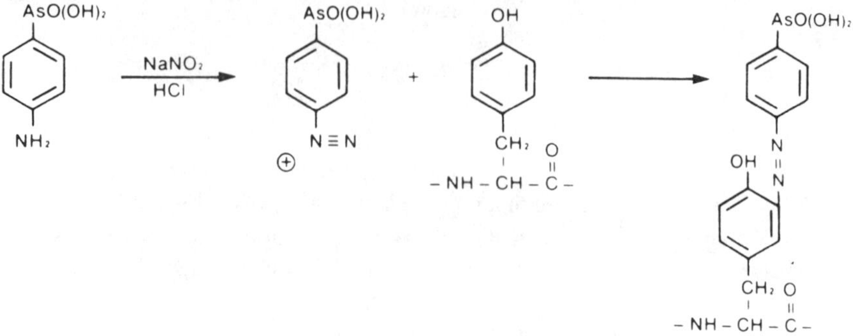

FIGURE 1. The formation of diazotized arsanilic acid and the modification of tyrosine with this reagent.

Amino acid analysis showed that only tyrosine is modified under these reaction conditions. Subsequent analysis showed that Tyr[146] is modified by each of the three reagents. The peptide with the modified tyrosine residue (possessing a yellow color) absorbs to the gel filtration matrix (G-10 equilibrated with 0.001 *M* HCl) and is eluted with 50% acetic acid. (This phenomenon is somewhat similar to that observed with tryptophan-containing peptides which have been modified with 2-hydroxy-5-nitrobenzyl bromide.[14])

Pancreatic ribonuclease has been modified by a diazonium salt derivative of uridine 2′(3′)5′-diphosphate.[15] Modification occurs at a specific tyrosine residue (Tyr[73]). Modification of ribonuclease with 5′-(4-diazophenyl phosphoryl)-uridine-2′(3′)-phosphate was accomplished by *in situ* generation of the diazonium salt from the corresponding amine by $NaNO_2$/HCl in the cold. The pH was then adjusted to pH 8.4 (NaOH); the solution was added to ribonuclease in 0.1 *M* borate, pH 8.4, and the reaction was allowed to proceed for 1 h at ambient temperature. The reaction was terminated by gel filtration (G-25 Sephadex) in 0.1 *M* acetic acid. The extent of modification was determined by spectral and amino acid analysis. Tyrosine was the only amino acid residue modified. Although it is relatively easy to assess the loss of tyrosyl residues, precise determination of diazotization can be obtained only after reduction to the corresponding amine with sodium sulfite. These investigators also examined the reaction of ribonuclease with *p*-diazophenylphosphate under the same conditions of solvent and temperature. Reaction with this reagent was far less specific, with losses of lysine, histidine, and tyrosine (3 mol/mol ribonuclease).

Reaction of bovine carboxypeptidase A with various diazonium salts has been explored in greater detail than that of the above proteins. Vallee and coworkers[16,17] reported on the reaction of bovine carboxypeptidase A crystals with diazotized *p*-arsanilic acid (conditions not specified) and obtained specific modification of Tyr[248]. Purification of the peptide containing the modified tyrosine residue was achieved by using antibody directed against the arsaniloazotyrosyl group. The antibodies were obtained from rabbits using

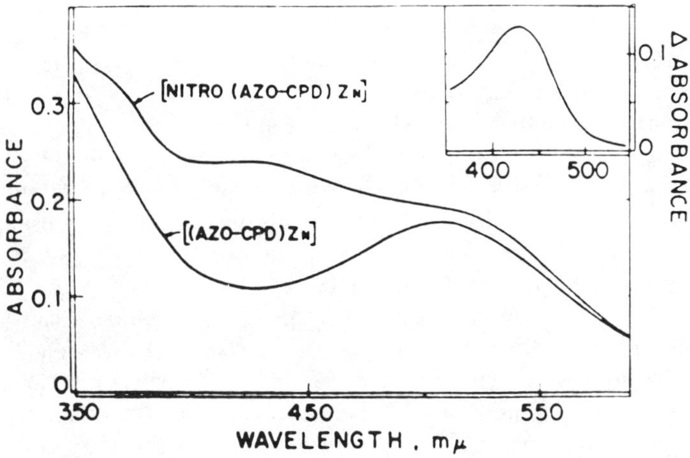

FIGURE 2. The UV absorption spectra of azocarboxypeptidase and nitroazocarboxypeptidase. Azocarboxypeptidase was obtained by the reaction of carboxypeptidase A with a sevenfold molar excess of diazonium-1H-tetrazole in 1.0 *M* NaCl-0.067 *M* potassium bicarbonate/carbonate, pH 8.8 at 0 to 4°C; the reaction was quenched after 30 min by the addition of excess Tris-Cl, pH 8.0 and excess reagent removed by dialysis. Nitroazocarboxypeptidase was obtained by the reaction of tetranitromethane with carboxypeptidase previously modified with a sevenfold molar excess of diazonium-1H-tetrazole as described above. Shown is the absorption spectra of azocarboxypeptidase [(AZO CPD)ZN] and nitroazocarboxypeptidase [NITRO(AZO CPN)ZN] in 0.1 *N* NaOH, both at 100 μ*M*. The inset represents the difference spectrum of nitroazocarboxypeptidase minus azocarboxypeptidase. (From Riordan, J. F., Sokolovsky, M., and Vallee, B. L., *Biochemistry,* 6, 3609, 1967. With permission.)

arsaniloazovalbumin and arsaniloazobovine γ-globulin as antigen. The reaction of bovine carboxypeptidase A with diazotized 5-amino-1H-tetrazole has been reported.[18] Diazotized 5-amino-1H-tetrazole also specifically reacts with Tyr^{248} in bovine carboxypeptidase A (in 0.67 *M* potassium bicarbonate/carbonate, 1.0 *M* NaCl, pH 8.8). A sevenfold molar excess of reagent was used, and the reaction was terminated after 30 min by the addition of Tris buffer. The extent of modification of tyrosine to tetrazolylazotyrosine is determined by absorbance at 483 nm (Figure 2) ($\varepsilon = 8.7 \times 10^3\ M^{-1}\ cm^{-1}$). Modification of Tyr^{248} in carboxypeptidase A by this reagent permits the subsequent modification of Tyr^{198} by tetranitromethane.

Iodination is used for the modification of tyrosyl residues in proteins.[19] The reaction is still of considerable value since the process of the radiolabeling of proteins with either of the iodine radioisotopes (^{125}I, ^{131}I) primarily involves the modification of tyrosine residues in proteins. It is, of course, of critical importance to appreciate the strength of the elemental halides as oxidizing agents.

Iodination has been utilized to study the reactivity of tyrosyl residues in cytochrome b_5.[20] Iodination is accomplished with a tenfold molar excess of I_2 (15 m*M* I_2 in 30 m*M* KI) in 0.025 *M* sodium borate, pH 9.8. Iodination with limiting amounts of iodine is accomplished with a two- to sixfold molar excess of iodine in 0.020 *M* potassium phosphate, pH 7.5 at 0°C. Monoiodination and diiodination of tyrosyl residues is observed. Iodination with a tenfold molar excess of I_2 results in the formation of 3 mol of diiodotyrosine per mole of cytochrome c.[21] The fourth tyrosyl residue is modified only in the presence of 4.0 *M* urea. Iodination of tyrosine results in a decrease in the pKa of the phenolic hydroxyl groups. Iodination with a limiting amount of iodine as described above results first in the formation of 2 mol of monoidotyrosine, then 1 mol of diiodotyrosine, and 1 mol of monoiodotyrosine. Tyrosyl residues which can be iodinated are also available for *O*-acetylation with acetic anhydride [0.1 *M* potassium phosphate, pH 7.5; acetic anhydride added in two portions over 1 h at 0°C; maintained at pH 7.8 with NaOH (1 *M*) addition].

The modification of tyrosyl residues in phosphoglucomutase by iodination has been reported.[22] Modification is achieved by reaction in 0.1 *M* borate, pH 9.5 with 1 m*M* I_2 (obtained by an appropriate dilution of a stock iodine/iodide solution, 0.05 *M* I_2 in 0.24 *M* KI) at 0°C for 10 min. Complete loss of enzymatic activity was observed with these reaction conditions, but the stoichiometry of modification was not established. Nitration of 7 of 20 tyrosyl residues resulted in an 83% loss of catalytic activity. These investigators also studied the reaction of phosphoglucomutase with diazotized sulfanilic acid and *N*-acetylimidazole.

The above modifications utilize the reaction of tyrosyl residues in proteins with iodine/iodide solutions at alkaline pH. Iodination of tyrosyl residues can also be accomplished with iodine monochloride (ICl) at mildly alkaline pH. One such study explores the modification of galactosyltransferase.[23] The modification is accomplished by reaction in 0.2 *M* sodium borate, pH 8.0. The reaction is initiated by a desired amount of a stock solution of ICl.[24] A stock solution of 0.02 *M* ICl is prepared by adding 21 ml 11.8 *M* HCl (stock concentrated HCl) to approximately 150 ml of H_2O containing 0.555 g KCl, 0.3567 g KIO_3, and 29.23 g NaCl. The solvent is taken to a final volume of 250 ml with H_2O. Free iodine is then extracted with CCl_4, if necessary, and the solution is aerated to remove trace amounts of CCl_4. The resulting solution of ICl is stable for an indefinite period of time under ambient conditions. Reaction proceeds for 1 min at ambient temperature and is terminated by the addition of a 1:6 volume of 0.5 *M* $Na_2S_2O_3$ (50 μl for a 0.300-ml reaction mixture). Radiolabeled sodium iodide ($Na^{125}I$) is included to provide a mechanism for establishing the stoichiometry of the reaction. The reaction mixture, after the addition of $Na_2S_2O_3$, is subjected to gel filtration on Bio-Gel P-10 (BioRad Laboratories, Richmond, Connecticut) in 0.1 *M* Tris, pH 7.4. In experiments designed to assess the relationships between reagent (ICl) concentration and the extent of modification, a maximum of 10 g-atom of iodine were incorporated into galactosyltransferase at a 40-fold molar excess of reagent. Incorporation of iodine is linear up to this excess of reagent and slowly declines at higher concentrations of ICl. Modification of tryptophanyl residues was excluded by direct analysis, and the only iodinated amino acids obtained from the modified protein were monoiodotyrosine and diiodotyrosine. Modification of

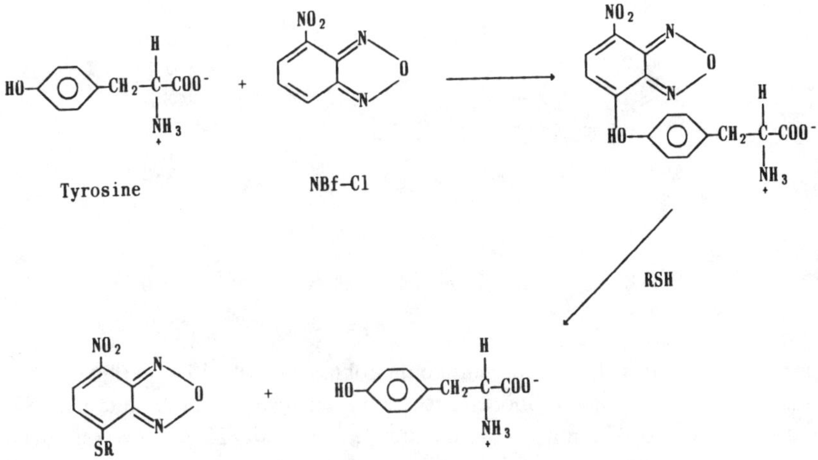

FIGURE 3. The structure of 7-chloro-4-nitrobenzo-2-oxa-1,3-diazole (Nbf-Cl), including a scheme for reaction with tyrosine and reversal with a thiol.

other residues such as histidine and methionine by oxidation without incorporation of iodine was not excluded.

Iodination can also be accomplished by peroxidase per H_2O_2 per NaI. A recent procedure was described for the modification of tyrosyl residues in insulin.[25] In these studies, 20 mg of porcine insulin in 20 ml 0.4 *M* sodium phosphate, 6.0 *M* urea, pH 7.8 was combined with 10 ml $Na^{125}I$ (1 mCi) and 3.6 mg urea, and H_2O_2 (5 μl 0.3 m*M* solution) and peroxidase (Sigma Chemical Company, 0.2 mg/ml; 5 μl) was added. The preparative reaction was terminated by dilution with an equal volume of 40% (w/v) sucrose. It is of interest to note that the enzyme-catalyzed iodination proceeds with efficiency in 6.0 *M* urea. Iodination of tyrosyl residues in peptides and proteins can also be accomplished with chloramine T.[26,27] The solution structure of insulin-like growth factor was investigated by iodination of tyrosyl residues mediated by either chloramine T or lactoperoxidase.[28] Chloramine T was more effective than lactoperoxidase.

Tyrosyl residues in proteins are also modified by reaction with cyanuric fluoride.[29,30] The reaction proceeds at alkaline pH (9.1) via modification of the phenolic hydroxyl group with a change in the spectral properties of tyrosine. The phenolic hydroxyl groups must be ionized (phenoxide ion) for reaction with cyanuric fluoride. The modification of tyrosyl residues in elastase[31] and yeast hexokinase[32] with cyanuric fluoride has been reported.

Modification of tyrosyl residues can occur as an interesting side reaction with other residue-specific reagents such as 7-chloro-4-nitrobenzo-2-oxa-1,3-diazole (Nbf-Cl).[33,34] Nbf-Cl reacts primarily with amino groups and sulfydryl groups in proteins. A scheme for the reaction of Nbf-Cl with tyrosine and subsequent reversal by a thiol compound is shown in Figure 3. The reaction

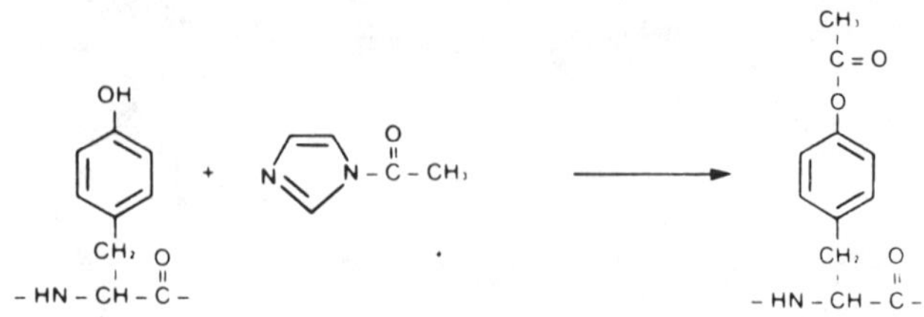

FIGURE 4. A scheme for the reaction of tyrosyl residues with *N*-acetylimidazole.

product obtained with tyrosine, unlike that obtained with either amino groups or sulfydryl groups, is not fluorescent and has an absorption maximum at 385 nm compared to 475 nm for amino derivatives and 425 nm for sulfydryl derivatives.[33] Transfer of the Nbf moiety from the phenolic hydroxyl group of tyrosine to a nitrogen nucleophile in a protein has been observed.[34] Modification of the phenolic hydroxyl group with 2,4-dinitrofluorobenzene has also been reported.[35]

Although the above reagents have proved useful for the selective modification of tyrosyl residues in proteins, the majority of the studies in this area have utilized reaction with either *N*-acetylimidazole and/or tetranitromethane.

The development of *N*-acetylimidazole as a reagent (Figure 4) for the selective modification of tyrosyl residues can, in part, be traced to the early observations[36–38] that *N*-acetylimidazole is, in fact, an energy-rich compound. The preparation of *O*-acyl derivatives via the action of carboxylic acid anhydrides (i.e., acetic anhydride) has been used for some time, but it is very difficult to obtain selective modification of tyrosine as these reagents readily react with primary amines to form stable *N*-acyl derivatives.[39,40] It is, however, possible to obtain the selective modification of tyrosine with acetic anhydride by reaction at mildly acidic pH (1.0 *M* acetate, pH 5.8 at 25°C), approximately 20,000-fold molar excess of acetic anhydride (5.1×10^{-2} *M* acetic anhydride, 2.9×10^{-6} *M* enzyme).[41] Bernad and colleagues[42] have reported on an extensive study comparing the modification of lysyl and tyrosyl residues in lysozyme with dicarboxylic acid anhydrides. In 50 m*M* HEPES, 1.25 *M* NaCl, pH 8.2, amino groups (primarily lysine residues) were far more reactive than hydroxyl groups (including tyrosine, serine, and threonine).

N-acetylimidazole was first used as a reagent for the modification of tyrosyl residues in bovine pancreatic carboxypeptidase A.[43] This same group of investigators subsequently reported on the use of *N*-acetylimidazole for the determination of "free" tyrosyl residues in proteins[44] as opposed to "buried" residues. This has not necessarily proved to be the case.[45] *N*-acetylimidazole is commercially available, but it can also be easily synthesized.[38] Our laboratory generally synthesizes the reagent, and always subjects reagent obtained from a commercial source to recrystallization from benzene after drying with sodium sulfate. It should be noted that, as with many reagents, *N*-acetylimidazole is hygro-

Table 1
REACTION OF PROTEINS WITH *N*-ACETYLIMIDAZOLE

Protein	Solvent/temp	Reagent excess[a]	*O*-AcTyr/Tyr[b]	Ref.
Carboxypeptidase[c]	0.02 *M* sodium barbital, 2.0 *M* NaCl, pH 7.5 at 23°C	60	4.3/19[d]	1
Pepsinogen	0.02 *M* sodium Veronal, 2.0 *M* NaCl, pH 7.5 at 25°C	60	7/16[e]	2
Pepsin	2.0 *M* NaCl, pH 5.8 at 25°C[f]	60	9/15[g]	2
Trypsin	0.01 *M* sodium borate, 0.01 *M* $CaCl_2$, pH 7.6 at 0°C	30	1.7/10[h]	3
Trypsin	0.01 *M* sodium borate,[i] 0.01 *M* $CaCl_2$, pH 7.6 at 0°C	465	3.0/10[j]	3
α-Amylase[k]	0.02 *M* Tris·Cl, pH 7.5 at 25°C	500	3.5/12[l]	4
Subtilisin novo	0.016 *M* barbital, pH 7.5	100	7/10[m]	5
Subtilisin Carlsberg	0.016 *M* barbital, pH 7.5	130	8.4/13[m]	5
Hemerythrin	0.05 *M* sodium borate, 0.05 *M* Tris, pH 7.5 at 0°C	800	—[n]	6
Thrombin	0.02 *M* Tris, 0.02 *M* imidazole, 0.02 *M* acetate, pH 7.5 at 23°C	300	4.4/12	7
Fructose diphosphatase	0.050 *M* sodium borate, pH 7.5	—	—	8
Erythrocyte ATPase				
Stroma	0.010 *M* Tris, pH 7.4 at 23°C[o]	—	—	9
Intact cells	0.010 *M* Tris, 0.140 *M* NaCl, pH 7.4 at 23°C[p]	—	—	9
α-Lactalbumin		200	2/5[q]	10,11
Pancreatic colipase	—	—	—	12,13
Pancreatic α-amylase	0.01 *M* phosphate, pH 7.5,[r] 0.1 m*M* CaCl at 25°C	120	5.9/18	14
Sweet potato α-amylase	0.01 *M* acetate, pH 7.5 at 25°C[r]	120	5.3/17	14
Aspergillus niger glucamylase	0.01 *M* acetate, pH 7.5 at 25°C	120	11.3/33	14
Emulsin β-D-glucosidase	0.01 *M* phosphate, pH 6.1 at 25°C[r]	300	—	15
Human placental taurine transporter	10 m*M* HEPES-Tris, pH 7.4 with 100 m*M* K_2SO_4 at 22°C	—	—	16
Renal Na,K ATPase	50 m*M* sodium borate with 2 m*M* EDTA, pH 7.5 at 20°C	—	—	17

[a] Moles *N*-acetylimidazole per mole of protein.

[b] Moles *O*-acetyltyrosine per moles of tyrosine in modified protein.

[c] Bovine pancreatic carboxypeptidase A-Anson.

[d] Changes in catalytic activity reversed by treatment with 0.01 *M* hydroxylamine, pH 7.5 at 23°C. Primary amino groups were not acetylated under those reaction conditions.

[e] Five out of ten lysine residues modified.

[f] pH maintained by NaOH from pH-stat.

[g] Lysine not acetylated under these conditions. Reaction with 1.0 *M* hydroxylamine, pH 5.8 (60 min; 37°C) reversed changes in catalytic activity produced on reaction with *N*-acetylimidazole and presumably deacetylated *O*-acetyl tyrosyl residues.

[h] Also 1.0 serine and 0.3 lysine.

[i] Also used Tris, TES, HEPES, and barbital buffers without any significant difference in nature of the reaction.

Table 1 (continued)
REACTION OF PROTEINS WITH *N*-ACETYLIMIDAZOLE

[j] Also 1.7 (probably serine and histidine) and 2.5 lysine residues modified.

[k] From *Bacillus subtilis*.

[l] Approximately two lysine residues modified under these conditions. Only a single tyrosine residue is modified with tetranitromethane. Either reagent (tetranitromethane or *N*-acetylimidazole) led to a 70 to 80% loss of catalytic activity.

[m] The reaction with *N*-acetylimidazole was performed with subtilisin preparation previously treated with phenylmethanesulfonyl fluoride. The active enzyme catalyzes the rapid hydrolysis of *N*-acetylimidazole under reaction conditions.

[n] Reaction performed on protein where lysine residue had been previously blocked by reaction with ethyl acetimidate. *N*-acetylimidazole was added in four 200-fold molar excess portion at 2-h intervals.

[o] Reaction for 1 h at ambient temperature with amount of *N*-acetylimidazole equivalent (weight/weight basis) to stroma. The reaction mixture was washed with distilled water to remove *N*-acetylimidazole.

[p] Reaction for 1 h at ambient temperature. The quantity of *N*-acetylimidazole used is not given. It is stated that this reagent should readily pass across the cell membrane, but this conclusion is based on analogy with acetic anhydride.

[q] Extensive modification of amino groups was reported.

[r] *N*-acetylimidazole added as a solid; pH maintained at 7.5 with pH-stat.

References for Table 1

1. **Simpson, R. T., Riordan, J. F., and Vallee, B. L.,** Functional tyrosyl residues in the active center of bovine pancreatic carboxypeptidase A, *Biochemistry,* 2, 616, 1963.
2. **Perlmann, G. E.,** Acetylation of pepsin and pepsinogen, *J. Biol. Chem.,* 241, 153, 1966.
3. **Houston, L. L. and Walsh, K. A.,** The transient inactivation of trypsin by mild acetylation with *N*-acetylimidazole, *Biochemistry,* 9, 156, 1970.
4. **Connellan, J. M. and Shaw, D. C.,** The inactivation of *Bacillus subtilis* α-amylase by *N*-acetylimidazole and tetranitromethane. Reaction of tyrosyl residues, *J. Biol. Chem.,* 245, 2845, 1970.
5. **Myers, B., II and Glazer, A. N.,** Spectroscopic studies of the exposure of tyrosine residues in proteins with special reference to the subtilisins, *J. Biol. Chem.,* 246, 412, 1971.
6. **Fan, C. C. and York, J. L.,** The role of tyrosine in the hemerythrin active site, *Biochem. Biophys. Res. Commun.,* 47, 472, 1972.
7. **Lundblad, R. L., Harrison, J. H., and Mann, K. G.,** On the reaction of purified bovine thrombin with *N*-acetylimidazole, *Biochemistry,* 12, 409, 1973.
8. **Kirtley, M. E. and Dix, J. C.,** Effects of acetylimidazole on the hydrolysis of fructose diphosphate and *p*-nitrophenyl phosphate by liver fructose diphosphatase, *Biochemistry,* 13, 4469, 1974.
9. **Masiak, S. J. and D'Angelo, G.,** Effects of *N*-acetylimidazole on human erythrocyte ATPase activity. Evidence for a tyrosyl residue at the ATP binding site of the (Na^+, K^+)-dependent ATPase, *Biochim. Biophys. Acta,* 382, 83, 1975.
10. **Kronman, M. J., Hoffman, W. B., Jeroszko, J., and Sage, G. W.,** Inter and intramolecular interactions of α-lactalbumin. XI. Comparison of the "exposure" of tyrosyl, tryptophyl and lysyl side chains in the goat and bovine proteins, *Biochim. Biophys. Acta,* 285, 124, 1972.
11. **Holohan, P., Hoffman, W. B., and Kronman, M. J.,** Chemical modification of tyrosyl and lysyl residues in goat alpha lactalbumin and the effect on the interaction with the galactosyl transferase, *Biochim. Biophys. Acta,* 621, 333, 1980.

References for Table 1 (continued)

12. **Erlanson, C., Barrowman, J. A., and Borgström, B.,** Chemical modifications of pancreatic colipase, *Biochim. Biophys. Acta,* 489, 150, 1977.
13. **Erlanson-Albertsson, C.,** The importance of the tyrosine residues in pancreatic colipase for its activity, *FEBS Lett.,* 117, 295, 1980.
14. **Hoschke, A., Laszlo, E., and Hollo, J.,** A study of the role of tyrosine groups at the active centre of amylolytic enzymes, *Carbohydrate Res.,* 81, 157, 1981.
15. **Kiss, L., Korodi, I., and Nanasi, P.,** Study on the role of tyrosine side-chains at the active center of emulsin β-D-glucosidase, *Biochim. Biophys. Acta,* 662, 308, 1981.
16. **Kulanthaivel, P., Leibach, F. H., Mahesh, V. B., and Ganapathy, V.,** Tyrosine residues are essential for the activity of the human placental taurine transporter, *Biochim. Biophys. Acta,* 985, 139, 1989.
17. **Arguello, J. M. and Kaplan, J. H.,** *N*-Acetylimidazole inactivates renal Na, K-ATPase by disrupting ATP binding at the catalytic site, *Biochemistry,* 29, 5775, 1990.

scopic and should be stored in a container, preferably a vacuum desiccator, over a suitable desiccant. A partial listing of proteins which have been modified with *N*-acetylimidazole is presented in Table 1. A stock solution of reagent is prepared in *dry* benzene (this stock solution is relatively stable for 2 to 4 weeks at 4°C), and a portion containing the desired amount of reagent is introduced to the reaction vessel. The solvent (benzene) is removed by a stream of *dry* air or *dry* nitrogen. The reaction is initiated by the addition of the protein solution to be modified to the residue of reagent. The reaction is usually performed at pH 7.0 to 7.5. A wide variety of buffers has been used for the study of the reaction of *N*-acetylimidazole. A high concentration of nucleophilic species such as Tris should be avoided because of reagent instability.[43] Likewise, although the modification occurs more rapidly at pH values more alkaline than 7.5, reagent and product (*O*-acetyl tyrosine) stability become a significant problem.

There are several approaches to the determination of the extent of tyrosine modification by *N*-acetylimidazole. The amount of acetylhydroxamate produced by the reaction of hydroxylamine can be determined.[46] The procedure described by these investigators involves the addition of 0.25 ml of a hydroxylamine solution (4 *M* $NH_2OH{\cdot}HCl$/3.5 *M* NaOH/0.001 *M* EDTA; 1:2:1) to 1.0 ml of the acetylated protein sample. After 1 min, 0.5 ml 25% trichloroacetic acid and 0.5 ml 20% $FeCl_2$, 6 H_2O in 2.5 *M* HCl is added, and the absorbance of the supernatant fraction is determined at 540 nm. We have found it convenient to use *p*-nitrophenyl acetate as the standard for this reaction. Second, *O*-acetylation of tyrosine produced a decrease in absorption at 278 nm. A $\Delta\varepsilon$ = 1160 M^{-1} cm^{-1} has been reported, while a subsequent study reported a $\Delta\varepsilon$ = 1210 M^{-1} cm^{-1}.[45] We have had more reliable results with the latter value in this laboratory. We have also found it more accurate to determine changes in absorbance at 278 nm as a function of time, taking into account spectral changes introduced by the addition of reagent to a solvent blank.[45] One of the major advantages of reaction with *N*-acetylimidazole is the ease of reversal of the reaction. The *O*-acetyl derivative of tyrosine is unstable under mildly

FIGURE 5. A scheme for the modification of tyrosyl residues with tetranitromethane.

alkaline conditions, and presence of a nucleophile such as Tris greatly decreases the stability of the *O*-acetyl derivative. Quantitative deacetylation occurs with hydroxylamine at pH 7.5. As would be expected, the rate of regeneration of free tyrosine is a function of hydroxylamine concentration. It should be noted that the primary side-reaction products of the reaction of *N*-acetylimidazole with proteins, ε-*N*-acetyllysines and *N*-acetyl amino-terminal amino acids, are stable to neutral or alkaline hydroxylamine. Assignment of changes in the biological activity of a protein on reaction with *N*-acetylimidazole to the *O*-acetylation of tyrosine can be verified by the reversibility of such changes in the presence of hydroxylamine.

The possible use of tetranitromethane for the modification of tyrosyl residues in proteins was advanced over 30 years ago.[47] However, it was not until some two decades later that the studies of Vallee, Riordan, Sokolovsky, and Harell established the specificity and characteristics of the reaction of tetranitromethane with proteins.[48,49]

The modification (Figure 5) proceeds optimally at alkaline pH. The rate of modification of *N*-acetyltyrosine is twice as rapid at pH 8.0 as at pH 7.0; it is approximately ten times as rapid at pH 9.5 as at pH 7.0. As shown above, the reaction of tetranitromethane with tyrosine produces 3-nitrotyrosine, nitroformate, and two protons. The spectral properties of nitroformate (ε at 350 nm = 14,400) suggested that monitoring the formation of this species would be a sensitive method for monitoring the time course of the reaction of tetranitromethane with tyrosyl residues.[48] Although determining the rate of nitroformate production appears to be effective in studying the reaction of tetranitromethane with model compounds such as *N*-acetyltyrosine (Figure 6), it has not proved useful with proteins.[49,50] Although the reaction of tetranitromethane with proteins is reasonably specific for tyrosine, oxidation of sulfhydryl groups has been reported[49,50] as has reaction with histidine,[49] methionine,[49] and tryptophan.[49,51]

The other potential problem associated with the use of tetranitromethane for the modification of tyrosyl residues in proteins is the covalent cross-linkage of tyrosyl residues resulting in inter- and intramolecular association. The magnitude of this problem is dependent on variables such as protein concentration and solvent conditions (i.e., pH). With respect to this latter consideration, it is noted that acidification of reaction mixtures tends to favor the cross-linkage reaction.[50] As would be expected, the extent of cross-linkage observed varies

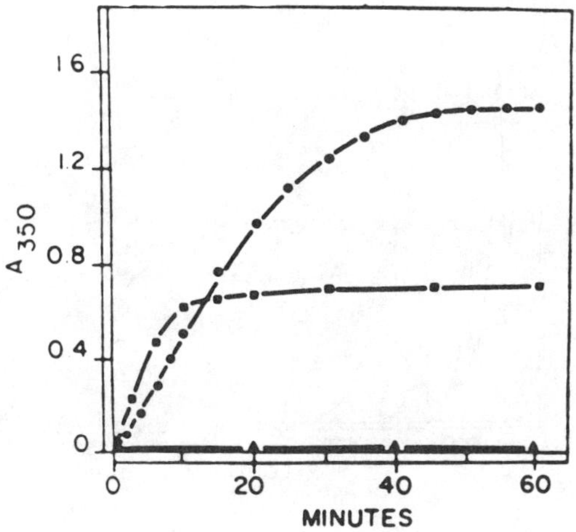

FIGURE 6. Time course of formation of nitroformate on reaction of tetranitromethane with several amino acid derivatives. Shown is the increase in absorbance at 350 nm on the reaction of tetranitromethane with *N*-acetyltyrosine (●), glutathione (■), and *N*-carbobenzoxyglycyl-L-trytophan, *N*-acetylhistidine, or *N*-carbobenzoxy-L-methionylglycine (▲), all at 0.1 m*M*. Tetranitromethane (5 μl, 42 μmol) was added to 3 ml 0.05 *M* Tris, pH 8.0, containing the amino acid derivative at 20°C. The data are corrected for the absorbance due to *N*-acetyl-3-nitrotyrosine. (From Sokolovsky, M., Riordan, J. F., and Vallee, B. L. *Biochemistry,* 5, 3582, 1966. With permission.)

with the protein being studied. For example, reaction of pancreatic deoxyribonuclease with tetranitromethane results in extensive formation of dimer.[52]

A related reaction is the free radical-induced cross-linking between tyrosyl residues and thymine providing a basis for the formation of nucleic acid-protein conjugates occurring as a result of ionizing radiation.[53] Finally, treatment of apoovotransferrin with periodate (50 m*M* HEPES, pH 7.4 with 5 m*M* sodium periodate) resulted in protein cross-linking via 3,3′-dityrosine.[54]

The extent of modification of tyrosyl residues by tetranitromethane in proteins can be assessed by either spectrophotometric means or by amino acid analysis. At alkaline pH (pH ≥8), 3-nitrotyrosine has an absorption maximum at 428 nm (Figure 7) with $\varepsilon = 4100\ M^{-1}\ cm^{-1}$; the absorption maximum of tyrosine at 275 nm increases from $\varepsilon = 1360$ to $4000\ M^{-1}\ cm^{-1}$. At acid pH (pH ≤6), the absorption maximum is shifted from 428 to 360 nm, with an isosbestic point at 381 nm ($\varepsilon = 2200\ M^{-1}\ cm^{-1}$) (see Figure 8). We have found it convenient to determine the A_{428} in 0.1 *M* NaOH. Amino acid analysis after acid hydrolysis has also proved to be a convenient method of assessing the extent of 3-nitrotyrosine formation. 3-Nitrotyrosine is stable to acid hydrolysis (6 *N* HCl, 105°C, 24 h). This approach has the added advantage that other modifications of tyrosine such as free radical-mediated cross-linkage can be either excluded or quantitatively determined. If nitration to form 3-nitrotyrosine is the only modification of tyrosyl residues in a protein occurring on reaction with tetranitromethane, the sum of 3-nitrotyrosine and tyrosine should be equivalent to the amount of tyrosine in the unmodified protein.

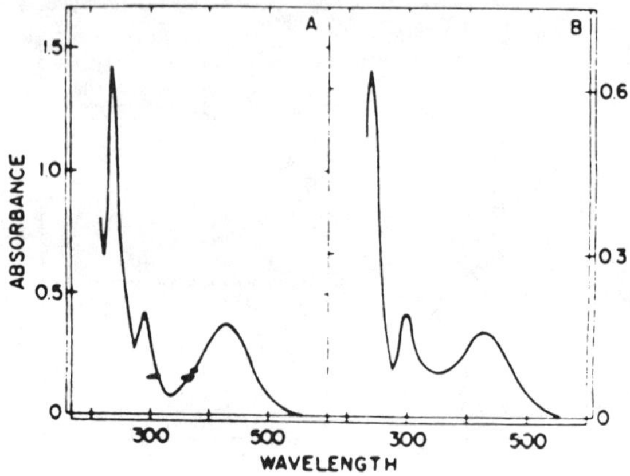

FIGURE 7. UV absorption spectra of *N*-acetyl-3-nitrotyrosine and nitrocarboxypeptidase. Shown is (A) the absorption spectrum of *N*-acetyl-3-nitrotyrosine minus *N*-acetyltyrosine, both at 0.100 m*M*, and (B) the absorption spectrum of nitrocarboxypeptidase minus carboxypeptidase, both at 17.6 μ*M*. Both spectra were obtained in 0.05 *M* Tris-1 *M* NaCl, pH 8.0 at ambient temperature. (From Riordan, J. F., Sokolovsky, M., and Vallee, B. L., *Biochemistry,* 6, 3609, 1967. With permission.)

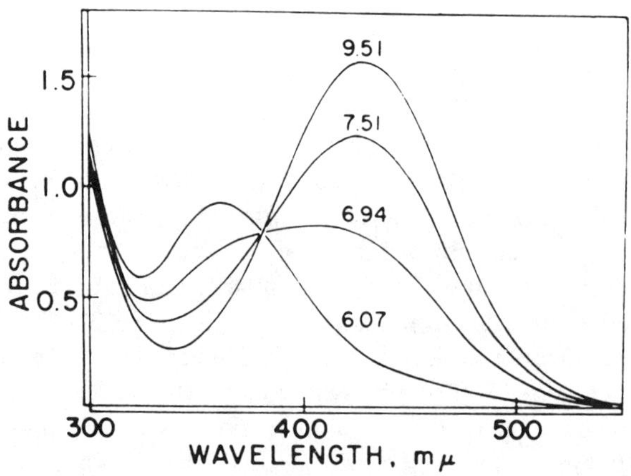

FIGURE 8. pH dependence of the UV absorption spectra of the 3-nitro derivative of tyrosine. Shown are the absorption spectra of *N*-acetyl-3-nitrotyrosine (250 μ*M*) in 0.2 *M* Tris, 0.2 *M* acetate, 0.5 *M* NaCl at the pH indicated. (From Riordan, J. F., Sokolovsky, M., and Vallee, B. L., *Biochemistry,* 6, 358, 1967. With permission.)

There are several consequences of the nitration of a tyrosyl residue. The most obvious is the placing of a somewhat bulky substituent (the nitro group) *ortho* to the phenolic hydroxyl function. The properties of the substituent nitro group "push" electrons into the benzene ring (inductive effect), lowering the pKa of the phenolic hydroxyl from approximately 10.3 to 7.3. This of course means that the phenolic hydroxyl of the nitrated tyrosyl residue will be in a

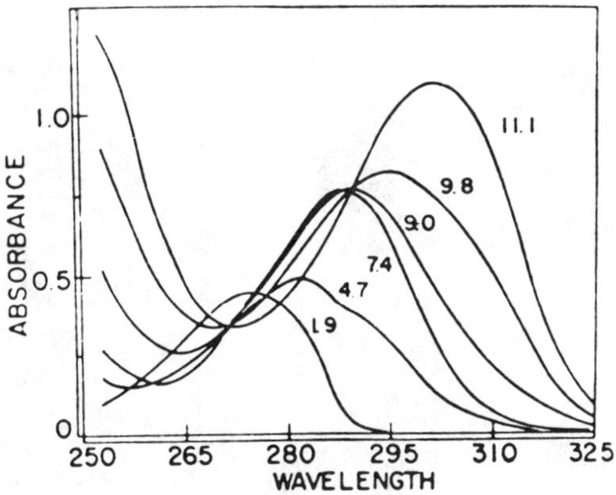

FIGURE 9. pH dependence of the UV absorption spectra of the 3-amino derivative of tyrosine. Shown are the absorption spectra of 267 μ*M* 3-aminotyrosine in 0.2 *M* Tris, 0.2 *M* acetate, 0.5 *M* NaCl at the pH value indicated. 3-Aminotyrosine was prepared by the reaction of 3-nitrotyrosine with a sixfold molar excess of sodium hydrosulfite ($Na_2S_2O_2$) in 0.05 *M* Tris, pH 8.0. (From Sokolovsky, M., Riordan, J. F., and Vallee, B. L., *Biochem. Biophys. Res. Commun.*, 27, 20, 1967. With permission.)

partially ionized state at physiological pH. The nitro function can be reduced to the corresponding amine under relatively mild conditions ($Na_2S_2O_4$, 0.05 *M* Tris, pH 8.0).[55] The conversion of 3-nitrotyrosine to 3-aminotyrosine is associated with the loss of the absorption maximum at 428 nm (Figure 9) and the change in the pKa of the phenolic hydroxyl group from approximately 7.0 to 10.0. The resultant amine function can be subsequently modified.[56]

In addition to changing the properties of a given tyrosyl residue, nitration also introduces a spectral probe which can be used to detect conformational change in the protein. The concept of using reagents to introduce probes with unique spectral and fluorescent properties has been introduced in Chapter 1. 3-Nitrotyrosine has an absorption maximum at 428 nm at alkaline pH. This spectral property was first used by Riordan and co-workers with studies on nitrated carboxypeptidase A[57] to study changes in the microenvironment around the modified residue. Addition of β-phenylpropionate, a competitive inhibitor of carboxypeptidase and nitrated carboxypeptidase, decreased the absorbance of mononitrocarboxypeptidase at 428 nm (Figure 10). This change is consistent with an increase in the hydrophobic quality of the microenvironment surrounding the modified tyrosyl residue.

Other uses of tetranitromethane in the study of the relationship between structure and function in various proteins are presented in Table 2.

Although the modification of tryptophanyl residues in proteins by tetranitromethane is a somewhat unusual reaction, there are a number of examples of the phenomena.[49,51,58–63] The modified protein is characterized by a broad absorption band with a maxima in the area of 340 to 360 nm. This absorbance,

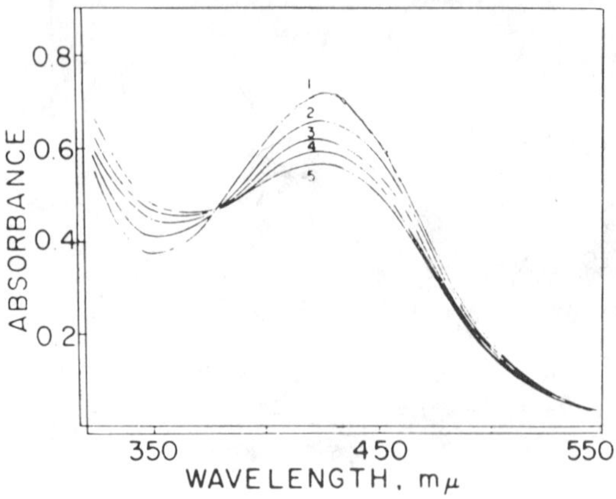

FIGURE 10. The effect of β-phenylpropionate, a competitive inhibitor of carboxypeptidase, on the UV absorption spectrum of nitrocarboxypeptidase. Spectra were measured in 0.2 *M* Tris, 0.2 *M* acetate, 0.5 *M* NaCl, pH 8.0. The concentrations of β-phenylpropionate were as follows: 1 (control), none; 2, 0.01 *M;* 3, 0.025 *M;* 4, 0.05 *M;* and 5, 0.1 *M*. (From Riordan, J. F., Sokolovsky, M., and Vallee, B. L., *Biochemistry,* 6, 358, 1967. With permission.)

unlike the absorbance of nitrotyrosine at 410 nm, is not pH dependent. The reader is recommended to the study of Cuatrecasas and co-workers[51] for additional details.

One of the early suggestions for the use of tetranitromethane concerned differentiation between "free" and "buried" tyrosyl residues.[46] Free residues were assumed to be readily accessible to solvent (e.g., on the surface of a globular protein), while buried residues were considered to be in the interior of the molecule (a more "hydrophobic" region). While this may be true in some cases, it is not likely to be a general occurrence. This argument has been addressed by Myers and Glazer.[45] The example of cytochrome c (horse) is worth consideration. Two of the four tyrosyl residues in this protein are converted to 3-nitrotyrosine on reaction with tetranitromethane at pH 8.0 (0.05 *M* Tris).[64] The two residues modified, Tyr-47 and Tyr-67, are located in the interior, while the two residues on the surface of the protein, Tyr-74 and Tyr-97, are not modified. Myers and Glazer emphasize the importance of viewing tetranitromethane (for example) as an organic compound which is more soluble in organic solvents (hydrophobic) than in water. Thus, it is not unreasonable to suggest that reaction at a given residue might, in fact, reflect selective partitioning of the reagent, in this case tetranitromethane, into the microenvironment around the given residue.

An interesting effect of substrate on the reaction of an enzyme with tetranitromethane has been reported by Christen and Riordan.[65] Generally, protection of an enzyme from loss of activity secondary to chemical modification with the concomitant lack of modification of an amino acid residue is observed in the presence of substrate. Christen and Riordan, however, ob-

Table 2
THE USE OF TETRANITROMETHANE TO MODIFY TYROSYL RESIDUES IN PROTEINS

Protein concentration	Solvent/temp	Molar excess TNM	Residues modified	Ref.
Carboxypeptidase A (10 mg/ml)	0.05 *M* Tris, 2 *M* NaCl at 20°C	4	1.2/18	1
Staphylococcal nuclease (2 mg/ml)	0.05 *M* Tris, pH 8.1 at 23°C	2	1.1/7	2
		4	1.6/7	
		8	3.4/7	
		12	3.7/7	
		16	4.2/7	
		20	4.7/7	
		30	5.0/7	
		60	4.8/7	
	4 *M* guanidine	60	6.4/7	
Horse heart cytochrome c (1 m*M*)	0.05 *M* Tris, pH 8.0	16	2/4	3
Aspartate aminotransferase (5 mg/ml)	0.05 *M* Tris, pH 7.5 at 22°C	30	—	4
Thrombin (0.06 mg/ml)	0.03 *M* sodium phosphate, pH 8.0 at 24°C	1000	4.9/12	5
2-Keto-4-hydroxyglutarate aldolase (1.2 mg/ml)	0.05 *M* Tris, pH 8.0			6
Porcine carboxypeptidase-β (5–7 mg/ml)	0.05 *M* Tris, pH 8.0 at 23°C	8	1.2/21	7
Bovine pituitary growth hormone (4 mg/ml)	0.05 *M* Tris, pH 8.0 at 0°C	30	2.7/6	8
Ovine pituitary growth hormone (4 mg/ml)	0.05 *M* Tris, pH 8.0 at 0°C	30	3.0/6	8
Aspartate transcarbamylase (catalytic subunit, 4 mg/ml)	0.1 *M* potassium phosphate, pH 6.7 at 23°C	2/3/8		9
Mucor miehei protease (1 mg/ml)	0.1 *M* phosphate, pH 8.0 at 15°C	20	2.8/21	10
Turnip yellow mosaic virus capsids (1.75 mg/ml)	0.05 *M* Tris, pH 8.0 at 22°C	50	3/3	11
α_1-Antiprotease inhibitor (1 mg/ml)	0.05 *M* Tris, pH 8.0 at 25°C	120	3/7	12
	5 *M* guanidine at 25°C	60	7/7	12
α_1-Acid glycoprotein	0.1 *M* Tris, pH 8.0 at 23°C	10	2.7/12	13
Carboxypeptidase A crystals (5 mg/ml)	0.05 *M* Tris, pH 8.0 at 20°C		1/9	14
Bovine growth hormone (2 mg/ml)	0.03 *M* Ringer phosphate at 25°C	12	3/7	15
Equine growth hormone (2 mg/ml)	0.03 *M* Ringer phosphate at 25°C	12	3/7	15
Lactose repressor protein	0.1 *M* Tris, 0.1 *M* mannose, pH 7.8 at 23°C	800	2.4/8	16
Aspartate transcarbamylase (8.3 mg/ml)	0.1 *M* Tris, acetate at 25°C	750	2.2/10	17

Table 2 (continued)
THE USE OF TETRANITROMETHANE TO MODIFY TYROSYL RESIDUES IN PROTEINS

Protein concentration	Solvent/temp	Molar excess TNM	Residues modified	Ref.
Aspartate transcarbamylase (5 mg/ml)	0.1 *M* Tris, pH 8.0 at 25°C		2.2/10	18
Human serum albumin	pH 8.0	80	9/18 1.2/18	19, 20
Prolactin (1 mg/ml)	0.05 *M* Tris, pH 8.0 at 23°C	175	1.9/7	21
Porcine pancreatic phospholipase (1 mg/ml)	0.05 *M* Tris, 0.1 *M* NaCl, 0.01 *M* $CaCl_2$, pH 8.0 at 30°C	10	—	22
Equine pancreatic phospholipase (1 mg/ml)	0.05 *M* Tris, 0.1 *M* NaCl, 0.01 *M* $CaCl_2$, pH 8.0 at 30°C	10	—	22
Bovine pancreatic phospholipase (1 mg/ml)	0.05 *M* Tris, 0.1 *M* NaCl, 0.01 *M* $CaCl_2$, pH 8.0 at 30°C	10	—	22
Troponin C (1 mg/ml)	0.05 *M* Tris, 0.002 *M* EGTA at 23°C	8	3/3	23
Mouse myeloma protein (5×10^{-5} *M*)	0.01 *M* Tris, pH 8.2 at 23°C	10		24
Escherichia coli elongation factor G (4–6 mg/ml)	0.1 *M* Tris, 0.01 *M* KCl, 5% glycerol, 0.2 m*M* EDTA, pH 8.0 at 25°C	250	4/20	25
Elapid venom cardiotoxins (7 mg/ml)	0.1 *M* Tris, pH 7.0 at 25°C or 0.05 *M* Tris, pH 8.0 at 25°C	—	—	26
Lactose repressor (0.1–1.0 mg/ml)	0.1 *M* Tris, pH 8.0 or 0.24 *M* potassium phosphate, 5% glucose, pH 8.0 at 23°C	50	—	27
L-Lactate monooxygenase (1.8 μ*M*)	0.05 *M* Tris, pH 8.0, 7.5 at 30°C			28
Tryptophanase apoenzyme (0.1 μ*M*)	0.05 *M* triethanolamine, pH 8.0 at 30°C	—	—	29
β-Lactamase (1.3 mg/ml)	0.05 *M* Tris, pH 8.0 at 25°C	5.20		30
Bacillus subtilis neutral protease	50 m*M* Tris-Cl, pH 8.0 with 5 m*M* $CaCl_2$ at 25°C	60	2	31
Fructose-1,6-bisphosphatase	50 m*M* Tris-Cl, pH 8.0	50	4	32
α-l-Anti-protease inhibitor	50 m*M* Tris-Cl, pH 8.0 at 22°C	105	3/6	33
Bovine thrombin	50 m*M* sodium phosphate, pH 8.0 with 100 m*M* NaCl or 50 m*M* Tris, pH 8.0 with 100 m*M* NaCl	50–200	1/12	34

References for Table 2

1. **Riordan, J. F., Sokolovsky, M., and Vallee, B. L.,** The functional tyrosyl residues of carboxypeptidase A. Nitration with tetranitromethane, *Biochemistry,* 6, 3609, 1967.
2. **Cuatrecasas, P., Fuchs, S., and Anfinsen, C. B.,** The tyrosyl residues at the active site of staphylococcal nuclease. Modifications by tetranitromethane, *J. Biol. Chem.,* 243, 4787, 1968.
3. **Skov, K., Hofmann, T., and Williams, G. R.,** The nitration of cytochrome c, *Can. J. Biochem.,* 47, 750, 1969.
4. **Christen, P. and Riordan, J. F.,** Syncatalytic modification of a functional tyrosyl residue in aspartate aminotransferase, *Biochemistry,* 9, 3025, 1970.
5. **Lundblad, R. L. and Harrison, J. H.,** The differential effect of tetranitromethane on the proteinase and esterase activity of bovine thrombin, *Biochem. Biophys. Res. Commun.,* 45, 1344, 1971.
6. **Lane, R. S. and Dekker, E. E.,** Oxidation of sulfhydryl groups of bovine liver 2-keto-4-hydroxyglutarate aldolase by tetranitromethane, *Biochemistry,* 11, 3295, 1972.
7. **Sokolovsky, M.,** Porcine carboxypeptidase B. Nitration of the functional tyrosyl residue with tetranitromethane, *Eur. J. Biochem.,* 25, 267, 1972.
8. **Glaser, C. B., Bewley, T. A., and Li, C. H.,** Reaction of bovine and ovine pituitary growth hormones with tetranitromethane, *Biochemistry,* 12, 3379, 1973.
9. **Kirschner, M. W. and Schachman, H. K.,** Conformational studies on the nitrated catalytic subunit of aspartate transcarbamylase, *Biochemistry,* 12, 2987, 1973.
10. **Rickert, W. S. and McBride-Warren, P. A.,** Structural and functional determinants of *Mucor miehei* protease. IV. Nitration and spectrophotometric titration of tyrosine residues, *Biochim. Biophys. Acta,* 371, 368, 1974.
11. **Re, G. G. and Kaper, J. M.,** Chemical accessibility of tyrosyl and lysyl residues in turnip yellow mosaic virus capsids, *Biochemistry,* 14, 4492, 1975.
12. **Busby, T. F. and Gan, J. C.,** The reaction of tetranitromethane with human plasma α_1-antitrypsin, *Int. J. Biochem.,* 6, 835, 1975.
13. **Kute, T. and Westphal, U.,** Steroid-protein interactions. XXXIV. Chemical modification of α_1-acid glycoprotein for characterization of the progesterone binding site, *Biochim. Biophys. Acta,* 420, 195, 1976.
14. **Muszynska, G. and Riordan, J. F.,** Chemical modification of carboxypeptidase A crystals. Nitration of tyrosine 248, *Biochemistry,* 15, 46, 1976.
15. **Daurat-Larroque, S. T., Portuguez, M. E. M., and Santome, J. A.,** Reaction of bovine and equine growth hormones with tetranitromethane, *Int. J. Pept. Protein Res.,* 9, 119, 1977.
16. **Alexander, M. E., Burgum, A. A., Noall, R. A., Shaw, M. D., and Matthews, K. S.,** Modification of tyrosine residues of the lactose repressor protein, *Biochim. Biophys. Acta,* 493, 367, 1977.
17. **Landfear, S. M., Lipscomb, W. N., and Evans, D. R.,** Functional modifications of aspartate transcarbamylase induced by nitration with tetranitromethane, *J. Biol. Chem.,* 253, 3988, 1978.
18. **Lauritzen, A. M., Landfear, S. M., and Lipscomb, W. N.,** Inactivation of the catalytic subunit of aspartate transcarbamylase by nitration with tetranitromethane, *J. Biol. Chem.,* 255, 602, 1980.
19. **Malan, P. G. and Edelhoch, H.,** Nitration of human serum albumin and bovine and human goiter thyroglobulins with tetranitromethane, *Biochemistry,* 9, 3205, 1970.
20. **Moravek, L., Saber, M. A., and Meloun, B.,** Steric accessibility of tyrosine residues in human serum albumin, *Collect. Czech. Chem. Commun.,* 44, 1657, 1979.
21. **Andersen, T. T., Zamierowski, M. M., and Ebner, K. E.,** Effect of nitration on prolactin activities, *Arch. Biochem. Biophys.,* 192, 112, 1979.
22. **Meyer, H., Verhoef, H., Hendriks, F. F. A., Slotboom, A. J., and de Haas, G. H.,** Comparative studies of tyrosine modification in pancreatic phospholipases. I. Reaction of tetranitromethane with pig, horse, and ox phospholipases A_2 and their zymogens, *Biochemistry,* 18, 3582, 1979.

References for Table 2 (continued)

23. **McCubbin, W. D., Hincke, M. T., and Kay, C. M.,** The utility of the nitrotyrosine chromophore as a spectroscopic probe in troponin C and modulator protein, *Can. J. Biochem.*, 57, 15, 1979.
24. **Gavish, M., Neriah, Y. B., Zakut, R., Givol, D., Dwek, R. A., and Jackson, W. R. C.,** On the role of Tyr 34_L in the antibody combining site of the dinitrophenyl binding protein 315, *Mol. Immunol.*, 16, 957, 1979.
25. **Alakhov, Y. B., Zalite, I. K., and Kashparov, I. A.,** Tyrosine residues in the C-terminal domain of the elongation factor G are essential for its interaction with the ribosome, *Eur. J. Biochem.*, 105, 531, 1980.
26. **Carlsson, F. H. H.,** The preparation of 3-nitrotyrosyl derivatives of three elapid venom cardiotoxins, *Biochim. Biophys. Acta*, 624, 460, 1980.
27. **Hsieh, W.-T. and Matthews, K. S.,** Tetranitromethane modification of the tyrosine residues of the lactose repressor, *J. Biol. Chem.*, 256, 4856, 1981.
28. **Durfor, C. N. and Cromartie, T. H.,** Inactivation of L-lactate monooxygenase by nitration with tetranitromethane, *Arch. Biochem. Biophys.*, 210, 710, 1981.
29. **Nihira, T., Toraya, T., and Fukui, S.,** Modification of tryptophanase with tetranitromethane, *Eur. J. Biochem.*, 119, 273, 1981.
30. **Wolozin, B. L., Myerowitz, R., and Pratt, R. F.,** Specific chemical modification of the readily nitrated tyrosine of the R_{TEM} β-lactamase and of *Bacillus cereus* β-lactamase. I. The role of this tyrosine in β-lactamase catalysis, *Biochim. Biophys. Acta*, 701, 153, 1982.
31. **Kobayashi, R., Kanatani, A., Yoshimoto, T., and Tsuru, D.,** Chemical modification of neutral protease from *Bacillus subtilis* var. *amylosaccharitícus* with tetranitromethane: assignment of tyrosyl residues nitrated, *J. Biochem. (Tokyo)*, 106, 1110, 1989.
32. **Liu, F. and Fromm, H. J.,** Investigation of the relationship between tyrosyl residues and the adenosine 5′-monophosphate binding site of rabbit liver fructose-1,6-bisphosphatase as studied by chemical modification and nuclear magnetic resonance spectroscopy, *J. Biol. Chem.*, 264, 18320, 1989.
33. **Mierzwa, S. and Chan, S. K.,** Chemical modification of human alpha-1-proteinase inhibitor by tetranitromethane. Structure-function relationship, *Biochem. J.*, 246, 37, 1987.
34. **Lundblad, R. L., Noyes, C. M., Featherstone, G. L., Harrison, J. H., and Jenzano, J. W.,** The reaction of bovine alpha-thrombin with tetranitromethane. Characterization of the modified protein, *J. Biol. Chem.*, 263, 3729, 1988.

served that aspartate aminotransferase was readily inactivated by tetranitromethane only in the presence of both substrates, glutamate and α-keto-glutarate ("syncatalytic" modification). This inactivation was associated with the modification of an additional tyrosyl residue.

Kirschner and Schachman[66] have reported some interesting spectral studies on the catalytic subunit of aspartate transcarbamylase modified by tetranitromethane (0.1 *M* potassium phosphate, pH 6.7). The addition of both substrates, carbamyl phosphate and succinate, resulted in a decrease in absorbance at 430 nm (Figure 11). This is consistent with the increase in the pKa of the phenolic hydroxyl (6.25 to 6.62). It should be noted that these investigators found that for the 3-nitrotyrosyl group in the modified aspartate transcarbamylase, $\varepsilon_{430} = 4.0 \times 10^{-3}$ cm^{-1} M^{-1} and $\varepsilon_{390} = 2.8 \times 10^{-3}$ cm^{-1} M^{-1}(390 nm is the isosbestic point).

p-Nitrobenzenesulfonyl fluoride (NBSF) is another environmentally sensitive reagent which can be used for the modification of tyrosyl residues in

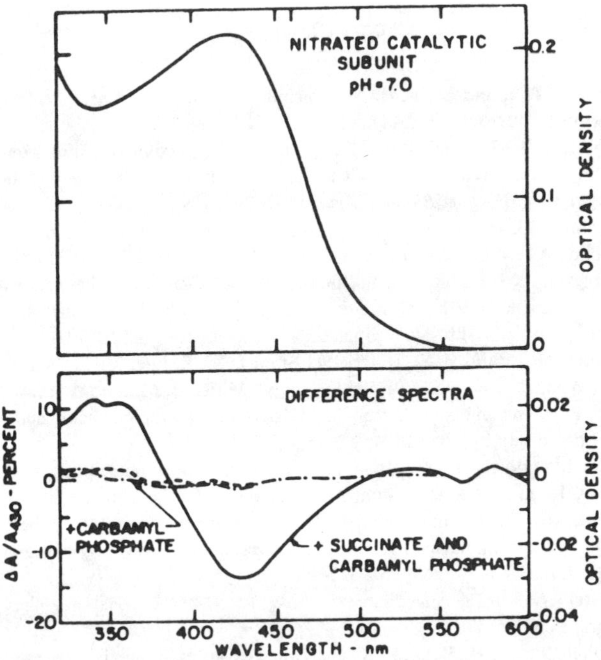

FIGURE 11. UV absorption spectrum and difference spectrum of the catalytic subunit of aspartate transcarbamylase modified with tetranitromethane. The top panel represents the absorption spectrum of a nitrated catalytic subunit containing 0.7 nitrotyrosine/polypeptide chain at a concentration of 3.0 mg/ml in 0.04 *M* potassium phosphate buffer (pH 7.0). The bottom panel represents the difference spectra of the same sample. The protein vs. protein spectrum is represented by (- - -), which superimposes with the carbamyl phosphate (2 m*M*) difference spectrum represented by (—·—·—·) above 450 nm. The solid line represents the difference spectrum obtained when succinate (2 m*M*) and carbamyl phosphate (4 m*M*) were added to one sample. (From Kirschner, M. W. and Schachman, H. K., *Biochemistry,* 12, 2987, 1973. With permission.)

proteins. This reagent was developed by Liao and co-workers for the selective modification of tyrosyl residues pancreatic DNase.[67] The modification reaction with NBSF can be performed in solvents (i.e., 0.1 *M* Tris-Cl, pH 8.0; 0.1 *M N*-ethylmorpholine acetate, pH 8.0) typically used for the modification of tyrosine residues by other reagents such as tetranitromethane or *N*-acetylimidazole. The rate of reagent hydrolysis is substantial and increases with increasing pH. In a recent study, NBSF has been used to characterize tyrosyl residues in NAD(P)H:quinone reductase.[68] Analysis of the product of the reaction showed that NBSF-modified tyrosyl residues were located in hydrophobic regions of the protein. In another study,[69] the reaction of NBSF with tyrosyl residues in human placental taurine transporter was compared with modification observed with tetranitromethane, *N*-acetylimidazole, and Nbf-Cl. Tetranitromethane and Nbf-Cl were the most effective reagents. NBSF was an order of magnitude less effective, while *N*-acetylimidazole was 500 times less potent.

REFERENCES

1. **Tsomides, T. J. and Eisen, H. N.,** Stoichiometric labeling of peptides by iodination on tyrosyl or histidyl residues, *Anal. Biochem.,* 210, 129, 1993.
2. **Skawinski, W. J., Adebodum, F., Cheng, J. T., Jordan, F., and Mendelsohn, R.,** Labeling of tyrosines in proteins with [15N]tetranitromethane, a new NMR reporter for nitrotyrosines, *Biochim. Biophys. Acta,* 1162, 297, 1993.
3. **Arguello, J. M. and Kaplan, J. H.,** *N*-Acetylimidazole inactivates renal Na, K-ATPase by disrupting ATP binding to the catalytic site, *Biochemistry,* 29, 5775, 1990.
4. **Kobayashi, R., Kanatani, A., Yoshimoto, T., and Tsuru, D.,** Chemical modification of neutral protease from *Bacillus subtilis* var. *amylosacchariticus* with tetranitromethane: assignment of tyrosyl residues nitrated, *J. Biochem.,* 106, 1110, 1993.
5. **Rosenfeld, R., Philo, J. S., Haniu, M., Stoney, K., Rohde, M. F., Wu, G.-M., Narhi, L. O., Wong, C., Boone, T., Hawkins, N. N., Miller, J. M., and Arakawa, T.,** Sites of iodination in recombinant human brain-derived neurotrophic factor and its effect on neurotrophic activity, *Protein Sci.,* 2, 1664, 1993.
6. **de Almeida Olivera, M. G., Rogana, E., Rose, J. C., Reinhold, B. B., Andrade, M. H., Greene, L. J., and Mares-Guia, M.,** Tyrosine 151 is part of the substrate activation binding site of bovine trypsin. Identification by covalent labeling with *p*-diazoniumbenzamidine and kinetic characterization of TYR-151-(*p*-benzamidino)-azo-β-trypsin, *J. Biol. Chem.,* 268, 26893, 1993.
7. **Landsteiner, K.,** *The Specificity of Serological Reactions,* Harvard University Press, Cambridge, 1945.
8. **Fraenkel-Conrat, H., Bean, R. S., and Lineweaver, H.,** Essential groups for the interaction of ovomucoid (egg white trypsin inhibitor) and trypsin, and for tryptic activity, *J. Biol. Chem.,* 177, 385, 1949.
9. **Riordan, J. F. and Vallee, B. L.,** Diazonium salts as specific reagents and probes of protein conformation, *Meth. Enzymol.,* 25, 521, 1972.
10. **Tabachnick, M. and Sobotka, H.,** Azoproteins. I. Spectrophotometric studies of amino acid azo derivatives, *J. Biol. Chem.,* 234, 1726, 1959.
11. **Tabachnick, M. and Sobotka, H.,** Azoproteins. II. A spectrophotometric study of the coupling of diazotized arsanilic acid with proteins, *J. Biol. Chem.,* 235, 1051, 1960.
12. **Fairclough, G. F., Jr. and Vallee, B. L.,** Arsanilazochymotrypsinogen. The extrinsic Cotton effects of an arsanilazotyrosyl chromophore as a conformation probe of zymogen activation, *Biochemistry,* 10, 2470, 1971.
13. **Gorecki, M., Wilchek, M., and Blumberg, S.,** Modulation of the catalytic properties of α-chymotrypsin by chemical modification at Tyr 146, *Biochim. Biophys. Acta,* 535, 90, 1978.
14. **Robinson, G. W.,** Reaction of a specific tryptophan residue in streptococcal proteinase with 2-hydroxy-5-nitrobenzyl bromide, *J. Biol. Chem.,* 245, 4832, 1970.
15. **Gorecki, M. and Wilchek, M.,** Modification of a specific tyrosine residue of ribonuclease A with a diazonium inhibitor analog, *Biochim. Biophys. Acta,* 532, 81, 1978.
16. **Johansen, J. T., Livingston, D. M., and Vallee, B. L.,** Chemical modification of carboxypeptidase A crystals. Azo coupling with tyrosine-248, *Biochemistry,* 11, 2584, 1972.
17. **Harrison, L. W. and Vallee, B. L.,** Kinetics of substrate and product interactions with arsanilazotyrosine-248 carboxypeptidase A, *Biochemistry,* 17, 4359, 1978.
18. **Cueni, L. and Riordan, J. F.,** Functional tyrosyl residues of carboxypeptidase A. The effect of protein structure on the reactivity of tyrosine-198, *Biochemistry,* 17, 1834, 1978.
19. **Roholt, O. A. and Pressman, D.,** Iodination-isolation of peptides from the active site, *Meth. Enzymol.,* 25, 438, 1972.
20. **Huntley, T. E. and Strittmatter, P.,** The reactivity of the tyrosyl residues of cytochrome b_5, *J. Biol. Chem.,* 247, 4648, 1972.

21. **McGowan, E. B. and Stellwagen, E.,** Reactivity of individual tyrosyl residues of horse heart ferricytochrome c toward iodination, *Biochemistry,* 9, 3047, 1970.
22. **Layne, P. P. and Najjar, V. A.,** Evidence for a tyrosine residue at the active site of phosphoglucomutase and its interaction with vanadate, *Proc. Natl. Acad. Sci. U.S.A.,* 76, 5010, 1979.
23. **Silva, J. S. and Ebner, K. E.,** Protection by substrates and α-lactalbumin against inactivation of galactosyltransferase by iodine monochloride, *J. Biol. Chem.,* 255, 11262, 1980.
24. **Izzo, J. L., Bale, W. F., Izzo, M. J., and Roncone, A.,** High specific activity labeling of insulin with ^{131}I, *J. Biol. Chem.,* 239, 3743, 1964.
25. **Linde, S., Sonne, O., Hansen, B., and Gliemann, J.,** Monoiodoinsulin labelled in tyrosine residue 16 or 26 of the insulin B-chain. Preparation and characterization of some binding properties, *Hoppe-Seyler's Z. Physiol. Chem.,* 362, 573, 1981.
26. **Hunter, W. M. and Greenwood, F. C.,** Preparation of iodine-131 labelled human growth hormone of high specific activity, *Nature (London),* 194, 495, 1962.
27. **Heber, D., Odell, W. D., Schedewie, H., and Wolfsen, A. R.,** Improved iodination of peptides for radioimmunoassay and membrane radioreceptor assay, *Clin. Chem.,* 24, 796, 1978.
28. **Maly, P. and Lüthi, C.,** The binding sites of insulin-like growth factor I (IGF I) to type I IGF receptor and to a monoclonal antibody. Mapping by chemical modification of tyrosine residues, *J. Biol. Chem.,* 263, 7068, 1988.
29. **Kurihara, K., Horinishi, H., and Shibata, K.,** Reaction of cyanuric halides with proteins. I. Bound tyrosine residues of insulin and lysozyme as identified with cyanuric fluoride, *Biochim. Biophys. Acta,* 74, 678, 1963.
30. **Gorbunoff, M. J.,** Cyanuration, *Meth. Enzymol.,* 25, 506, 1972.
31. **Gorbunoff, M. J. and Timasheff, S. N.,** The role of tyrosines in elastase, *Arch. Biochem. Biophys.,* 152, 413, 1972.
32. **Coffe, G. and Pudles, J.,** Chemical reactivity of the tyrosyl residues in yeast hexokinase. Properties of the nitroenzyme, *Biochim. Biophys. Acta,* 484, 322, 1977.
33. **Ferguson, S. J., Lloyd, W. J., Lyons, M. H., and Radda, G. K.,** The mitochondrial ATPase. Evidence for a single essential tyrosine residue, *Eur. J. Biochem.,* 54, 117, 1975.
34. **Ferguson, S. J., Lloyd, W. J., and Radda, G. K.,** The mitochondrial ATPase. Selective modification of a nitrogen residue in the β subunit, *Eur. J. Biochem.,* 54, 127, 1975.
35. **Andrews, W. W. and Allison, W. S.,** 1-Fluoro-2,4-dinitrobenzene modifies a tyrosine residue when it inactivates the bovine mitochondrial F_1-ATPase, *Biochem. Biophys. Res. Commun.,* 99, 813, 1981.
36. **Wieland, T. and Schneider, G.,** *N*-Acylimidazoles as acyl derivatives of high energy, *Ann. Chem. Justus Liebigs,* 580, 159, 1953.
37. **Stadtman, E. R. and White, F. H., Jr.,** The enzymic synthesis of *N*-acetylimidazole, *J. Am. Chem. Soc.,* 75, 2022, 1953.
38. **Stadtman, E. R.,** On the energy-rich nature of acetyl imidazole, an enzymatically active compound, in *A Symposium on the Mechanism of Enzyme Action,* McElroy, W. D. and Glass, B., Eds., Johns Hopkins Press, Baltimore, MD, 1954, 581.
39. **Riordan, J. F. and Vallee, B. L.,** Acetylation, *Meth. Enzymol.,* 11, 565, 1967.
40. **Karibian, D., Jones, C., Gertler, A., Dorrington, K. J., and Hofmann, T.,** On the reaction of acetic and maleic anhydrides with elastase. Evidence for a role of the NH_2-terminal valine, *Biochemistry,* 13, 2891, 1974.
41. **Ohnishi, M., Suganuma, T., and Hiromi, K.,** The role of a tyrosine residue of bacterial liquefying α-amylase in the enzymatic hydrolysis of linear substrates as studied by chemical modification with acetic anhydride, *J. Biochem. (Tokyo),* 76, 7, 1974.
42. **Bernad, A., Nieto, M. A., Vioque, A., and Palacian, E.,** Modification of the amino groups and hydroxyl groups of lysozyme with carboxylic acid anhydrides: a comparative study, *Biochim. Biophys. Acta,* 873, 350, 1986.

43. **Simpson, R. T., Riordan, J. F., and Vallee, B. L.,** Functional tyrosyl residues in the active center of bovine pancreatic carboxypeptidase A, *Biochemistry,* 2, 616, 1963.
44. **Riordan, J. F., Wacker, W. E. C., and Vallee, B. L.,** *N*-Acetylimidazole: a reagent for determination of "free" tyrosyl residues of proteins, *Biochemistry,* 4, 1758, 1965.
45. **Myers, B., II and Glazer, A. N.,** Spectroscopic studies of the exposure of tyrosine residues in proteins with special reference to the subtilisins, *J. Biol. Chem.,* 246, 412, 1971.
46. **Tildon, J. T. and Ogilvie, J. W.,** The esterase activity of bovine mercaptalbumin. The reaction of the protein with *p*-nitrophenyl acetate, *J. Biol. Chem.,* 247, 1265, 1972.
47. **Herriott, R. M.,** Reactions of native proteins with chemical reagents, *Adv. Protein Chem.,* 3, 169, 1947.
48. **Riordan, J. F., Sokolovsky, M., and Vallee, B. L.,** Tetranitromethane. A reagent for the nitration of tyrosine and tyrosyl residues in proteins, *J. Am. Chem. Soc.,* 88, 4104, 1966.
49. **Sokolovsky, M., Harell, D., and Riordan, J. F.,** Reaction of tetranitromethane with sulfhydryl groups in proteins, *Biochemistry,* 8, 4740, 1969.
50. **Riordan, J. F. and Vallee, B. L.,** Nitration with tetranitromethane, *Meth. Enzymol.,* 25, 515, 1972.
51. **Cuatrecasas, P., Fuchs, S., and Anfinsen, C. B.,** The tyrosyl residues at the active site of staphylococcal nuclease. Modifications by tetranitromethane, *J. Biol. Chem.,* 243, 4787, 1968.
52. **Hugli, T. E. and Stein, W. H.,** Involvement of a tyrosine residue in the activity of bovine pancreatic deoxyribonuclease A, *J. Biol. Chem.,* 246, 7191, 1971.
53. **Margolis, S. A., Coxon, B., Gajewski, E., and Dizdaroglu, M.,** Structure of a hydroxyl radical induced cross-link of thymine and tyrosine, *Biochemistry,* 27, 6353, 1988.
54. **Hsuan, J. J.,** The cross-linking of tyrosine residues in apo-ovotransferrin by treatment with periodate anions, *Biochem. J.,* 247, 467, 1987.
55. **Sokolovsky, M., Riordan, J. F., and Vallee, B. L.,** Conversion of 3-nitrotyrosine to 3-aminotyrosine in peptides and proteins, *Biochem. Biophys. Res. Commun.,* 27, 20, 1967.
56. **Riordan, J. F., Sokolovsky, M., and Vallee, B. L.,** Environmentally sensitive tyrosyl residues. Nitration with tetranitromethane, *Biochemistry,* 6, 358, 1967.
57. **Riordan, J. F., Sokolovsky, M., and Vallee, B. L.,** The functional tyrosyl residues of carboxypeptidase A. Nitration with tetranitromethane, *Biochemistry,* 6, 3609, 1967.
58. **Riggle, W. L., Long, J. A., and Borders, C. L., Jr.,** Reaction of turkey egg-white lysozyme with tetranitromethane. Modification of tyrosine and tryptophan, *Can. J. Biochem.,* 51, 1433, 1973.
59. **Sokolovsky, M., Fuchs, M., and Riordan, J. F.,** Reaction of tetranitromethane with tryptophan and related compounds, *FEBS Lett.,* 7, 167, 1970.
60. **Teuwissen, G., Masson, P. L., Osinski, P., and Heremans, J. F.,** Metal-combining properties of human lactoferrin. The effect of nitration of lactoferrin with tetranitromethane, *Eur. J. Biochem.,* 35, 366, 1973.
61. **Katsura, T., Lam, E., Packer, L., and Seltzer, S.,** Light dependent modification of bacteriorhodopsin by tetranitromethane. Interaction of a tyrosine and a tryptophan residue with bound retinal, *Biochem. Int.,* 5, 445, 1982.
62. **Atassi, M. Z. and Habeeb, A. F. S. A.,** Enzymic and immunochemical properties of lysozyme. I. Derivatives modified at tyrosine. Influence of nature of modification on activity, *Biochemistry,* 8, 1385, 1969.
63. **Lundblad, R. L., Featherstone, G. L., Jenzano, J. W., Church, F. C., and Roberts, H. R.,** The modification of tryptophanyl residues in human prothrombin and thrombin with tetranitromethane, manuscript in preparation, 1994.
64. **Skov, K., Hofmann, T., and Williams, G. R.,** The nitration of cytochrome c, *Can. J. Biochem.,* 47, 750, 1969.
65. **Christen, P. and Riordan, J. F.,** Syncatalytic modification of a functional tyrosyl residue in aspartate aminotransferase, *Biochemistry,* 9, 3025, 1970.

66. **Kirschner, M. W. and Schachman, H. K.,** Conformational studies on the nitrated catalytic subunit of aspartate transcarbamylase, *Biochemistry,* 12, 2987, 1973.
67. **Liao, T.-H., Ting, R. S., and Young, J. E.,** Reactivity of tyrosine in bovine pancreatic deoxyribonuclease with *p*-nitrobenzenesulfonyl fluoride, *J. Biol. Chem.,* 257, 5637, 1982.
68. **Haniu, M., Yuan, H., Chen, S., Iyanagi, T., Lee, T. D., and Shively, J. E.,** Structure-function relationship of NAD(P)H:quinone reductase: characterization of NH_2-terminal blocking group and essential tyrosine and lysine residues, *Biochemistry,* 27, 6877, 1988.
69. **Kulanthaivel, P., Leibach, F. H., Mahesh, V. B., and Ganapathy, V.,** Tyrosine residues are essential for the activity of the human placental taurine transporter, *Biochim. Biophys. Acta,* 985, 139, 1989.

Chapter 14

THE MODIFICATION OF CARBOXYL GROUPS

The site-specific modification of carboxyl groups in proteins is somewhat difficult to achieve, as is differentiation between aspartyl residues and glutamyl residues. Most of the current work utilizes water-soluble carbodiimides as described below. The reader is recommended to several recent articles addressing the modification of carboxyl groups in proteins.[1–4]

Parsons and co-workers used triethyloxonium fluoroborate (Figure 1) to modify the β-carboxyl groups of an aspartic residue essential for the enzymatic activity of lysozyme.[5,6] Paterson and Knowles[7] used trimethyloxonium fluoroborate to determine the number of carboxyl groups in pepsin which are essential for catalytic activity. This article discusses in some depth the rigorous precautions necessary for the preparation of this reagent. This reagent is highly reactive and considerable care is required for its introduction into the reaction mixture containing protein. The reaction is performed at pH 5.0 (0.020 *M* sodium citrate, pH maintained at 5.0 with 2.5 *M* NaOH). These investigators also report the preparation of the ^{14}C-labeled reagent from sodium methoxide and [^{14}C]-methyliodide.

Woodward and co-workers[8,9] developed *N*-ethyl-5-phenylisoxazolium-3′-sulfonate (Woodward's Reagent K) and various other *N*-alkyl-5-phenylisoxazolium fluoroborates as reagents for the "activation" of carboxyl groups for synthetic purposes (Figure 2). Bodlaender and co-workers[10] used *N*-ethyl-5-phenylisoxazolium-3-sulfonate, the *N*-methyl and *N*-ethyl derivatives of 5-phenylisoxazolium fluoroborate or *N*-methylbenzisoxazolium fluoroborate (Figure 3) to activate carboxyl groups on trypsin for subsequent modification with methylamine or ethylamine. The extent of modification obtained ranged from approximately 3 residues modified (*N*-methyl-5-phenylisoxazolium fluoroborate or *N*-ethyl-5-phenylisoxazolium fluoroborate, pH 3.80, 20°C, 80 min) to approximately 11 residues modified (*N*-methyl-5-phenylisoxazolium fluoroborate, pH 6.0, 20°C, 10 min). Reagent decomposition occurs quite rapidly, even at ice-bath temperature (2°C). The modification appears fairly selective for carboxyl groups, although some modification of lysine was observed under conditions where extensive modification was obtained (250-fold molar excess of *N*-methyl-5-phenylisoxazolium fluoroborate, pH 4.75, 72 min, 20°C, methylamine as the attacking nucleophile).

Saini and Van Etten[11] reported on the reaction of *N*-ethyl-5-phenylisoxazolium-3′-sulfonate with human prostatic acid phosphatase. The modification was performed with a 4000- to 10,000-fold molar excess of reagent in 0.020 *M* pyridinesulfonic acid, pH 3.6 at 25°C. Ethylamine was utilized as the attacking nucleophile to determine the extent of modification. A substantial number of carboxyl groups in the protein were modified under these experimental conditions. Arana and Vallejos[12] have compared the reaction of chlo-

Structure	Nomenclature	Abbreviation
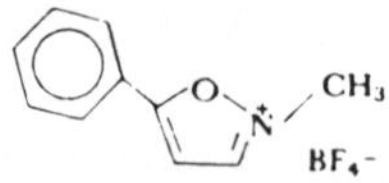	*N*-Methyl-5-phenylisoxazolium fluoroborate	MPI
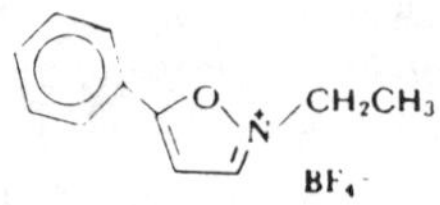	*N*-Ethyl-5-phenylisoxazolium fluoroborate	EPI
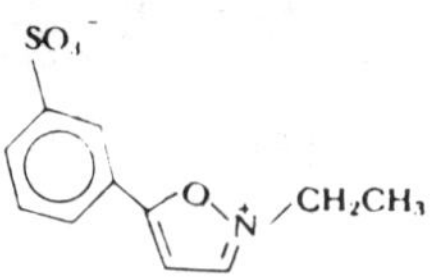	*N*-Ethyl-5-phenylisoxazolium-3′-sulfonate Woodwards K reagent)	K
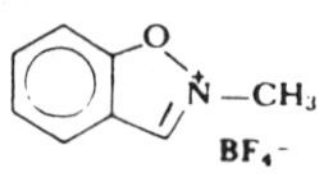	*N*-Methyl-benzisoxazolium fluoroborate	MBI

FIGURE 1. The structures of some isoxazolium salts. (From Bodlaender, P., Feinstein, G., and Shaw, E., *Biochemistry*, 8, 4941, 1969. With permission.)

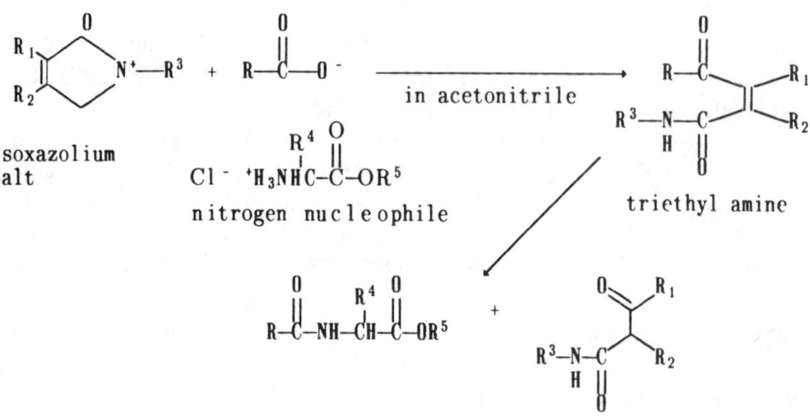

FIGURE 2. Modification of a protein carboxyl group with a nitrogen nucleophile mediated with an isoxazolium salt.

roplast coupling factor with *N*-ethyl-5-phenylisoxazolium-3′-sulfonate (Woodward's Reagent K) and dicyclohexylcarbodiimide. Reaction with Woodward's Reagent K was accomplished at 25°C in 0.040 *M* Tricine, pH 7.9, while reaction with dicyclohexylcarbodiimide was accomplished at 30°C in 0.040 *M*

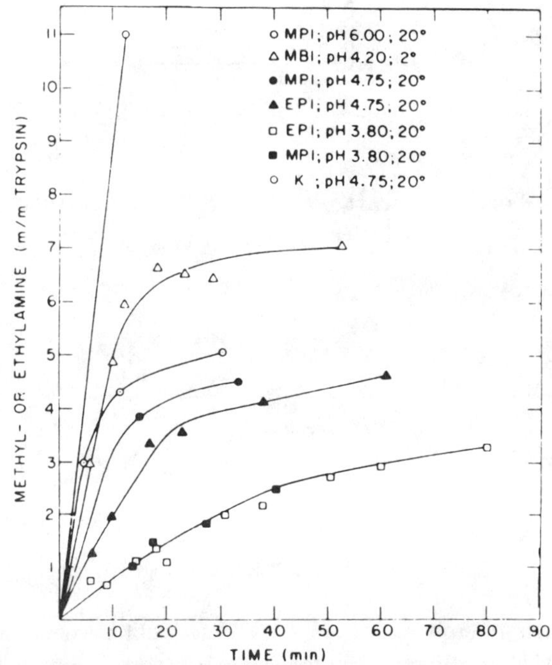

FIGURE 3. The reaction of isoxazolium salts with β-trypsin at various pH values. The isoxazolium salts were MPI, *N*-methyl-5-phenylisoxazolium fluoroborate; EPI, *N*-ethyl-5-phenylisoxazolium fluoroborate; K, *N*-ethyl-5-phenylisoxazolium-3′-sulfonate (Woodward's Reagent K); and MBI, *N*-methyl-benzisoxazolium fluoroborate. In all reactions, the initial reagent and protein concentrations were 10 and 5 mg/ml, respectively, except for the reaction with Reagent K in which the respective concentrations were 5 and 1.5 mg/ml. Each point was obtained by removing a portion from the reaction mixture and adjusting its pH to 2.5 with formic acid at the indicated times. Amino acid analyses were performed on the gel-filtered samples to determine methyl- or ethylamine content per mole of trypsin. (From Bodlaender, P., Feinstein, G., and Shaw, E., *Biochemistry,* 8, 4941, 1969. With permission.)

MOPS, pH 7.4. ATP and derivatives such as ADP and inorganic phosphate protect against the loss of activity occurring upon reaction with Woodward's Reagent K, but they do not have any effect on inactivation by dicyclohexylcarbodiimide. The reverse was seen with divalent cations such as Ca^{2+}. The modification of an essential carboxyl group in pancreatic phospholipase A_2 by 5-ethyl-5-phenylisoxazolium-3-sulfonate has been reported.[13] The reaction was performed in 0.01 *M* sodium phosphate, pH 4.75 (pH stat) at 25°C. A second-order rate constant of $k_2 = 25.5\ M^{-1}\ min^{-1}$ was obtained for the loss of catalytic activity. This rate inactivation is increased more than twofold in the presence of 30 m*M* $CaCl_2$ (69.3 $M^{-1}\ min^{-1}$). Quantitative information on the extent of modification is obtained with [^{14}C]-glycine ethyl ester. It is of interest that treatment with a water-soluble carbodiimide, 1-(3′-dimethyl-aminopropyl)-3-ethylcarbodiimide, results in the loss of catalytic activity in a reaction with characteristics different from those seen with Woodward's Reagent K. Kooistra and Sluyterman[14] modified guanidated mercuripapain with *N*-ethylbenzio-

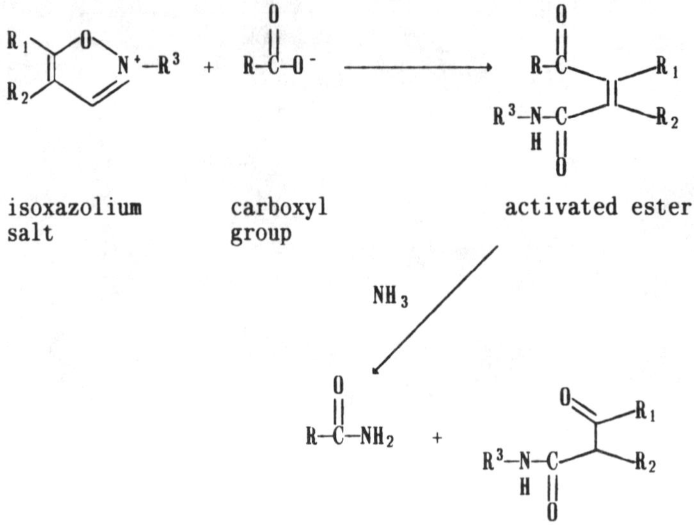

FIGURE 4. Isosteric modification of a carboxyl group.

soxazolium tetrafluoroborate at pH 4.2, 0°C to yield a protein modified with *N*-ethylsalicylamide esters. When the active ester groups on the protein were allowed to undergo amminolysis (2.0 *M* ammonium acetate, pH 9.2) and the esters were converted to the corresponding amides (an isosteric modification, see Figure 4), the conversion to the amides resulted in the loss of a bulky substituent group with an increase in activity.

The use of carbodiimide-mediated modification (Figure 5) of carboxyl functional groups in proteins is by far the most widely used method for the study of such functional groups (Figure 6). The most popular approach utilizes a water-soluble carbodiimide (Figure 7) as the activating agent as introduced by Hoare and Koshland.[15] These investigators show that virtually quantitative modification of the carboxyl groups in lysozyme (8.1:11), chymotrypsin (15.5:17), and trypsin (12.5:11) occurred with 0.1 *M* carbodiimide, pH 4.75 with 1.0 *M* glycine methyl ester in a pH-Stat at 25°C. The possibility of a side reaction was discussed with reference to the possible modification of the phenolic hydroxyl of tyrosine to form the *O*-arylisourea. The modification of the active-site cysteinyl group in papain has been reported[16] as occurring under conditions (pH 4.75, 25°C) where 6:14 carboxyl groups are modified together with 9/19 tyrosyl residues. Modification of the tyrosyl residues is reversed by 0.5 *M* hydroxylamine, pH 7.0 (5 h at 25°C) as first demonstrated by Carraway and Koshland.[17] Despite the problems with side reaction, modification of the carboxyl group in proteins with a water-soluble carbodiimide and an appropriate nucleophile (e.g., [^{14}C]-glycine ethyl ester, norleucine methyl ester — easily detected by amino acid analysis, aminomethylsulfonic acid) has proved extremely useful.[18] It should be noted that ammonium ions can be used as the

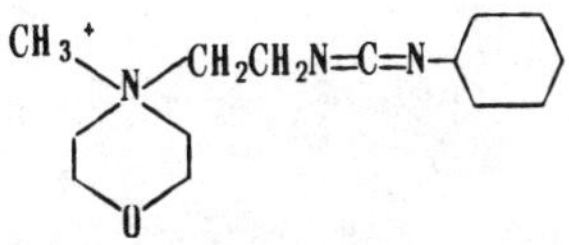

1–Cyclohexyl–3(2–morpholinoethyl)–carbodiimide
(relatively hydrophobic)

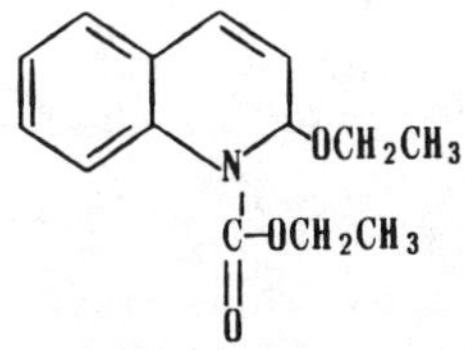

N–ethoxy–2–ethoxy–1,2–dihydroquinone
(relatively hydrophobic)

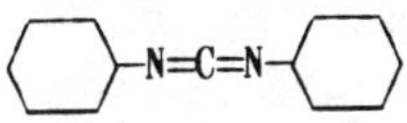

1,3–Dicyclohexylcarbodiimide
(relatively hydrophobic)

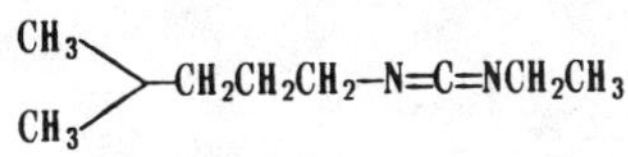

1–ethyl–3(3–dimethylaminopropyl)–carbodiimide
(relatively hydrophobic)

FIGURE 5. A comparison of some reagents used for the modification of carboxyl groups in proteins.

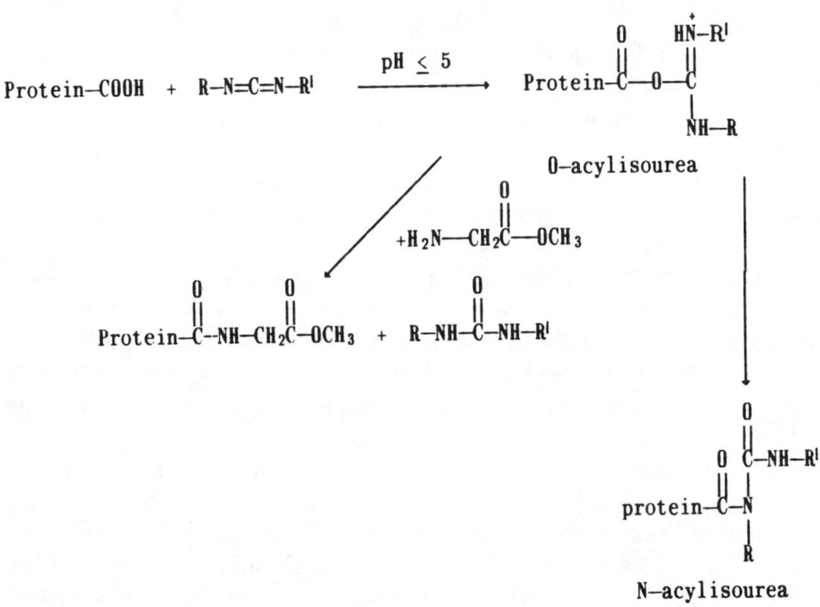

FIGURE 6. The reaction of carbodiimide with protein carboxyl groups.

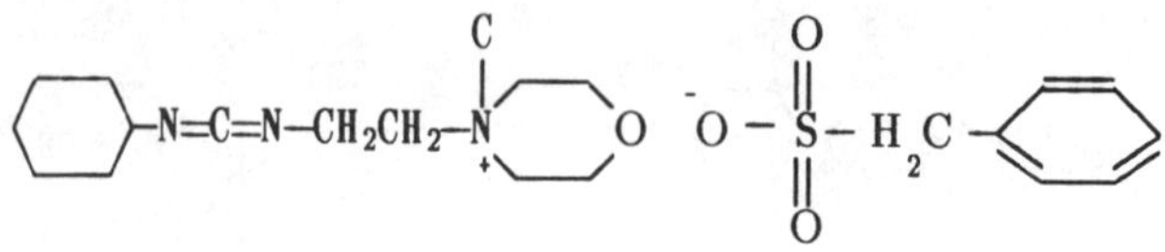

1–cyclohexyl–3–(2–morpholinoethyl)–carbodiimide metho–p–toluenesulfonate

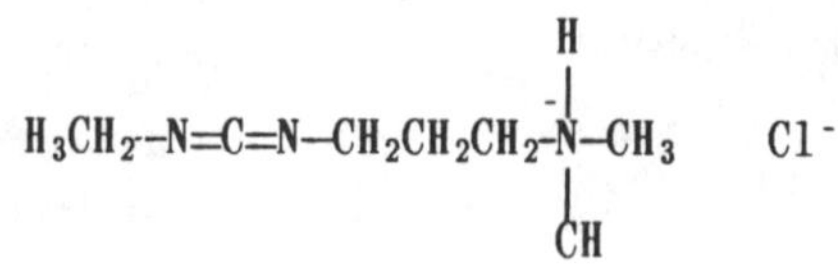

1–ethyl–3–(3–dimethylaminopropyl)–carbodiimide hydrochloride

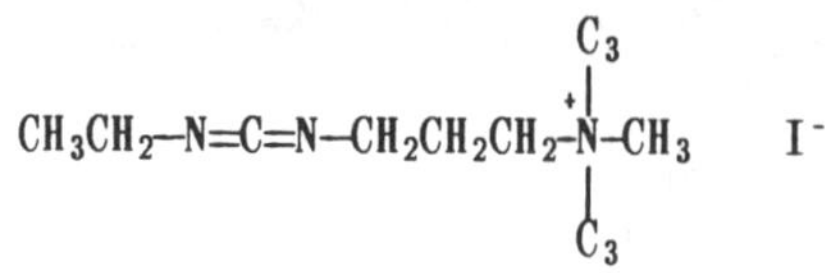

1–ethyl–3–(4–azonia–4,4–dimethyl–pentyl) carbodiimide iodide

FIGURE 7. Some examples of water-soluble carbodiimides.

attacking nucleophile to generate asparaginyl and glutaminyl residues from "exposed" carboxyl groups. The modification was accomplished in 5.5 *M* NH_4Cl, pH 4.75 for 3 h at 25°C. Under these conditions, approximately 11 of the 15 free carboxyl groups in chymotrypsinogen were converted to the corresponding amide.[19] 1,2-Diaminoethane or diaminomethane can be coupled to aspartic acid residues to produce a trypsin-sensitive bond.[20] Another example of using this chemistry to introduce a new functional group into a protein is provided by the work of Lin et al.[21] This procedure is outlined in Figure 8. The first step involves the water-soluble carbodiimide-mediated coupling of cystamine to protein carboxyl groups. Reduction of the coupled cystamine with dithiothreitol results in 2-aminothiol functional groups bound to protein carboxyl groups. Examples of the use of this approach are given in Table 1.

Another possible reaction occurring with a water-soluble carbodiimide and a protein in the absence of added nucleophile is zero-length cross-linking via isopeptide bond formation. This reaction is somewhat favored by participation of an unprotonated lysine residue and, hence, proceeds somewhat more rapidly at pH values greater than that generally used for protein modification with carbodiimide reagents.

There are examples of carboxyl group modification with reagents expected to react far more effectively with other nucleophiles. An example of this is the reaction of iodoacetamide with ribonuclease T_1 to form the glycolic acid derivative of the glutamic acid residue as elegantly shown by Takahashi and co-workers.[22] Another example is the modification of a specific carboxyl group in pepsin by *p*-bromophenacyl bromide[23] (the use of *p*-bromophenacyl bromide in the specific modification of proteins is not uncommon, but is generally associated with the modification of cysteine, histidine, or methionine). In the

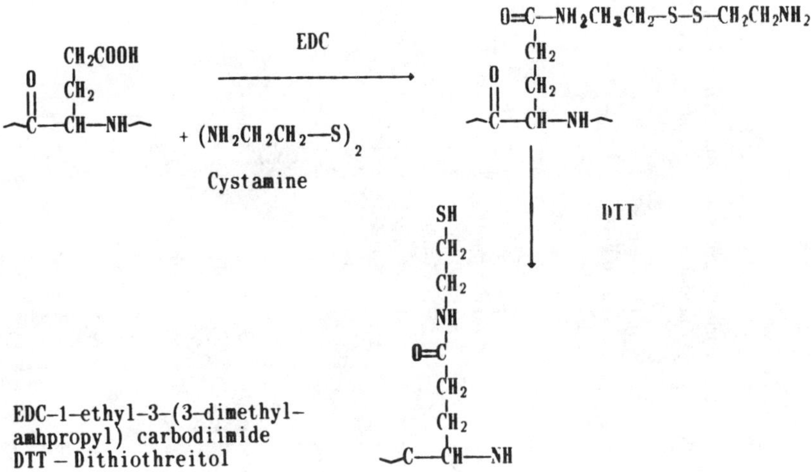

FIGURE 8. Introduction of a sulfydryl function (2-aminoethanethiol) via isopeptide linkage mediated by a water-soluble carbodiimide.

study of pepsin, optimal inactivation (approximately 12-fold molar excess of reagent, 3 h, 25°C) was obtained in the pH range of 1.5 to 4.0 with a rapid decrease in the extent of inactivation at pH 4.5 and above (the effect of pH greater than 5.5 to 6.0 on the modification of pepsin cannot be studied because of irreversible denaturation of pepsin at pH 6.0 and above). In studies with a 10% molar excess of *p*-bromophenacyl bromide, pH 2.8 at 37°C for 3 h, complete inactivation of the enzyme was obtained concomitant with the incorporation of 0.93 mol of reagent per mole of pepsin (assessed by bromide analysis). Attempts to reactivate the modified enzyme with a potent nucleophile such as hydroxylamine were unsuccessful, but reactivation could be obtained with sulfhydryl-containing reagents (i.e., β-mercaptoethanol, 2,3-dimercaptopropanol, thiophenol). It has been subsequently established that reaction occurs at the β-carboxy group of an aspartic acid residue (formation of 2-*p*-bromophenyl-1-ethyl-2-one β-aspartate).[24] These investigators noted that the reduction of enzyme under somewhat harsh conditions ($LiBH_4$ in tetrahydrofuran) resulted in the formation of homoserine.

Table 1
THE MODIFICATION OF CARBOXYL GROUPS IN PROTEINS

Protein	Reaction conditions	Carbodiimide	Molar excess	Nucleophile	Carboxyl groups modified	Other functional groups modified	Ref.
Lysozyme	pH 4.75 at 25°C, pH-stat,[a] 0.1 *M* carbodiimide	BDC[b]	140	1.0 *M* glycine methyl ester	2.1/11 (5 min)	—	1
					4.7/11 (60 min)	—	
					8.1/11 (5–6 h)	—	
Lysozyme			140	0.1 *M* nitro-tyrosine ethyl ester	1.4/11[c]	—	1
Trypsin			250	1.0 *M* glycine methyl ester	4.6/11 (5 min)	—	1
					8.8/11 (60 min)	—	
					12.5/11 (5–6 h)[d]	—	
Chymotrypsin			250	1.0 *M* glycine methyl ester	6.2/17 (5 min)	—	1
					11.8/17 (60 min)	—	
					15.5/17 (5–6 h)	—	
Trypsinogen	pH 4.5 at 25°C	EDC[f]	25	1.0 *M* glycine ethyl ester[e]	—	—	2
Chymo-trypsinogen	pH 4.0 at 25°C, pH-stat	EDC		1.0 *M* glycine ethyl ester	13/14[g]	—	3
	pH 6.0 at 25°C, pH-stat				10/14	—	3
	pH 8.0 at 25°C, pH-stat				3/14	—	3
Chymotrypsin	pH 4.75 at 25°C, pH-stat	EDC		1.0 *M* glycine methyl ester	12.7[h]	—	4
					15.6[i]		
					10.6		4
					13.5[i]		

Lysozyme	pH 4.75 at 25°C, pH-stat	BDC		0.25 *M* amino-methanesulfonic acid	8.5–9.5[j]	—	5
				10 *M* glycine methyl ester	6.5–7.5[j]	—	5
Trypsin	pH 4.75 at 25°C, pH-stat	EDC		1.0 *M* glycinamide	7.9[k] 1.7[k]	3–5 Tyr[l]	6
Albumin	pH 4.75	EDC		Glycine methyl ester or L-argininamide	1[m]	—	7
Chymotrypsin	pH 4.0[n]	EDC		Glycine ethyl ester	15/15	—	8
Papain	pH 4.75 at 25°C	EDC	300–1200	Glycine ethyl ester	6/14	6–10 Tyr,[o] active site cysteinyl residue	9
Yeast hexokinase	0.1 *M* phosphate, pH 6.0 at 20°C[p]	CMC[q]	500–3000	Nitrotyrosine[r] ethyl ester	1	Cysteinyl (2)	10
Phos-phorylase[b]	pH 5.1 at 25°C	CMC		Glycine[s] ethyl ester	3[t]	—	11
α-Manno-sidase	pH 4.2, 0.1 *M* MES, 1.0 *M* NaCl	EDC	2000	Glycine[g] ethyl ester	8	—	12
Yeast enolase	0.050 *M* MES, 1 m*M* $MgCl_2$, 0.01 m*M* EDTA, pH 6.1	EDC, CMC[u]	2000	—	—	—[v]	13

Table 1 (continued)
THE MODIFICATION OF CARBOXYL GROUPS IN PROTEINS

Protein	Reaction conditions	Carbodiimide	Molar excess	Nucleophile	Carboxyl groups modified	Other functional groups modified	Ref.
3-Phosphoglycerate kinase	0.1 *M* phosphate, pH 6.1 at 17°C	CMC	2000	Nitrotyrosine[v] ethyl ester	1	—	14
cAMP-dependent protein kinase	pH 6.5, 0.050 *M* MES, 23°C	EDC	—	Glycine ethyl ester	1.7[w]	—[x]	15
Human Fc fragment	pH 4.75	EDC	—	Glycine ethyl ester	25	—[y]	16
Pancreatic phospholipase A_2	0.25 *M* cacodylate, pH 5.5	EDC	—	Semicarbazide	13/15[z]	—[aa]	17
	pH 3.5	EDC	—	Semicarbazide	14/15[z]	—[aa]	17
Spinach plastocyanin	pH 6.0 at 23°C borate	EDC	—	Ethylenediamine	4.3/16	—	18
Mitochondrial F_1-ATPase	0.05 *M* triethanolamine H_2SO_4, pH 7.0	DCC[cc]	6	—	2[bb]	—	19
Mitochondrial transhydrogenase	1 m*M* Tricine, pH 7.0 with 0.1 *M* choline chloride and 2% MeOH	DCC	—	—	—	—[dd]	20

Lysozyme	pH 5.0	EDC	3.5[ee]	Ethylenediamine, ethanolamine, 4-(5)-(aminomethyl) imidazole, histamine, D-glucosamine, methylamine	1	—	21
Restriction endonuclease Eco R_1	0.1 *M* triethanolamine pH 7.0, 2.0 *M* KCl at 20°C	CMC	—	—[ff]	—	—	22
Thylakoid membrane proteins	pH 7.5 (HEPES)	DDC	—	Glycine ethyl ester	—	—	23
		EDC	—	Glycine ethyl ester	—	—	

[a] It is generally necessary to add dilute HCl (0.2 *M*) during the course of the reaction to maintain the pH at 4.75.

[b] BCD, *N*-benzyl-*N'*-(3-dimethylaminopropyl)carbodiimide.

[c] Reaction time not given.

[d] Excess may result from autolysis of trypsin preparations which would create "new" free carboxyl groups.

[e] Reactions are generally terminated by dilution into cold sodium acetate (1.0 *M*, pH 3.5–5.5).

[f] EDC, 1-ethyl-(3-dimethylaminopropyl)carbodiimide.

[g] Determined by incorporation of [^{14}C]-glycine ethyl ester.

[h] 1-H reaction terminated with 1.0 *M* acetate, pH 4.75.

[i] In 7.5 *M* urea.

[j] Note the interesting difference in the extent of modification which is dependent upon nucleophile used. There is also an interesting difference in the time course of modification.

[k] In the presence and absence of the competitive inhibitor, benzamidine.

[l] Tyrosyl residues regenerated in 0.5 *M* hydroxylamine, pH 7.1 with no effect on the EDC/glycinamide changes in catalytic activity.

[m] Complete modification of the carboxyl groups was achieved in 6.0 *M* guanidine with either L-argininamide or glycine methyl ester. After reduction and carboxymethylation, approximately 20% of the carboxyl groups are unreactive with either nucleophile. There is a further decrease in modification with the reduced and cyanoethylated derivative.

[n] Reaction at pH 4.0 results in apparent quantitative modification of carboxyl groups.

Table 1 (continued)
THE MODIFICATION OF CARBOXYL GROUPS IN PROTEINS

[o] Tyrosine modification is acid-stable, but reversed in 0.5 *M* hydroxylamine, pH 7.0 for 5 h at 25°C. "Activated" papain isirreversibly modified at active site cysteinyl residue (Cys-25) by EDC, while mercuripapain is not.

[p] Optimal inactivation occurred at pH 5.5–6.0 with marked decrease in extent of inactivation at more alkaline pH.

[q] CMC, 1-cylclohexyl-3-(2-morpholinoethyl)carbodiimide metho *p*-toluenesulfonate.

[r] Isolated peptide containing modified glutamic acid residue using affinity chromatography with antinitrotyrosyl γ-globulin.

[s] Inactivation is not dependent upon addition of nucleophile, but rate is greatly enhanced.

[t] Determined by incorporation of [^{14}C] from [Metho-^{14}C] CMC. Also, determined from the extent of incorporation of *N*-(2,4-dinitrophenyl)-ethylene diamine (spectrophotometry); $\varepsilon = 15{,}000\ M^{-1}\ cm^{-1}$. It was determined that the modification of one carboxyl group is critical for potential catalytic function.

[u] These investigators also used 1-ethyl-(4-azonia-4,4-dimethylpentyl)-carbodiimide (1-ethyl-3-(3-dimethylaminopropyl)-carbodiimide). This reagent was more effective than either EDC or CMC (least active) in the inactivation of the enzyme under these conditions.

[v] Reaction at cysteine and tyrosine excluded by amino acid analysis after acid hydrolysis.

[w] Determined γ-glu-gly after proteolysis with trypsin (2X), pronase, carboxypeptidases A and B, and leucine aminopeptidase.

[x] Direct determination not performed. Reaction at pH 8.0 (carbodiimide is more specific for phenolic hydroxyl at alkaline pH) did not result in loss of catalytic activity.

[y] Tyrosine modification excluded by amino acid analysis after acid hydrolysis.

[z] Radiolabeled semicarbazide (synthesized from [^{14}C] cyanate) incorporation.

[aa] Tyrosine modification did occur. Tyrosine regenerated with neutral hydroxylamine. These investigators stressed the need to keep hydroxylamine exposure as brief as possible to avoid side reactions such as peptide bond cleavage or deamidation.

[bb] From incorporation of [^{14}C]-carbodiimide.

[cc] DCC, dicyclohexylcarbodiimide.

[dd] Interchain cross-linking of transhydrogenase dimer occurred under these conditions.

[ee] Obtained specific modification of Asp-101 by using low molar excess of carbodiimide. Extent of modification is somewhat independent of amine (nucleophile) used. These investigators speculate that increased specificity is a reflection of binding of carbodiimide to substrate binding site close to Asp-101. These investigators did purify reaction products to obtain selectively modified protein derivatives.

[ff] Rate and extent of inactivation not changed by the addition of glycine ethyl ester.

References for Table 1

1. **Hoare, D. G. and Koshland, D. E., Jr.,** A procedure for the selective modification of carboxyl groups in proteins, *J. Am. Chem. Soc.*, 88, 2057, 1966.
2. **Radhakrishnan, T. M., Walsh, K. A., and Neurath, H.,** Relief by modification of carboxylate groups of the calcium requirement for the activation of trypsinogen, *J. Am. Chem. Soc.*, 89, 3059, 1967.
3. **Abita, J. P., Maroux, S., Delaage, M., and Lazdunski, M.,** The reactivity of carboxyl groups in chymotrypsinogen, *FEBS Lett.*, 4, 203, 1969.
4. **Carraway, K. L., Spoerl, P., and Koshland, D. E., Jr.,** Carboxyl group modification in chymotrypsin and chymotrypsinogen, *J. Mol. Biol.*, 42, 133, 1969.
5. **Lin, T.-Y. and Koshland, D. E., Jr.,** Carboxyl group modification and the activity of lysozyme, *J. Biol. Chem.*, 244, 505, 1969.
6. **Eyl, A. W., Jr. and Inagami, T.,** Identification of essential carboxyl groups in the specific binding site of bovine trypsin by chemical modification, *J. Biol. Chem.*, 246, 738, 1971.
7. **Frater, R.,** Reactivity of carboxyl groups in modified proteins, *FEBS Lett.*, 12, 186, 1971.
8. **Johnson, P. E., Stewart, J. A., and Allen, K. G. D.,** Specificity of α-chymotrypsin with exposed carboxyl groups blocked, *J. Biol. Chem.*, 251, 2353, 1976.
9. **Perfetti, R. B., Anderson, C. D., and Hall, P. L.,** The chemical modification of papain with 1-ethyl-3(3-dimethylaminopropyl) carbodiimide, *Biochemistry*, 15, 1735, 1976.
10. **Pho, D. B., Roustan, C., Tot, A. N. T., and Pradel, L.-A.,** Evidence for an essential glutamyl residue in yeast hexokinase, *Biochemistry*, 16, 4533, 1977.
11. **Ariki, M. and Fukui, T.,** Modification of rabbit muscle phosphorylase b by a water-soluble carbodiimide, *J. Biochem. (Tokyo)*, 83, 183, 1978.
12. **Paus, E.,** Reaction of α-mannosidase from *Phaseolus vulgaris* with group-specific reagents. Essential carboxyl groups, *Biochim. Biophys. Acta*, 526, 507, 1978.
13. **George, A. L., Jr. and Borders, C. L., Jr.,** Essential carboxyl residues in yeast enolase, *Biochem. Biophys. Res. Commun.*, 87, 59, 1979.
14. **Desvages, G., Roustan, C., Fattoum, A., and Pradel, L.-A.,** Structural studies on yeast 3-phosphoglycerate kinase. Identification by immunoaffinity chromatography of one glutamyl residue essential for 3-phosphoglycerate kinase activity. Its location in the primary structure, *Eur. J. Biochem.*, 105, 259, 1980.
15. **Matsuo, M., Huang, C.-H., and Huang, L. C.,** Modification and identification of glutamate residues at the arginine recognition site in the catalytic subunit of adenosine 3′,5′-cyclic monophosphate-dependent protein kinase of rabbit skeletal muscle, *Biochem. J.*, 187, 371, 1980.
16. **Vivanco-Martinez, F., Bragado, R., Albar, J. P., Juarez, C., and Ortiz-Masllorens, F.,** Chemical modification of carboxyl groups in human Fc_γ fragment: structural role and effect on the complement fixation, *Mol. Immunol.*, 17, 327, 1980.
17. **Fleer, E. A. M., Verheij, H. M., and de Haas, G. H.,** Modification of carboxylate groups in bovine pancreatic phospholipase A_2. Identification of aspartate-49 as Ca^{2+}-binding ligand, *Eur. J. Biochem.*, 113, 283, 1981.
18. **Burkey, K. O. and Gross, E. L.,** Effect of carboxyl group modification on redox properties and electron donation capability of spinach plastocyanin, *Biochemistry*, 20, 5495, 1981.
19. **Pennington, R. M. and Fisher, R. R.,** Dicyclohexylcarbodiimide modification of bovine heart mitochondrial transhydrogenase, *J. Biol. Chem.*, 256, 8963, 1981.
20. **Esch, F. S., Böhlen, P., Otsuka, A. S., Yoshida, M., and Allison, W. S.,** Inactivation of the bovine mitochondrial F_1-ATPase with dicyclohexyl [^{14}C] carbodiimide leads to the modification of a specific glutamic acid residue in the β-subunit, *J. Biol. Chem.*, 256, 9084, 1981.
21. **Yamada, H., Imoto, T., Fujita, K., Okazaki, K., and Motomura, M.,** Selective modification of aspartic acid-101 in lysozyme by carbodiimide reaction, *Biochemistry*, 20, 4836, 1981.
22. **Woodhead, J. L. and Malcolms, D. B.,** The essential carboxyl group in restriction endonuclease Eco RI, *Eur. J. Biochem.*, 120, 125, 1981.

REFERENCES

1. **Funane, K., Shiraiwa, M., Hashimoto, K., Ichishima, E., and Kobayashi, M.,** An active-site peptide containing the second essential carboxyl group of dextransucrase from *Leuconostoc mesenteroides* by chemical modifications, *Biochemistry,* 32, 13696, 1993.
2. **Turk, T. and Macek, P.,** The role of lysine, histidine and carboxyl residues in biological activity of equinatoxin II, a pore forming polypeptide from the sea anemone Actinia equina L, *Biochim. Biophys. Acta,* 1119, 5, 1992.
3. **Xia, C., Meyer, D. J., Chen, H., Reinemer, P., Huber, R., and Ketterer, B.,** Chemical modification of GSH transferase P1–1 confirms the presence of Arg-13, Lys-44 and one carboxylate group in the GSH-binding domain of the active site, *Biochem. J.,* 293, 357, 1993.
4. **Mattsson, P., Pohjalainen, T., and Korpela, T.,** Chemical modification of cyclomaltodextrin glucanotransferase from *Bacillus circulans* var. *alkalophilus, Biochim. Biophys. Acta,* 1122, 33, 1992.
5. **Parsons, S. M., Jao, L., Dahlquist, F. W., Borders, C. L., Jr., Groff, T., Racs, J., and Raftery, M. A.,** The nature of amino acid side chains which are critical for the activity of lysozyme, *Biochemistry,* 8, 700, 1969.
6. **Parsons, S. M. and Raftery, M. A.,** The identification of aspartic acid residue 52 as being critical to lysozyme activity, *Biochemistry,* 8, 4199, 1969.
7. **Paterson, A. K. and Knowles, J. R.,** The number of catalytically essential carboxyl groups in pepsin. Modification of the enzyme by trimethyloxonium fluoroborate, *Eur. J. Biochem.,* 31, 510, 1972.
8. **Woodward, R. B., Olofson, R. A., and Mayer, H.,** A new synthesis of peptides, *J. Am. Chem. Soc.,* 83, 1010, 1961.
9. **Woodward, R. B. and Olofson, R. A.,** The reaction of isoxazolium salts with nucleophiles, *Tetrahedron,* Suppl. 7, 415, 1966.
10. **Bodlaender, P., Feinstein, G., and Shaw, E.,** The use of isoxazolium salts for carboxyl group modification in proteins. Trypsin, *Biochemistry,* 8, 4941, 1969.
11. **Saini, M. S. and Van Etten, R. L.,** An essential carboxylic acid group in human prostate acid phosphatase, *Biochim. Biophys. Acta,* 568, 370, 1979.
12. **Arana, J. L. and Vallejos, R. H.,** Two different types of essential carboxyl groups in chloroplast coupling factor, *FEBS Lett.,* 123, 103, 1981.
13. **Dinur, D., Kantrowitz, E. R., and Hajdu, J.,** Reaction of Woodward's Reagent K with pancreatic porcine phospholipase A_2: modification of an essential carboxylate residue, *Biochem. Biophys. Res. Commun.,* 100, 785, 1981.
14. **Kooistra, C. and Sluyterman, L. A. A.,** Isosteric and non-isosteric modification of carboxyl groups of papain, *Biochim. Biophys. Acta,* 997, 115, 1989.
15. **Hoare, D. G. and Koshland, D. E., Jr.,** A procedure for the selective modification of carboxyl groups in proteins, *J. Am. Chem. Soc.,* 88, 2057, 1966.
16. **Perfetti, R. B., Anderson, C. D., and Hall, P. L.,** The chemical modification of papain with 1-ethyl-3(3-dimethylaminopropyl) carbodiimide, *Biochemistry,* 15, 1735, 1976.
17. **Carraway, K. L. and Koshland, D. E., Jr.,** Reaction of tyrosine residues in proteins with carbodiimide reagents, *Biochim. Biophys. Acta,* 160, 272, 1968.
18. **Carraway, K. L. and Koshland, D. E., Jr.,** Carbodiimide modification of proteins, *Meth. Enzymol.,* 25, 616, 1972.
19. **Lewis, S. D. and Shafer, J. A.,** Conversion of exposed aspartyl and glutamyl residues in proteins to asparaginyl and glutaminyl residues, *Biochim. Biophys. Acta,* 303, 284, 1973.
20. **Wang, T.-T. and Young, N. M.,** Modification of aspartic acid residues to induce trypsin cleavage, *Anal. Biochem.,* 91, 696, 1978.
21. **Lin, C., Mihal, K. A., and Krueger, R. J.,** Introduction of sulfydryl groups into proteins at carboxyl sites, *Biochim. Biophys. Acta,* 1038, 382, 1990.

22. **Takahashi, K., Stein, W. H., and Moore, S.,** The identification of a glutamic acid residue as part of the active site of ribonuclease T_1, *J. Biol. Chem.,* 242, 4682, 1967.
23. **Erlanger, B. F., Vratsanos, S. M., Wassermann, M., and Cooper, A. G.,** Specific and reversible inactivation of pepsin, *J. Biol. Chem.,* 240, PC3447, 1965.
24. **Gross, E. and Morell, J. L.,** Evidence for an active carboxyl group in pepsin, *J. Biol. Chem.,* 241, 3638, 1966.

Chapter 15

THE CHEMICAL CROSS-LINKING OF PEPTIDE CHAINS

The formation of either inter- or intramolecular covalent cross-links between specific amino acid residues in proteins is a valuable tool in protein chemistry. The major use has been in the study of protein-protein interaction.[1] As an example of this use, the recent work on the affinity labeling of the epidermal growth factor (EGF) receptor with EGF is cited.[2] A secondary use has been in increasing protein stability. The major purpose of this chapter is to introduce the basic concepts of this approach to the study of protein structure. The reader is directed to a recent review for a more comprehensive discussion of protein cross-linking.[3] Further information can be obtained from specialty chemical suppliers. The Pierce Chemical Company (Rockford, IL; see Appendix I) is a good source of reagents and information.

Zero-length cross-linking[4] is a procedure which joins peptide chains via existing functional groups such that a "spacer" group is not utilized. Examples include the covalent linkage of proteins to nucleic acids, the formation of 3,3′-dityrosine mediated by tetranitromethane (see Chapter 13), and isopeptide bond formation. Isopeptide bond formation (Figure 1) mediated by carbodiimide[5,6] (see Chapter 14) has been the most extensively used approach. This technique has been applied to the cross-linking of heavy meromyosin and F-actin[7,8] and components of the *Azotobacter vinelandii* nitrogenase complex[9]. A more recent modification of this technique[10] involves a two-step procedure where one protein is first incubated with a water-soluble carbodiimide and *N*-hydroxysuccinimide resulting in the formation of a *N*-oxysuccinimide ester. The second step involves condensation with lysine residues in the second component of the reaction.

An interesting pair of reagents (Figure 2) which cross-link proteins by consecutive Michael reactions have been described.[11] These investigators made the point that since the cross-linkage reaction is driven by consecutive Michael additions, eventually the most thermodynamically stable cross-link will be established, which can be subsequently stabilized by reduction of the nitro function with sodium dithionite. These investigators explored the reaction of pancreatic ribonuclease with 2-(*p*-nitrophenyl) allyl-4-nitro-3-caboxyphenyl sulfide (twofold molar excess with respect to ribonuclease in 0.1 *M* sodium phosphate, pH 10.5 at 37°C, 36 h, cross-link stabilized by sodium dithionite fivefold molar excess). No reaction occurred at pH 8.0. Analysis of the reaction mixture showed 61% monomer, 21% dimer, 10% trimer, and a trace of tetramer. The monomer fraction was characterized; the predominant cross-links occurred between lysine-7 and lysine-37 and between lysine-31 and lysine-41.

Care must be taken in the size analysis of intramolecularly linked species in any denaturing medium (i.e., sodium dodecylsulfate, guanidine hydrochloride, etc.) since, unless the cross-linking reagent is cleaved by reduction, the in-

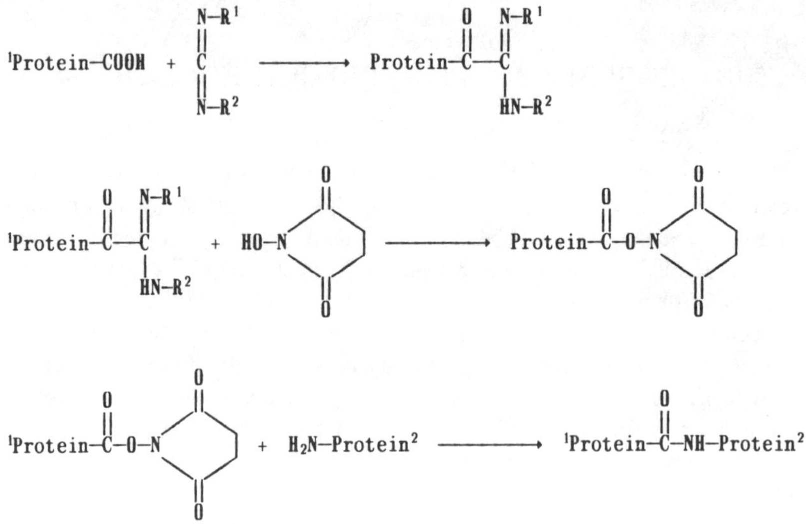

FIGURE 1. A scheme for zero-length cross-linking using a two-step process as described in Reference 10.

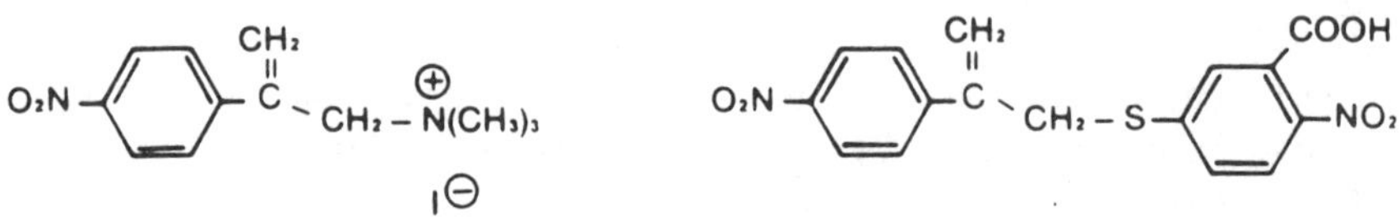

FIGURE 2. The structure of 2-(*p*-nitrophenyl)-allyltrimethylammonium iodide (on the left) and the structure of 2-(*p*-nitrophenyl)-allyl-4-nitro-3-carboxyphenyl sulfide (on the right).

tramolecularly cross-linked protein will not denature properly and will probably give falsely low molecular weight results.[12,13]

Intramolecular cross-linking can be enhanced by the following reaction conditions:

1. Low protein concentration (<0.1 mg/ml)
2. High net charge on protein
3. High ratio of protein reactive sites to reagent concentration

The remainder of our consideration will involve intermolecular cross-linking, which includes the cross-linking of identical promoters to form homopolymers (e.g., cross-linkage of identical subunits in an oligomeric protein).[14,15] Cross-linkage to form heteropolymers includes studies on protein-protein interactions (this could result in homopolymers in self-associating systems),[16,17] studies on multienzyme complexes,[18–20] and protein-ligand interactions with cell membrane receptors.[21–26]

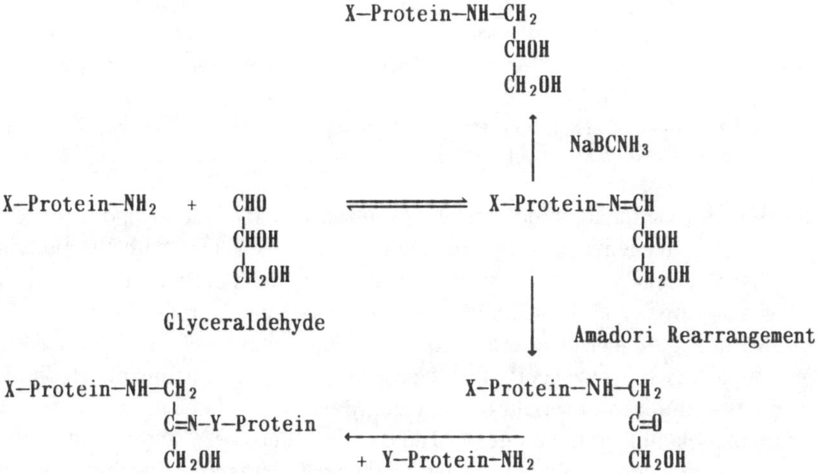

FIGURE 3. A scheme for the cross-linking of proteins using glyceraldehyde as described in Reference 35.

The following discussion will focus on the various reagents which have been used for intermolecular protein cross-linking studies.

Glutaraldehyde should cross-link proteins with the formation of α,ω-Schiff bases, which should be a readily reversible process in the absence of reduction of the Schiff base. This is not the case, as summarized by Richards and Knowles.[27] These investigators noted that the reaction of proteins with glutaraldehyde was essentially irreversible even without reduction and that, in the absence of reduction, there was a loss of lysine on amino acid analysis following acid hydrolysis. These investigators proposed the formation of a complex reagent resulting from aldol condensation of glutaraldehyde which would then react with the protein. It seems clear that the chemistry of the reaction of glutaraldehyde is complex. The use of glutaraldehyde has received considerable attention in the study of the properties of protein crystals in solution.[28–32] The rationale in these studies has been to show that the properties of a protein in crystalline form are similar to those for the protein in solution.[32] Although glutaraldehyde has a fair degree of specificity for the ε-amino group of lysine, reaction has also occurred with other nucleophilic functional groups in proteins such as the sulfhydryl group of cysteine, the imidazole ring of histidine, and the phenolic hydroxyl group of tyrosine.[33]

Manning and Manning have examined the related chemistry of protein cross-linking using glycoaldehyde.[34] Acharya and co-workers[35] have examined the cross-linking of proteins with glyceraldehyde in some detail. A scheme for the chemistry of this reaction is shown in Figure 3. These studies provide some insight into the cross-linking of proteins with other carbohydrate-like compounds.[36]

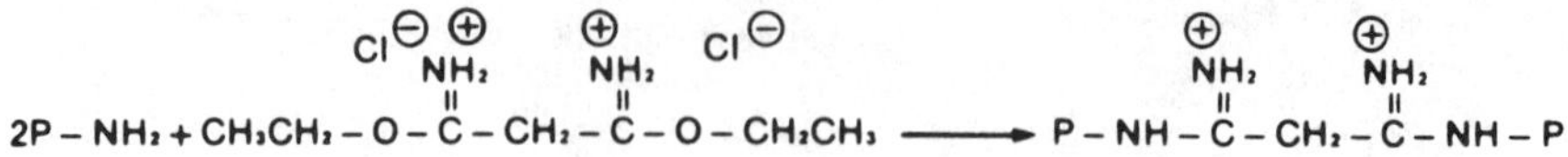

FIGURE 4. The reaction of a homobifunctional imidoester with the primary amino groups in a protein resulting in covalent cross-linking.

Homobifunctional imidoesters were introduced by Dutton and co-workers.[37] These reagents have the advantage that the reaction with the protein results in charge preservation of the lysine residue modified. This class of reagents is highly specific for primary amines in the following reactions as shown in Figure 4. Buffer effects on the reaction have not been extensively investigated, except to specify that the use of potential competing nucleophiles (e.g., Tris, imidazole) should be avoided. Most studies have used 0.02 to 0.1 *M* triethanolamine in the range of pH 8.0 to 9.0. It has been suggested that the amidation reaction is enhanced by the presence of triethanolamine in studies on the reaction of methyl-4-mercaptobutyrimidate.[24]

It is essential to know the effect of reaction with monofunctional reagents on biological activity before considering the effect of cross-linking on biological activity.[38] Sinha and Brew[39] developed a useful procedure employing the prior trace labeling of the protein with acetic anhydride. Since reaction with acetic anhydride and imidoester is mutually exclusive, fragmentation and subsequent determination of specific radioactivity at specific lysine residues allows the identification of the site(s) of reaction with bifunctional imidoesters.

bis-(Sulfosuccinimidyl)suberate has been used to assess the dimerization of EGF receptors in Triton X-100 extracts of A431 cells. The introduction of a cross-linking reagent that could be subsequently cleaved, 3,3′-dithiobis-(succinimidylpropionate),[40] has proved to be of considerable use in the study of proteins.[41–49] This reagent provides for the reversible formation of cross-links according to Figure 5.

The use of bifunctional maleimides is of considerable value in cross-linking between sulfhydryl groups. A particularly useful study in this area is the work of Heilmann and Holzner[50] in which the synthesis and use of a number of bifunctional maleimides for the elucidation of the structure of tryptophan synthetase is reported. One of the more useful derivatives is bis-*N*-maleimido-1,8-octane (Figure 6). A cleavable bifunctional dimaleimide has been recently reported.[51] The synthesis of the reagent, maleimidomethyl-3-maleimido propionate (Figure 7), is reported as is the use of the reagent to probe spatial relationships in the erythrocyte membrane. *p*-Phenylene dimaleimide (Figure 8) has been used to cross-link subunits of transducin and cGMP phosphodiesterase complexes in bovine rod photoreceptors.[52]

While homobifunctional reagents have been quite useful, the majority of work in this area is utilizing heterobifunctional reagents. Of particular interest has been the use of photoactivatable derivatives. An example of this type of

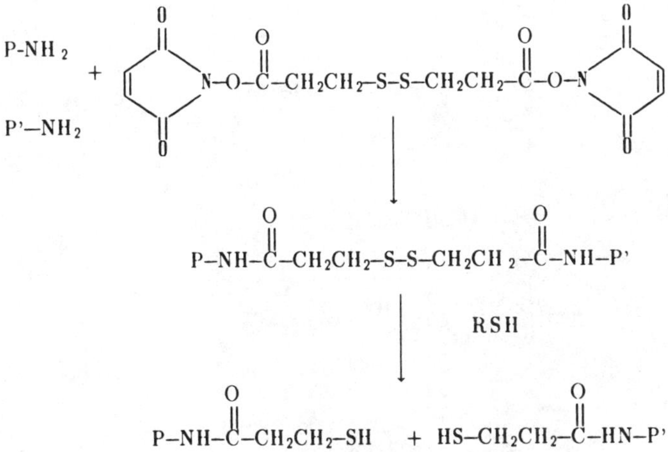

FIGURE 5. The structure of 3,3′-dithiobis-(succinimidylpropionate) and the formation of reversible cross-links in proteins via coupling with primary amino groups.

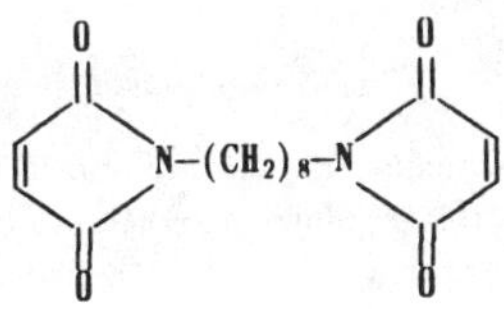

FIGURE 6. The structure of bis-*N*-maleimido-1,8-octane.

derivative is methyl-3-(*p*-azidophenyl)dithiopropioimidate[21] (Figure 9). In these experiments, epidermal growth factor was reacted in the dark at pH 8.5 (0.1 *M* triethanolamine, 0.2 *M* NaCl, pH 8.5) for reaction of the imido ester formation with lysine. This reaction is terminated by the addition of ammonium acetate. Photoactivation in the presence of mouse 3T3 cells resulted in the specific labeling of a cell surface protein. It is noted that this reagent can be cleaved by reduction, permitting isolation of such a cell surface protein free of ligand.

A series of homo- and heterobifunctional reagents which form acid-labile cross-links in proteins have been reported.[53] These are shown in Figure 10. The ability to be cleaved by mild acid (pH 5.0) is based on the presence of *ortho* ester, acetal, and ketal functional groups.

The synthesis of 4-(6-formyl-3-azidophenoxy)butyrimidate (Figure 11) has been reported.[54] The imido function can react with the amino groups on a protein (0.1 *M* sodium borate, pH 8.6). Cross-linking can occur either with nitrene formation from the azido functional group upon irradiation or via reductive alkylation utilizing the aldehyde formation. In either instance, radio-

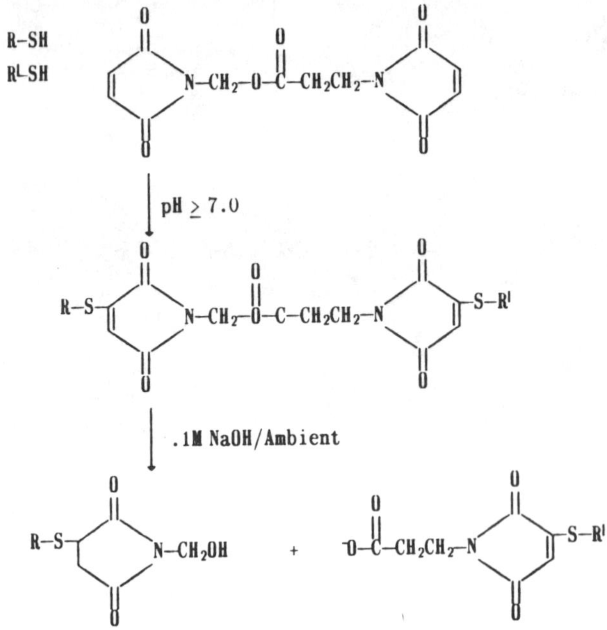

FIGURE 7. The structure of maleimidomethyl-3-maleimido propionate, the reaction of this reagent with protein sulfhydryl groups, and the subsequent cleavage of this cross-link with base.

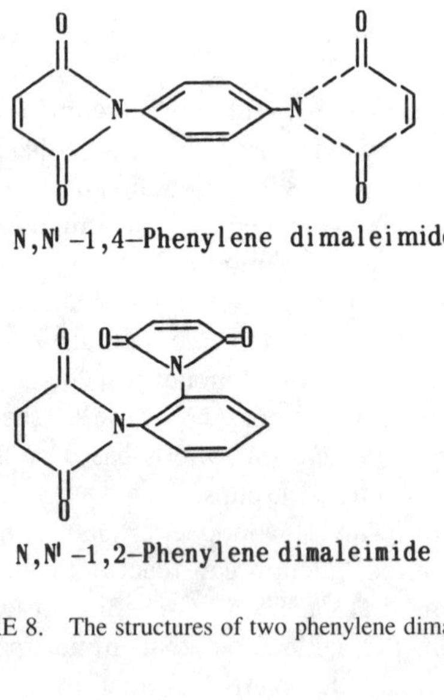

N,N' –1,4–Phenylene dimaleimide

N,N' –1,2–Phenylene dimaleimide

FIGURE 8. The structures of two phenylene dimaleimides.

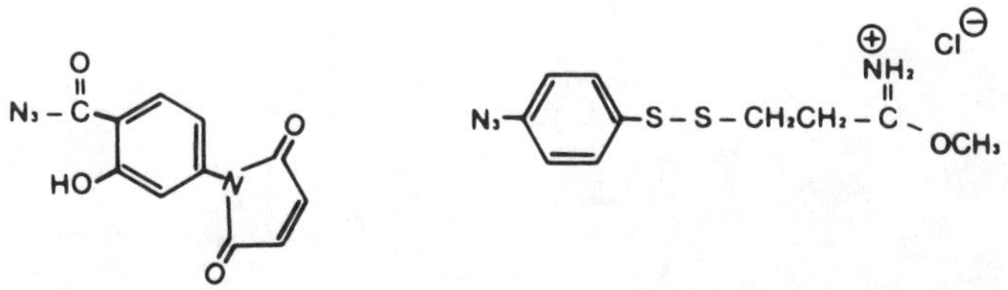

FIGURE 9. The structure of *N*-(4-azidocarbonyl-3-hydroxyphenyl)maleimide (on the left) and methyl-3-(*p*-azidophenyl)dithiopropioimidate (on the right).

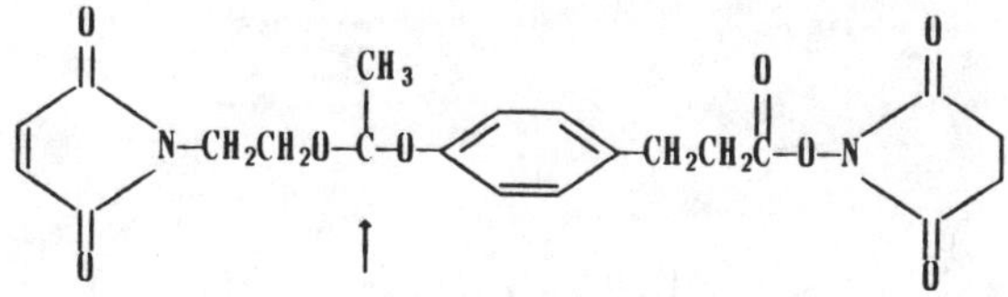

I

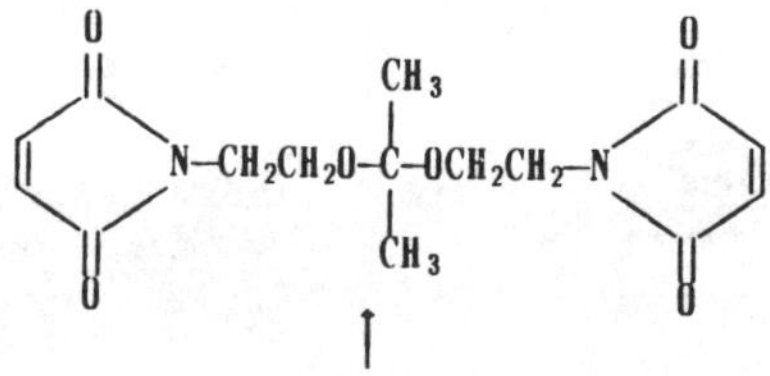

II

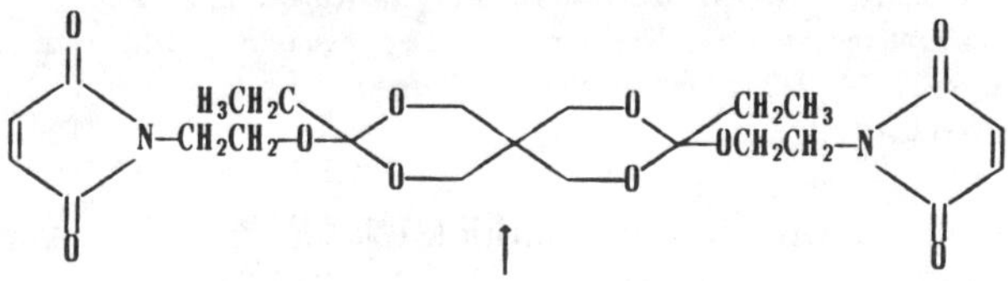

III

FIGURE 10. The structures of some acid-labile cross-linking reagents. Shown are I, a heterobifunctional acetal cross-linking agent; II, a maleimido ketal cross-linking agent; and III, an *ortho* ester cross-linking agent.

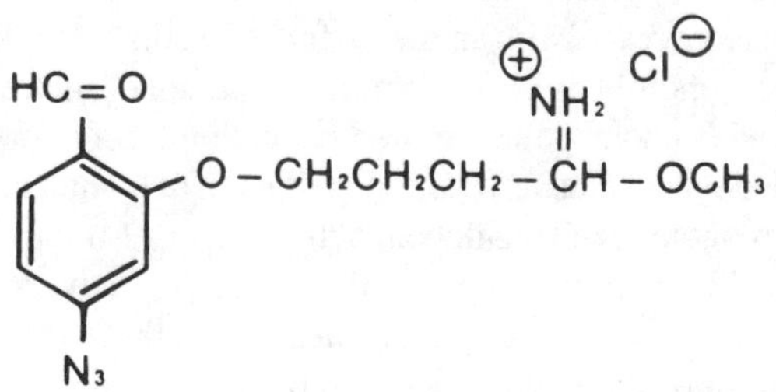

FIGURE 11. The structure of 4-(6-formyl-3-azidophenoxy)butyrimidate.

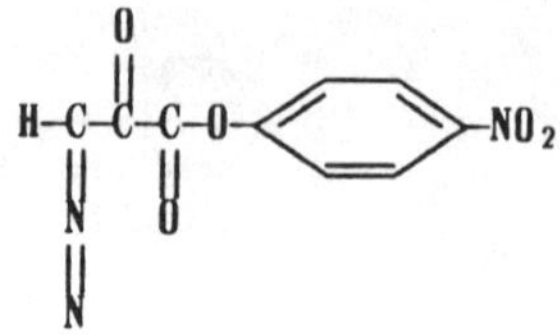

FIGURE 12. The structure of *p*-nitrophenyl-3-diazopyruvate.

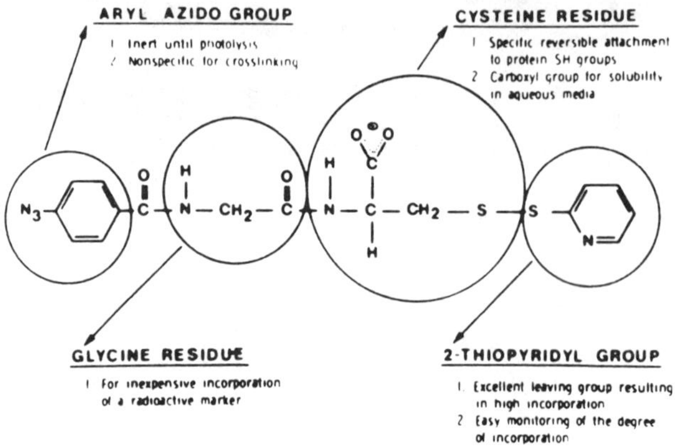

FIGURE 13. The design of a heterobifunctional cross-linking reagent for the study of biological interactions between proteins. Shown is the rationale for the design of *N*-(4-azidobenzoylglycyl)-*S*-(2-thiopyridyl)-cysteine (ACTC). (From Chong, P. C. S. and Hodges, R. S., *J. Biol. Chem.*, 256, 5064, 1981. With permission.)

label can be introduced with sodium borotritiide ($NaB[^3T]_4$) with conversion of the free aldehyde to an alcohol or by reducing the Schiff base formed during reductive alkylation. Sulfosuccinimidyl-2-(*p*-azidosalicylamido)-1,3-dithiopropionate has been used to identify the receptor for phytohemaglutinin in mononuclear cells.[55]

A related approach involves the use of *p*-nitrophenyl-3-diazopyruvate (Figure 12).[56] This reagent reacts with amines to form the corresponding pyruvamide derivatives. Upon photolysis at 300 nm, a ketene amide is formed, which is highly reactive with nucleophiles to produce malonic acid derivatives.

Chong and Hodges[57,58] have reported studies of a complex hetero-bifunctional affinity reagent, *N*-(4-azidobenzoylglycyl)-5-(2-thiopyridyl)-cysteine (AGTC) (Figure 13). Figure 14 describes strategy in the use of this reagent.

One of the uses of cross-linking reagents such as those described above is the characterization of the interaction(s) between biologically active peptides and proteins and cell surface receptors. An excellent example of this approach is provided by the work of Ji and co-workers.[22,59] Some reagents developed for this purpose[60] are shown in Figure 15.

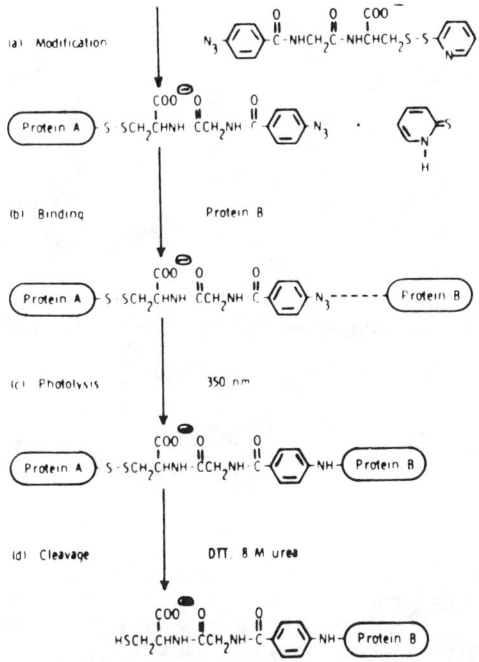

FIGURE 14. The use of a heterobifunctional cross-linking reagent for the study of the biological interactions between proteins. Shown is a scheme for the radiolabeling of a binding site on one protein which is in the vicinity of an SH-group on another protein. The SH-groups are modified with the heterobifunctional photoaffinity probe *N*-(4-azidobenzoyl[2-^{3}H]glycyl)-*S*-(2-thiopyridyl)-cysteine. Part a, the introduction of the arylazide structure into a thioprotein A by thiol-disulfide interchange; b, noncovalent binding of modified protein A to protein B; c, covalent cross-linking of the proteins by photolysis of the arylazide moiety; and d, the cleavage of the disulfide bridge linking the two proteins by dithiothreitol, thus completing the transfer of radiolabel from protein A to protein B. (From Chong, P. C. S. and Hodges, R. S., *J. Biol. Chem.*, 256, 5071, 1981. With permission.)

bis-Pyridoxal polyphosphates (Figure 16) have been described by Benesch and Kwong[61] as a class of specific intramolecular cross-linking agents for hemoglobin. While specifically designed for hemoglobin, this reagent would likely be useful for other site-specific cross-link reagents. In another study on hemoglobin, Hosmane and Bertha[62] describe the use of 2,2′-sulfonylbis[3-methoxy-(E,E)-2-propenitrile] (Figure 17) as a cross-linking reagent. Lee et al.[63] described an interesting heterobifunctional reagent (Figure 18) designed for coupling glycopeptides to proteins.

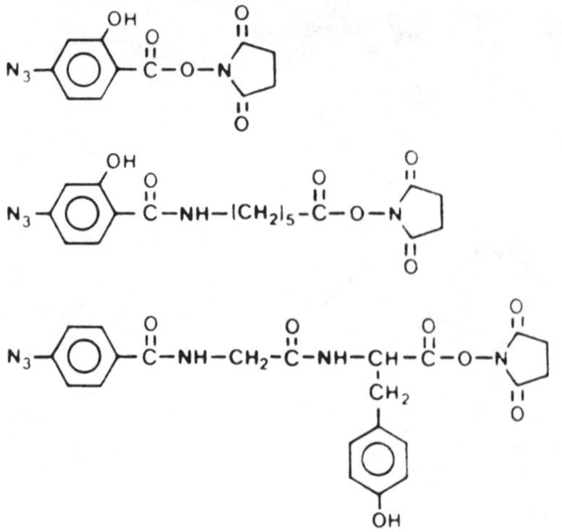

FIGURE 15. Structures of some cross-linking reagents which have proved useful in the study of the interaction of peptide hormones with cell surface receptors. Shown is NHS-ASA, the *N*-hydroxysuccinimide ester of 4-azido-salicylic acid; NHS-ABGT, the *N*-hydroxysuccinimide ester of 4-azidobenzoylglycyltyrosine; and HNS-ASC, the *N*-hydroxysuccinimide ester of *N*-(4-azidosalicyl)-6-aminocaproic acid. (From Ji, T. H. and Ji, I., *Anal. Biochem.*, 121, 286, 1982. With permission.)

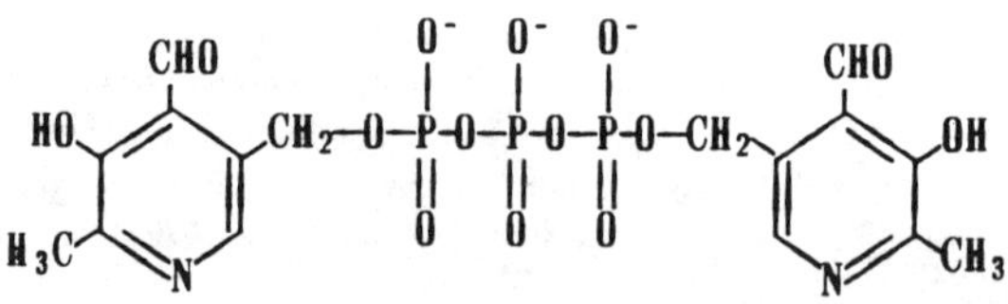

FIGURE 16. The structure of a bis-pyridoxal polyphosphate.

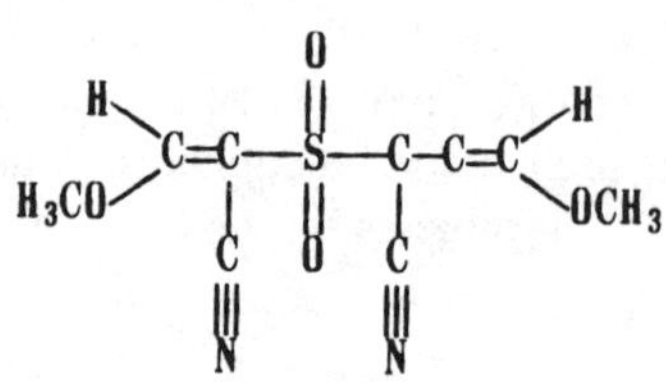

FIGURE 17. The structure of 2,2′-sulfonylbis[3-methoxy-(E,E)-2-propenenitrile].

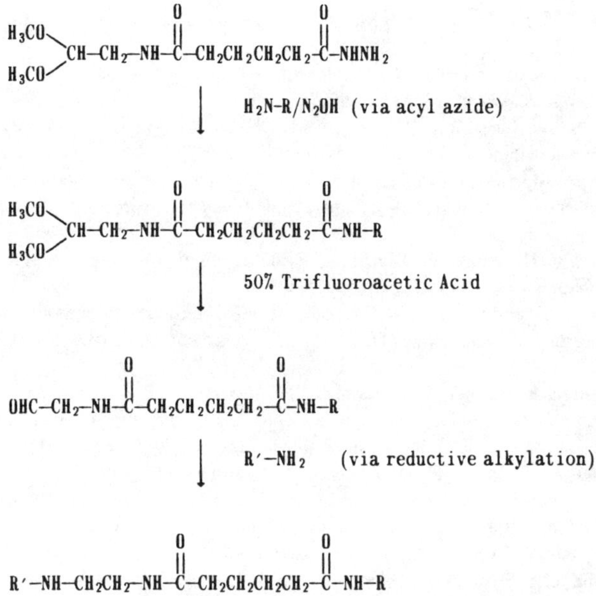

FIGURE 18. A scheme for cross-linking a peptide to a protein.

REFERENCES

1. **Brinkley, M.,** A brief survey of methods for preparing protein conjugates with dyes, haptens, and cross-linking reagents, *Bioconjugate Chem.,* 3, 2, 1992.
2. **Woltjer, R. L., Weclas-Henderson, L., Papyannopoulos, J. A., and Staros, J. V.,** High-yield covalent attachment of epidermal growth factor to its receptor by kinetically controlled, stepwise affinity cross-linking, *Biochemistry,* 31, 7341, 1992.
3. **Wong, S. S.,** *Chemistry of Protein Conjugation and Cross-Linking,* CRC Press, Boca Raton, FL, 1991.
4. **Kunkel, G. R., Mehrabian, M., and Martinson, H. G.,** Contact-site cross-linking agents, *Mol. Cell. Biochem.,* 34, 3, 1981.
5. **Sheehan, J. C. and Hlavka, J. J.,** The use of water-soluble and basic carbodiimides in peptide synthesis, *J. Org. Chem.,* 21, 439, 1956.
6. **Sheehan, J. C. and Hlavka, J. J.,** The cross-linking of gelatin using a water-soluble carbodiimide, *J. Am. Chem. Soc.,* 79, 4528, 1957.
7. **Onishi, H., Maita, T., Matsuda, G., and Fujiwara, K.,** Evidence for the association between two myosin heads in rigor acto-smooth muscle heavy meromyosin, *Biochemistry,* 28, 1898, 1989.
8. **Onishi, H., Maita, T., Matsuda, G., and Fujiwara, K.,** Carbodiimide-catalyzed cross-linking sites in the heads of gizzard heavy meromyosin attached to F-actin, *Biochemistry,* 28, 1905, 1989.

9. **Willing, A. and Howard, J. B.,** Cross-linking site in *Azotobacter vinelandii* complex, *J. Biol. Chem.,* 265, 6596, 1990.
10. **Grabarek, Z. and Gergely, J.,** Zero-length cross-linking procedure with the use of active esters, *Anal. Biochem.,* 185, 131, 1990.
11. **Mitra, S. and Lawton, R. G.,** Reagents for the cross-linking of proteins by equilibrium transfer alkylation, *J. Am. Chem. Soc.,* 101, 3097, 1979.
12. **Ruoho, A., Bartlett, P. A., Dutton, A., and Singer, S. J.,** A disulfide-bridge bifunctional imidoester as a reversible cross-linking reagent, *Biochem. Biophys. Res. Commun.,* 63, 417, 1975.
13. **Steele, J. C. H., Jr. and Nielson, T. B.,** Evidence of cross-linked polypeptides in SDS gel electrophoresis, *Anal. Biochem.,* 84, 218, 1978.
14. **Davies, G. E. and Stark, G. R.,** Use of dimethyl suberimidate, a cross-linking reagent, in studying the subunit structure of oligomeric proteins, *Proc. Natl. Acad. Sci. U.S.A.,* 66, 651, 1970.
15. **Carpenter, F. H. and Harrington, K. T.,** Intermolecular cross-linking of monomeric proteins and cross-linking of oligomeric proteins as a probe of quaternary structure. Application to leucine aminopeptidase (bovine lens), *J. Biol. Chem.,* 247, 5580, 1972.
16. **Tarvers, R. C., Noyes, C. M., Roberts, H. R., and Lundblad, R. L.,** Influence of metal ions on prothrombin self-association. Demonstration of dimer formation by intermolecular cross-linking with dithiobis(succinimidyl propionate), *J. Biol. Chem.,* 257, 10708, 1982.
17. **Lewis, R. V., Roberts, M. F., Dennis, E. A., and Allison, W. S.,** Photoactivated heterobifunctional cross-linking reagents which demonstrate the aggregation state of phospholipase A_2, *Biochemistry,* 16, 5650, 1977.
18. **DeAbreu, R. A., DeVries, J., DeKok, A., and Veeger, C.,** Cross-linking studies with the pyruvate dehydrogenase complexes from *Azotobacter vinelandii* and *Escherichia coli, Eur. J. Biochem.,* 97, 379, 1979.
19. **Monneron, A. and d'Alayer, J.,** Effects of cross-linking agents on adenylate cyclase regulation, *FEBS Lett.,* 109, 75, 1980.
20. **Baskin, L. S. and Yang, C. S.,** Cross-linking studies of cytochrome P-450 and reduced nicotinamide adenine dinucleotide phosphate-cytochrome P-450 reductase, *Biochemistry,* 19, 2260, 1980.
21. **Das, M., Miyakawa, T., Fox, C. F., Pruss, R. M., Aharonov, A., and Herschman, H. R.,** Specific radiolabeling of a cell surface receptor for epidermal growth factor, *Proc. Natl. Acad. Sci. U.S.A.,* 74, 2790, 1977.
22. **Ji, T. H.,** A novel approach to the identification of surface receptors. The use of photosensitive heterobifunctional cross-linking reagent, *J. Biol. Chem.,* 252, 1566, 1977.
23. **Pilch, P. R. and Czech, M. P.,** Interaction of cross-linking agents with the insulin effector system of isolated fat cells. Covalent linkage of ^{125}I-insulin to a plasma membrane receptor protein of 140,000 daltons, *J. Biol. Chem.,* 254, 3375, 1979.
24. **Birnbaumer, M. F., Schrader, W. T., and O'Malley, B. W.,** Chemical cross-linking of chick oviduct progesterone-receptor subunits by using a reversible bifunctional cross-linking agent, *Biochem. J.,* 181, 201, 1979.
25. **Kasuga, M., Van Obberghen, E., Nissley, S. P., and Rechler, M. M.,** Demonstration of two subtypes of insulin-like growth factor receptors by affinity cross-linking, *J. Biol. Chem.,* 256, 5305, 1981.
26. **Rebois, R. V., Omedeo-Sale, F., Brady, R. O., and Fishman, P. H.,** Covalent cross-linking of human chorionic gonadotropin to its receptor in rat testes, *Proc. Natl. Acad. Sci. U.S.A.,* 78, 2086, 1981.
27. **Richards, F. M. and Knowles, J. R.,** Glutaraldehyde as a protein cross-linking reagent, *J. Mol. Biol.,* 37, 231, 1968.
28. **Quiocho, F. A. and Richards, F. M.,** Intermolecular cross-linking of a protein in the crystalline state: carboxypeptidase A, *Proc. Natl. Acad. Sci. U.S.A.,* 52, 833, 1964.

29. **Wong, C., Lee, T. J., Lee, T. Y., Lu, T. H., and Hung, C. S.,** Intermolecular cross-linking of a protein crystal — acid protease from *Endothia parasitica* — in 2.7 *M* ammonium sulfate solution, *Biochem. Biophys. Res. Commun.,* 80, 886, 1978.
30. **Spillburg, C. A., Bethune, J. L., and Vallee, B. L.,** Kinetic properties of crystalline enzymes. Carboxypeptidase A, *Biochemistry,* 16, 1142, 1977.
31. **Wong, C., Lee, T. J., Lee, T. Y., Lu, T. H., and Hung, C. S.,** The structure of acid protease from *Endothia parasitica* in cross-linked form at 3.5 Å resolution, *Biochem. Biophys. Res. Commun.,* 80, 891, 1978.
32. **Tüchsen, E., Hvidt, A., and Ottesen, M.,** Enzymes immobilized as crystals. Hydrogen isotope exchange of crystalline lysozyme, *Biochimie,* 62, 563, 1980.
33. **Habeeb, A. F. S. A. and Hiramoto, R.,** Reaction of proteins with glutaraldehyde, *Arch. Biochem. Biophys.,* 126, 16, 1968.
34. **Manning, L. R. and Manning, J. M.,** Influence of ligation state and concentration of hemoglobin A on its cross-linking by glycolaldehyde: functional properties of cross-linked, carboxymethylated hemoglobin, *Biochemistry,* 27, 6640, 1988.
35. **Acharya, A. S., Cho, Y. J., and Manjula, B. N.,** Cross-linking of proteins by aldotriose: reaction of the carbonyl function of the keto amines generated in situ with amino groups, *Biochemistry,* 27, 4522, 1988.
36. **Harding, J. J.,** Nonenzymatic covalent posttranslational modification of proteins *in vivo, Adv. Enzymol.,* 37, 247, 1985.
37. **Dutton, A., Adams, M., and Singer, S. J.,** Bifunctional imidoesters as cross-linking reagents, *Biochem. Biophys. Res. Commun.,* 23, 730, 1966.
38. **Monneron, A. and d'Alayer, J.,** Effects of imido-esters on membrane-bound adenylate cyclase, *FEBS Lett.,* 122, 241, 1980.
39. **Sinha, S. K. and Brew, K.,** A label selection procedure for determining the location of protein-protein interaction sites by cross-linking with bisimidoesters. Application to lactose synthase, *J. Biol. Chem.,* 256, 4193, 1981.
40. **Lomant, A. J. and Fairbanks, G.,** Chemical probes of extended biological structures: synthesis and properties of the cleavable protein cross-linking reagent [^{35}S] dithiobis(succinimidyl propionate), *J. Mol. Biol.,* 104, 243, 1976.
41. **Tarvers, R. C., Noyes, C. M., Roberts, H. R., and Lundblad, R. L.,** Influence of metal ions on prothrombin self-association. Demonstration of dimer formation by intermolecular cross-linking with dithiobis(succinimidylpropionate), *J. Biol. Chem.,* 257, 10708, 1982.
42. **Tarvers, R. C., Roberts, H. R., and Lundblad, R. L.,** Self-association of bovine prothrombin fragment 1 in the presence of metal ions. Use of a covalent cross-linking agent to study the reaction, *J. Biol. Chem.,* 259, 1944, 1984.
43. **Tarvers, R. C., Noyes, C. M., Tarvers, J. K., and Lundblad, R. L.,** Mechanism of the calcium dependent self-association of bovine prothrombin. Use of a covalent cross-linking reagent to study the reaction, *J. Biol. Chem.,* 261, 4855, 1986.
44. **Tarvers, R. C., Roberts, H. R., Straight, D. L., Featherstone, G. L., and Lundblad, R. L.,** Homo- and heterodimer formation with prothrombin and prothrombin fragment 1 in the presence of calcium ions, *Arch. Biochem. Biophys.,* 257, 439, 1987.
45. **Hantula, J. and Bamford, D. H.,** Chemical cross-linking of bacteriophage 6 nucleocapsid proteins, *Virology,* 165, 482, 1988.
46. **Cornell, R.,** Chemical cross-linking reveals a dimeric structure for CTP: phosphocholine cytidylyltransferase, *J. Biol. Chem.,* 264, 9077, 1989.
47. **Wiland, E., Siemieniako, B., and Trzeciak, W. H.,** Binding of low mobility group protein from rat liver chromatin with histones studied by chemical cross-linking, *Biochem. Biophys. Res. Commun.,* 166, 11, 1990.
48. **D'Souza, S. E., Ginsberg, M. H., Lam, S. C.-T., and Plow, E. F.,** Chemical cross-linking of the arginyl-glycyl-aspartic acid peptides to an adhesion receptor on platelets, *J. Biol. Chem.,* 263, 3943, 1988.

49. **Rexin, M., Busch, W., and Gehring, U.,** Chemical cross-linking of heteromeric glucocorticoid receptors, *Biochemistry,* 27, 5593, 1988.
50. **Heilmann, H. D. and Holzner, M.,** The spatial organization of the active sites of the bifunctional oligomeric enzyme tryptophan synthetase: cross-linking by a novel method, *Biochem. Biophys. Res. Commun.,* 99, 1146, 1981.
51. **Sato, S. and Nakao, M.,** Cross-linking of intact erythrocyte membrane with a newly synthesized cleavable bifunctional reagent, *J. Biochem.,* 90, 1177, 1981.
52. **Fanger, B. O., Stephens, J. E., and Staros, J. V.,** High-yield trapping of EGF-induced receptor dimers by chemical cross-linking, *FASEB J.,* 3, 71, 1989.
53. **Srinivasachar, K. and Neville, D. M., Jr.,** New protein cross-linking reagents that are cleaved by mild acid, *Biochemistry,* 28, 2501, 1989.
54. **Maassen, J. A.,** Cross-linking of ribosomal proteins by 4-(6-formyl-3-azidophenoxy) butyrimidate. A heterobifunctional cleavable cross-linker, *Biochemistry,* 18, 1288, 1979.
55. **Shephard, E. G., de Beer, F. C., von Holt, C., and Hapgood, J. P.,** The use of *S*-sulfosuccinimidyl-2-(*p*-azidosalicylamido)-1,3′-dithiopropionate as a cross-linking reagent to identify cell surface receptors, *Anal. Biochem.,* 168, 306, 1988.
56. **Goodfellow, V. S., Settineri, M., and Lawton, R. G.,** *p*-Nitrophenyl 3-diazopyruvate and diazopyruvamides, a new family of photoactivatable cross-linking bioprobes, *Biochemistry,* 28, 6346, 1989.
57. **Chong, P. C. S. and Hodges, R. S.,** A new heterobifunctional cross-linking reagent for the study of biological interactions between proteins. I. Design, synthesis and characterization, *J. Biol. Chem.,* 256, 5064, 1981.
58. **Chong, P. C. S. and Hodges, R. S.,** A new heterobifunctional cross-linking reagent for the study of biological interactions between proteins. II. Application to the troponin C-troponin I interaction, *J. Biol. Chem.,* 256, 5071, 1981.
59. **Ji, I., Yoo, B. Y., Kaltenbach, C., and Ji, T. H.,** Structure of the lutropin receptor on granulosa cells. Photoaffinity labeling with the α-subunit in human choriogonadotropin, *J. Biol. Chem.,* 256, 10853, 1981.
60. **Ji, T. H. and Ji, I.,** Macromolecular photoaffinity labeling with radioactive photoactiveable heterobifunctional reagents, *Anal. Biochem.,* 121, 286, 1982.
61. **Benesch, R. E. and Kwong, S.,** Bis-pyridoxal polyphosphates: a new class of specific intramolecular crosslinking agents for hemoglobin, *Biochem. Biophys. Res. Commun.,* 156, 9, 1988.
62. **Hosmane, R. S. and Bertha, C. M.,** A versatile, highly reactive, cross-linking reagent: 2,2′-sulfonylbis[3-methoxy-(E,E)-2-propenenitrile], *Biochem. Biophys. Res. Commun.,* 166, 567, 1990.
63. **Lee, R. T., Wong, T.-C., Lee, R., Yue, L., and Lee, Y. C.,** Efficient coupling of glycopeptides to proteins with a heterobifunctional reagent, *Biochemistry,* 28, 1856, 1989.

Appendix I

SOME COMMONLY USED REAGENTS FOR SITE-SPECIFIC MODIFICATION IN PROTEINS[a]

Compound[b]	Amino acid modified[c]	M.W.[d]	Source[e]
Acetaldehyde	Lysine, α-amino groups	44.05	SP, F, A, S
Acetic acid	Solvent	60.05	SP, F, A, S, P
Acetic anhydride	Lysine, tyrosine	102.1	SP, F, A, S
Acetone (2-propanone; dimethylketone)	Solvent	58.1	SP, F, A, S
Acetonitrile	Solvent	41.1	SP, F, A, S, P
N-acetyl-imidazole (1-acetylimidazole)	Tyrosine, lysine	110.1	A, S
Arsanlic acid (4-aminophenyl arsonic acid)	Tyrosine	217.0	A, S
Benzene	Organic solvent	78.1	SP, F, A, P, S
Bolton-Hunter Reagent [*N*-succinimidyl-3-(4-hydroxyphenyl) propionate; 3-*p*-(hydroxyphenyl)-propionic acid, *N*-hydroxysuccinide ester]	Iodination reagent	263.6	A, S, P
BNPS-Skatole [2-(2-Nitrophenylsulfenyl)-3-methyl-3-bromoindolenine]	Tryptophan, tyrosine	363.2	P
Bromoacetamide	Cysteine, histidine, lysine	138	A
Bromoacetic acid	Cysteine, histidine	139	SP, A, S
N-Bromosuccimide	Tryptophan, cysteine methionine	178	SP, F, A, S
2,3-Butanedione	Arginine, also can partipate in photoinactivation	86.1	A, S
Chloramine T (*N*-Chloro-4-methyl benzenesulfonamide, sodium salt)	Oxidizing agent, catalyzes iodination of proteins	227.7	SP, F, A, P, S
Chloroacetamide	Cysteine	93.5	A, S
Chloroacetic acid	Cysteine	94.5	SP, F, A, S
N-Chlorosuccinimide	Tryptophan, peptide bond cleavage	133.5	A, S
Citraconic anhydride (methylmaleic anhydride)	Lysine, α-amino groups	112.1	A, S, P
Cyanogen bromide	Methionine, peptide bond cleavage	105.9	SP, A, S, P
1,2-Cyclohexanedione	Arginine	112.1	A, S
1-Cyclohexyl-3-(2-morpholinyl-4-ethyl) carbodiimide, *p*-toluene-sulfonate (CMC metho-*p*-toluenesulfonate, morpho CDI)	Carboxyl groups	423.8	A, S

Compound[b]	Amino acid modified[c]	M.W.[d]	Source[e]
Cystamine (2-mercaptoethylamine; 2-aminoethanethiol)	Coupled to carboxyl groups to yield a sulfydryl function; also used as antioxidant	152.3	CA, A, S
Cysteine	Reducing agent	121.1	SP, F, CA, A, S, P
Dabsyl chloride [4-(dimethylamino)-azobenzene-4′-sulfonyl chloride]	Lysine, α-amino groups	323.8	A, S, P
Dansyl chloride [5-dimethylamino-1-naphthalene-sulfonyl chloride]	Lysine, α–amino groups	269.7	A, S, P
N,N′-Dicyclohexyl carbodiimide	Carboxyl groups	206.3	SP, A, P
Diethyl pyrocarbonate (diethyl oxydiformate; ethoxyformic anhydride)	Histidine, tyrosine, lysine	162.1	SP, CA, A, S
Dimethyl formamide	Solvent	73.1	SP, F, CA, A, S, P
Dimethyl sulfoxide	Organic solvent	78.1	SP, F, CA, S, P
5,5′-Dithiobis-(2-nitrobenzoic acid) Ellman's Reagent; DTNB)	Sulfydryl groups	396.4	CA, A, S, P
p-Dioxane (1,4-dioxane)	Solvent	88.1	SP, F, S
2,2′-Dithiopyridine also includes 2,2′-dithio-bis-(5-nitropyridine)	Sulfydryl groups	293.2	A, S
1,4-Dithiothreitol (Cleland's Reagent)	Reducing agents	154	SP, CA, A, S, P
Ethanolamine (2-aminoethanol; monoethanolamine)	Buffer	61.1	SP, F, A, S
1-Ethyl-3-(3-dimethyl aminoprophyl) carbodiimide hydrochloride	Carboxyl groups	191.8	CA, A, S, P
Ethyl acetamide	Lysine, α-amino groups	87.2	A, S
Ethyl acetate	Organic solvent	88.1	SP, F, A, S, P
Ethyl alcohol (ethanol)	Organic solvent	46.1	SP, F, A, S, P
N-Ethylmaleimide	Sulfydryl groups	125.1	SP, F, CA, A, S, P
1-Fluoro-2,4-dinitrobenzene (2,4-dinitrofluorobenzene; Sanger's Reagent)	Lysine, α-amino groups	186.1	SP, A, S
Glutaraldehyde (1,5-pentanedial)	Lysine	100.1	SP, F, A, S
Glyceraldehyde (2,3-dihydroxy propanal)	Lysine, α-amino groups	90.1	A, S

Compound[b]	Amino acid modified[c]	M.W.[d]	Source[e]
Guanidine hydrochloride	Denaturing agent	95.5	SP, F, CA, A, S, P
Guanidine thiocyanate	Denaturing agent	181.2	SP, F, CA, A, S
Hydrogen chloride (muriatic acid)	Mineral acid	36.51	SP, F, A, S, P
Hydrogen peroxide	Oxidizing agent, cysteine, methionine	34.0	SP, F, CA, A, S
Hydroxylamine	Peptide bond cleavage, also hydrolysis of *O*-acetyl tyrosine	33.1	SP, F, A, S
2-Hydroxy-5-nitrobenzyl bromide (Koshland's Reagent)	Tryptophan, cysteine	232.1	A, S
N-Hydroxysuccinimide	Lysine, α-amino groups	115.1	A, S, P
Imidazole (1,3-diaza-2,4-cyclopentanediene)	Buffer	68.1	SP, F, CA, A,S
Iodine(I_2)	Tyrosine	253.8	SP, F, A, S, P
Iodoacetamide	Cysteine, histidine	185	CA, A, S
Iodoacetic acid	Cysteine, histidine	186	SP, CA, A, S, P
Maleic anhydride	Lysine	98.1	SP, A, S
2-Mercaptoethanol (β-mercaptoethanol)	Reducing agent	78.1	SP, F, A, S, P
Methanol (methyl alcohol)	Organic solvent	32	SP, F, A, S, P
O-Methylisourea sulfate	Lysine	172.2	A, S
Methylene blue (3,7-bis(dimethyl-amino) phenothiazine-5-onium chloride	Photooxidizing reagent	319.9	SP, F, A, S
Ninhydrin (1, 2, 3-triketohydrindene monohydrate)	Arginine, also used as reagent for amino acid detection	178.1	SP, F, A, S, P
2-Nitrophenyl-sulfenyl chloride (2-nitrobenzene-sulfenyl chloride; *o*-nitrophenyl-sulfenyl chloride)	Tryptophan	189.6	A, S
Phenylglyoxal	Arginine, also potential for photooxidation	113	A, S, P
Pyridoxal-5′-phosphate	Lysine, α-amino groups	247.2	A, S
Sodium borohydride	Reducing agent for Schiff bases, also reduces disulfide bonds	37.8	SP, F, A, S
Sodium cyanate	Lysine, α-amino groups	65.0	A
Sodium cyanide	Cysteine, peptide bond cleavage	49.0	SP, F, A, S
Sodium cyanoborohydride	Reducing agent for Schiff bases	62.8	SP, F, A, S
Sodium periodate	Oxidizing agent	213.9	SP, F, A, S

Compound[b]	Amino acid modified[c]	M.W.[d]	Source[e]
Sodium sulfite	Cystine, cleavage of disulfide bonds	126.0	SP, F, A, S
Sodium tetrathionate	Cystine, cysteine	270.3	SP, A, S
Succinic anhydride	Lysine, α-amino groups, protein cross-linking	100.1	A, S
Tetranitromethane	Tyrosine, cysteine, tryptophan	196.0	A, S
Tri-*n*-butyl phosphine	Reducing agent for disulfide bonds	266.3	A, S
Trichloroacetic acid	Protein precipitant	163.4	SP, F, CA, A, S
Triethanolamine [tris-(2-hydroxyethyl)amine]	Buffer	149.2	SP, F, CA, A, S
Trifluoroacetic acid	HPLC solvent, counter-ion	114.0	SP, F, A, S, P
Tris-(hydroxymethyl) aminomethane		121.1	SP, F, CA, A, S, P
2,4,6-Trinitro benzenesulfonic acid(TNBS) (picryl sulfonic acid)	Lysine, α-amino groups	293.2	S, P

[a] This list does not include cross-linking reagents, active-site directed reagents, or suicide inhibitors. The reader is recommended to specific reviews for more information on these types of reagents. Also, see Pierce Chemical Company literature.

[b] The most commonly used compound name. This is frequently incorrect according to precise chemical nomenclature. Synonyms, when useful, are given.

[c] The most commonly modified amino acids. Solvents are indicated.

[d] M.W., molar weight. These are values which are useful, but are not precisely accurate.

[e] The most common sources from the author's experience are presented. With the most common chemicals, any large laboratory supply house will be a useful source, while it may be somewhat difficult to obtain the more complicated reagents. Although not referenced above, Molecular Probes (P.O. Box 22010, Eugene, OR, 97402-0414; 503-465-8300) is an excellent source of fluorescent-labeled probes.

Scientific Products (SP)
1430 Waukegan Road
McGaw Park, IL 60085-6787
708-689-8410/1-800-444-0880

Fischer Scientific (F)
711 Forbes Avenue
Pittsburgh, PA 16219-4785
(412) 562-8300/1-800-766-7000

Pierce Chemical Company (P)
3747 N. Meridan Road
P.O. Box 117
Rockford, IL 61105
815-968-0747/1-800-874-3723

Calbiochem-Novabiochem Corp. (CA)
10394 Pacific Center Court
San Diego, CA 92121
619-450-9600/1-800-854-3417

Aldrich Chemical Company (A)
1001 West Saint Paul Avenue
Milwaukee, WI 53233
414-273-3850/1-800-558-9160

Sigma Chemical Company (S)
P.O. 14508
St. Louis, MO 63178
314-777-5750/1-800-325-3010

The reader is also directed to several directories that appear on a annual basis. These include *Linscott's Guide and Chemical Cyclopedia* (American Chemical Society, Washington, D.C.)

Appendix II

ASSAY OF PROTEIN CONCENTRATION IN SOLUTION

The determination of the concentration of chemically modified proteins is critical for the characterization of such molecules. The modification process can create difficulties with the use of certain techniques such as absorbance at 280 nm. For example, the modification of tryptophan or tyrosine can markedly change the absorption characteristics, as these residues are the primary contributors to absorbance at 280 nm (phenylalanine and cystine make only minor contributions to such absorbance). The reader is recommended to a recent review of protein assay methods.[1]

The purpose of this short section is to describe some commonly used techniques for the determination of protein concentration. Care must be taken with the use of these techniques several of the more frequently used techniques depend on protein quality as well as quantity.

MICROPLATE BIURET ASSAY

This procedure is based on the macroprocedure originally developed by Gornall et al.,[2] and modified by Jenzano et al.[3]

Reagent Preparation — Dissolve 90.0 g NaOH in 300 ml deionized H_2O (this reaction is exothermic, so cooling with an ice bath is recommended). While this solution is cooling after dissolving NaOH, dissolve 1.5 g $CuSO_4 \cdot 5 H_2O$ and 6.0 g sodium potassium tartrate in 500 ml deionized H_2O. After the NaOH solution has cooled to ambient (room) temperature, combine the two solutions and take to a final volume of 1.0 l with deionized H_2O.

Standard — In general, a standard solution of bovine serum albumin (3.0 mg/ml established by spectroscopy) in 0.15 *M* NaCl is used; on occasion, we have used the Pierce Albumin Standard (Cat. No. 23210; 2.0 mg/ml in 0.9% NaCl with sodium azide as stabilizer). The choice of solvent is critical and should reflect the solvent used for the sample and reaction blank. There are significant solvent effects in this reaction. Tris-based buffers and other amino and guanidino compounds do react with the biuret reagent.

Blank — In general, we use 0.15 *M* NaCl as the blank, but the above comments regarding solvent selection must be considered. Unless there is an exceptional situation, we will only use a reaction blank and not an optical blank.

Procedure — Add 160 μl biuret reagent to a 40-μl sample, standard or blank (see attached format sheet and system sheet). Allow to stand at ambient temperature for 20 min and then read on microplate reader at 540 nm.

Comments — While this technique is generally quite reliable and is not significantly influenced by protein quality, it is insensitive when compared with other techniques (c.f. Reference 2). Selected references on the use of this method with various proteins are provided on the following pages.[3–9]

BCA (BICINCHONINIC ACID) ASSAY FOR PROTEIN

This procedure can be used for dye-binding assays based on the original procedure of Smith et al.,[10] as illustrated by Jenzano et al.[3] This procedure is a modification of the Lowry et al.[11] reaction, but it is significantly easier and somewhat more sensitive (see Reference 10).

Reagent Preparation — We use the commercial reagent as supplied by Pierce Chemical Company (Rockford, IL). Specifically, the BCA protein assay reagent is obtained (Cat. No. 23225). For the assay, one part reagent B is mixed with 50 parts of reagent A. This reagent (working reagent) is stable for at least 4 h at room temperature.

Standard — A commercially available bovine serum albumin standard is used. The use of Pierce Albumin Standard (Cat. No. 23210; 2.0 mg/ml in 0.9% NaCl with sodium azide as stabilizer) is recommended. Appropriate dilutions of this standard are used as indicated below. It may be necessary to alter these standards, but this is easily done with the BioRad Microplate Reader or other similar instrumentation. As indicated above in the discussion of the biuret reaction, solvent selection is critical. Specific examples of solvent effects are provided in the reference section.

Blank — In general, we use 0.15 *M* NaCl as the blank, but solvent selection should reflect the solvent used for the sample.

Procedure — Add 100 μl reagent to a 5-μl sample, standard or blank (see attached sheet for system and formal sheets), allow to stand for 30 min at 37°C (incubator), and then read on microplate reader at 570 nm (562 nm is recommended by manufacturer, but 570 nm is effective).

Comments — This reaction is quite sensitive, but it does reflect qualitative differences in proteins. As a reflection of the dependence on protein quality,[2] it is critical to select a standard that is qualitatively similar, if not identical, with the samples. This is obviously difficult when the assay is used with heterogeneous mixtures such as saliva or serum. Selected references to the use of this method are given below.[12–20]

DYE-BINDING ASSAY FOR PROTEIN USING COUMASSIE BRILLIANT BLUE G-250

The dye-binding assay for proteins using Coumassie Brilliant Blue G-250 is likely the most sensitive and most extensively used protein assay at this time. It is also extremely easy to perform. The technique, as noted below, is extremely dependent on the quality of the protein. The procedure given below is based on the procedure of Bradford[21] as illustrated by Jenzano et al.[3]

Reagent Preparation — For most procedures we use the commercial reagent as supplied by Pierce Chemical Company (Rockford, IL). Variations are in accordance with respective modification of procedure.

Standard — In general, a standard solution of bovine serum albumin (3.0 mg/ml established by spectroscopy) in 0.15 *M* NaCl is used. Solvent choice must reflect the solvent used for the sample and blank. The choice of a standard is critical because of the dependence of the spectral shift on the quality of the protein (see Reference 3).

Blank — As with the other assays, we use 0.15 *M* NaCl as the blank, but solvent choice must reflect the solvent used with the sample. If this is not possible, it is necessary to include sample solvent as a sample and correct final absorbance readings for any contribution by solvent.

Procedure — Add 195 µl reagent to a 5-µl sample, standard or blank (see attached sheet for system and format sheets), allow to stand for 15 min at 37°C (incubator), and then read on microplate reader at 595 nm.

Comments — As noted above, this assay technique is likely the most sensitive and facile of the currently available procedures. Rigorous application of the dye-binding assay to the quantitative determination of a broad spectrum of proteins is difficult because of the marked influence of protein quality on the reaction. This is reflected by various studies attempting to modify the assay system to eliminate dependence on the quality of the protein.[22–27] Literature citations are provided showing a broad application of this technique.[28–53]

REFERENCES

1. **Dawnay, A. B. StJ., Hirst, A. D., Perry, D. E., and Chambers, R. E.,** A critical assessment of current analytical methods for the routine assay of serum total protein and recommendations for their improvement, *Ann. Clin. Biochem.,* 28, 556, 1991.
2. **Gornall, A. G., Bardawill, C. J., and David, M. M.,** Determination of serum proteins by means of the biuret reaction, *J. Biol. Chem.,* 177, 751, 1949.
3. **Jenzano, J. W., Hogan, S. L., Noyes, C. M., Featherstone, G. L., and Lundblad, R. L.,** Comparison of five techniques for the determination of protein content in mixed human saliva, *Anal. Biochem.,* 159, 370, 1986.
4. **Arneberg, P.,** Quantitative determination of albumin, lysozyme, and amylase. A comparison of seven analytical methods, *Scand. J. Dent. Res.,* 78, 435, 1970.
5. **Wolf, R. O. and Taylor, L. L.,** A comparative study of saliva protein analysis, *Arch. Oral Biol.,* 9, 135, 1964.
6. **Skurk, A., Mlynski, G., and Fendel, K.,** Methoden der Quantitativen Eiweiszbestimmung im Menschlichen Parotisspeichel, *Acta Oto-Laryngol.,* 71, 71, 1971.
7. **Itzhaki, R. F. and Gill, D. M.,** A micro-biuret method for estimating proteins, *Anal. Biochem.,* 9, 401, 1964.
8. **Eckfeldt, J. H., Kershaw, M. J., and Dahl, I. I.,** Direct analysis for urinary protein with biuret reagent, with use of urine ultrafiltrate blanking: comparison with a manual biuret method involving trichloroacetic acid precipitation, *Clin. Chem.,* 30, 443, 1984.
9. **Schlabach, T. D.,** Postcolumn detection of serum proteins with the biuret and lowry reactions, *Anal. Biochem.,* 139, 309, 1984.
10. **Smith, P. K., Krohn, R. I., Hermanson, G. T., Mallia, A. K., Gartner, F. H., Provenzano, M. D., Fujimoto, E. K., Goeke, N. M., Olson, B. J., and Klenk, D. C.,** Measurement of protein using bicinchoninic acid, *Anal. Biochem.,* 150, 76, 1985.
11. **Lowry, O. H., Rosebrough, N. J., Farr, A. L., and Randall, R. J.,** Protein measurement with the folin phenol reagent, *J. Biol. Chem.,* 193, 265, 1951.

12. **Kirazov, L. P., Venkov, L. G., and Kirazov, E. P.,** Comparison of the Lowry and the Bradford assays as applied for protein estimation of membrane-containing fractions, *Anal. Biochem.,* 208, 44, 1993.
13. **Zhang, J.-X. and Halling, P. J.,** pH and buffering in the bicinchoninic acid (4,4′-dicarboxy- 2,2′-biquinoline) protein assay, *Anal. Biochem.,* 188, 9, 1990.
14. **Stich, T. M.,** Determination of protein covalently bound to agarose supports using bicinchoninic acid, *Anal. Biochem.,* 191, 343, 1990.
15. **Baker, W. L.,** Potential interference of hydrogen peroxide in the 2,2′-bicinchoninic acid protein assay, *Anal. Biochem.,* 192, 212, 1991.
16. **Hill, H. D. and Straka, J. G.,** Protein determination using bicinchoninic acid in the presence of sulfydryl reagents, *Anal. Biochem.,* 170, 203, 1988.
17. **Wiechelman, K. J., Braun, R. D., and Fitzpatrick, J. D.,** Investigation of the bicinchoninic acid protein assay: identification of the groups responsible for color formation, *Anal. Biochem.,* 175, 231, 1988.
18. **Goldschmidt, R. C. and Kimelberg, H. K.,** Protein analysis of mammalian cells in monolayer culture using the bicinchoninic assay, *Anal. Biochem.,* 177, 41, 1989.
19. **Brown, R. E., Jarvis, K. L., and Hyland, K. J.,** Protein measurement using bicinchoninic acid: elimination of interfering substances, *Anal. Biochem.,* 180, 136, 1989.
20. **Tuszynski, G. P. and Murphy, A.,** Spectrophotometric cell-adhesion assay with bicinchoninic acid, *Anal. Biochem.,* 184, 189, 1990.
21. **Bradford, M. M.,** A rapid and sensitive method for the determination of microgram quantities of protein utilizing the principle of protein-dye binding, *Anal. Biochem.,* 72, 248, 1976.
22. **Wilmsatt, D. K. and Lott, J. A.,** Improved measurement of urinary total protein (including light-chain proteins) with a Coomassie Brilliant Blue G-250-sodium dodecyl sulfate reagent, *Clin. Chem.,* 33, 2100, 1987.
23. **Tal, M., Silberstein, A., and Nusser, E.,** Why does Coomassie Brilliant Blue R interact differently with different proteins, *J. Biol. Chem.,* 260 (18), 9976, 1985.
24. **Pierce, J. and Suelter, C. H.,** An evaluation of the Coomassie Brilliant Blue G-250 dye-binding method for quantitative protein determination, *Anal. Biochem.,* 81, 478, 1977.
25. **Read, S. M. and Northcote, D. H.,** Minization of variation in the response to different proteins of the coomassie blue G dye-binding assay for protein, *Anal. Biochem.,* 116, 53, 1981.
26. **Sedmak, J. J. and Grossberg, S. F.,** A rapid, sensitive, and versatile assay for protein using coomassie brilliant blue G250, *Anal. Biochem.,* 79, 544, 1977.
27. **Stoscheck, C. M.,** Increased uniformity in the response of the Coomassie blue G protein assay to different proteins, *Anal. Biochem.,* 184, 111, 1990.
28. **Pollard, H. B., Menard, R., Brant, H. A., Pazoles, C. J., Creutz, C. E., and Ramu, A.,** Applications of Bradford's protein assay to adrenal gland subcellular fractions, *Anal. Biochem.,* 86, 761, 1978.
29. **Khan, M. Y. and Newman, S. A.,** An assay for heparin by decrease in color yield (DECOY) of a protein-dye-binding reaction, *Anal. Biochem.,* 187, 124, 1990.
30. **Gotham, S. M., Fryer, P. J., and Paterson, W. R.,** The measurement of insoluble proteins using a modified Bradford assay, *Anal. Biochem.,* 173, 353, 1988.
31. **Spector, T.,** Refinement of the Coomassie Blue Method of protein quantitation. A simple and linear spectophotometric assay for <0.5 to 50 µg of protein, *Anal. Biochem.,* 86, 142, 1978.
32. **Hanggi, D. and Carr, P.,** Analytical evaluation of the purity of commercial preparations of cibacron blue F3GA and related dyes, *Anal. Biochem.,* 149, 91, 1985.
33. **Compton, S. J. and Jones, C. G.,** Mechanism of dye response and interference in the Bradford protein assay, *Anal. Biochem.,* 151, 369, 1985.
34. **Lea, M. A., Grasso, S. V., Hu, J., and Seilder, N.,** Factors affecting the assay of histone H1 and polylysine by binding of coomassie blue G, *Anal. Biochem.,* 141, 390, 1984.

35. **Wise, B. L.,** The quantitiation and fractionantion of proteins in cerebrospinal fluid, *Am. J. Med. Technol.,* 48, 821, 1982.

36. **Pollard, H. B., Menard, R., Brant, H. A., Pazoles, C. J., Creutz, C. E., and Ramu, A.,** Applications of Bradford's protein assay to adrenal gland subcellular fractions, *Anal. Biochem.,* 86, 761, 1978.

37. **Almog, R. and Berns, D.,** A sensitive assay for proteins and biliproteins, *Anal. Biochem.,* 114, 336, 1981.

38. **Meola, J. H., Vargas, M. A., and Brown, H. H.,** Simple procedure for measuring total protein in urine, *Clin. Chem.,* 23, 975, 1977.

39. **Zaman, Z. and Verwilghen, R. L.,** Dye-binding assay for protein solubilized in the presence or absence of sodium dodecyl sulfate, *Anal. Biochem.,* 109, 454, 1980.

40. **Rubin, R. W. and Warren, R. W.,** Quantitation of microgram amounts of protein in SDS-mercaptoethanol-Tris electrophoresis sample buffer, *Anal. Biochem.,* 83, 773, 1977.

41. **Perini, J. M., Mizon, C., Dehon, B., Fialdes, P., Lefebvre, I., Mizon, J., and Roussel, P.,** Urinary protein determination using Coomassie Brilliant Blue in the presence of sodium dodecyl sulphate, *Clin. Chim. Acta,* 143, 321, 1984.

42. **Kain, S. R. and Henry, H. L.,** Quantitation of proteins bound to polyvinylidene difluoride membranes by elution of Coomassie brilliant blue R-250, *Anal. Biochem.,* 189, 169, 1990.

43. **Minamide, L. S. and Bamburg, J. R.,** A filter paper dye-binding assay for quantitative determination of protein without interference from reducing agents or detergents, *Anal. Biochem.,* 190, 66, 1990.

44. **Said-Fernández, S., González-Garza, M. T., Mata-Cárdenas, B. D., and Navarro-Marmolejo, L.,** A multipurpose solid-phase method for protein determination with Coomassie brilliant blue G-250, *Anal. Biochem.,* 191, 119, 1990.

45. **Kirazov, L. P., Venkov, L. G., and Kirazov, E. P.,** Comparison of the Lowry and the Bradford assays as applied for protein estimation of membrane-containing fractions, *Anal. Biochem.,* 208, 44, 1993.

46. **Gotham, S. M., Fryer, P. J., and Paterson, W. R.,** The measurement of insoluble proteins using a modified Bradford assay, *Anal. Biochem.,* 173, 353, 1988.

47. **Goldschmidt, R. C. and Kimelberg, H. K.,** Protein analysis of mammalian cells in monolayer culture using the bicinchoninic assay, *Anal. Biochem.,* 177, 41, 1989.

48. **Löffler, B.-M. and Kunze, H.,** Refinement of the Coomassie brilliant blue G assay for quantitative protein determination, *Anal. Biochem.,* 177, 100, 1989.

49. **Ball, E. H.,** Quantitation of proteins by elution of Coomassie brilliant blue R from stained bands after sodium dodecyl sulfate-polacrylamide gel electrophoresis, *Anal. Biochem.,* 155, 23, 1986.

50. **Vincent, R. and Nadeau, D.,** A micromethod for the quantitation of cellular proteins in percoll with the coomassie brilliant blue dye-binding assay, *Anal. Biochem.,* 135, 355, 1983.

51. **Rubin, R. W. and Warren, R. W.,** Quantitation of microgram amounts of protein in SDS-mercaptoethanol-Tris electrophoresis sample buffer, *Anal. Biochem.,* 83, 773, 1977.

52. **Friedenauer, S. and Berlet, H. H.,** Sensitivity and variability of the Bradford protein assay in the presence of detergents, *Anal. Biochem.,* 178, 263, 1989.

53. **Splittgerber, A. G. and Sohl, J.,** Nonlinearity in protein assays by the coomassie blue dye-binding method, *Anal. Biochem.,* 179, 198, 1989.

INDEX

A

B

C

D

E

F

G

H

I

K

P

R

S

T

U

V

X

Y

Z

THE ECONOMICS OF EDUCATION